ANGIOGENESIS

FROM THE MOLECULAR TO INTEGRATIVE PHARMACOLOGY

ADVANCES IN EXPERIMENTAL MEDICINE AND BIOLOGY

Recent Volumes in this Series

Volume 467
TRYPTOPHAN, SEROTONIN, AND MELATONIN: Basic Aspects and Applications
Edited by Gerald Huether, Walter Kochen, Thomas J. Simat, and Hans Steinhart

Volume 468
THE FUNCTIONAL ROLES OF GLIAL CELLS IN HEALTH AND DISEASE: Dialogue between Glia and Neurons
Edited by Rebecca Matsas and Marco Tsacopoulos

Volume 469
EICOSANOIDS AND OTHER BIOACTIVE LIPIDS IN CANCER, INFLAMMATION, AND RADIATION INJURY, 4
Edited by Kenneth V. Honn, Lawrence J. Marnett, and Santosh Nigam

Volume 470
COLON CANCER PREVENTION: Dietary Modulation of Cellular and Molecular Mechanisms
Edited under the auspices of the American Institute for Cancer Research

Volume 471
OXYGEN TRANSPORT TO TISSUE XXI
Edited by Andras Eke and David T. Delpy

Volume 472
ADVANCES IN NUTRITION AND CANCER 2
Edited by Vincenzo Zappia, Fulvio Della Ragione, Alfonso Barbarisi, Gian Luigi Russo, and Rossano Dello Iacovo

Volume 473
MECHANISMS IN THE PATHOGENESIS OF ENTERIC DISEASES 2
Edited by Prem S. Paul and David H. Francis

Volume 474
HYPOXIA: Into the Next Millennium
Edited by Robert C. Roach, Peter D. Wagner, and Peter H. Hackett

Volume 475
OXYGEN SENSING: Molecule to Man
Edited by Sukhamay Lahiri, Naduri R. Prabhakar, and Robert E. Forster, II

Volume 476
ANGIOGENESIS: From the Molecular to Integrative Pharmacology
Edited by Michael E. Maragoudakis

CONTENTS

ANGIOGENIC FACTORS AND THEIR RECEPTORS

REGULATION OF ANGIOGENESIS AND TRANSDUCTION MECHANISMS INVOLVED

ROLE OF EXTRACELLULAR MATRIX AND ADHESION MOLECULES IN ANGIOGENESIS

INHIBITORS OF ANGIOGENESIS AND THEIR MECHANISMS OF ACTION

ANGIOGENESIS AND MALIGNANT TRANSFORMATION

PRECLINICAL DEVELOPMENTS OF ANGIOGENESIS INHIBITORS

CLINICAL APPLICATIONS

INTRODUCTORY COMMENTS

Michael E. Maragoudakis
University of Patras Medical School, Department of Pharmacology, 261 10 Rio, Patras, GREECE

The complexity of the angiogenic cascade has attracted the interest of basic science such as embryology, physiology, anatomy, biochemistry molecular and cellular biology and pharmacology. With their multidisciplinary approach and the powerful new techniques that have been available the progress has been impressive indeed as evidenced by the logarithmic increase in publications. It is estimated that 3000 laboratories and 120 bio-pharmaceutical companies are actively involved in angiogenesis research.

Many endogenous modulators of angiogenesis have been characterized. A large number of cytokines that modulate angiogenesis have been studied and many of them have been sequenced and cloned. The list of endogenous modulators of angiogenesis is increasing. Hundreds of inhibitors of angiogenesis have been reported and more than 150 medical centres are evaluating in the clinic drug candidates. This explosive clinical and commercial interest is understandable based on the prevailing view that angiogenesis-based therapy is possible in variety of clinical problems.

Therapeutic suppression of angiogenesis is possible in: cancer, ocular neovascularization, child hemangiomas, rheumatoid arthritis, Kaposis sarcoma, etc. On the other hand therapeutic stimulation of angiogenesis in conditions such as: ischemic heart disease, peripheral vascular disease, non-healing diabetic ulcers, peptic ulcer diseases, etc.

Angiogenesis: From the Molecular to Integrative Pharmacology
Edited by Maragoudakis, Kluwer Academic / Plenum Publishers, New York, 2000

Angiogenesis can also serve as a tool to prognosis and guide to therapy in cancer patients, using angiogenic markers such as microvessel count and angiogenic cytokines

However, the concept of angiogenesis-based therapy is based on certain assumptions that are not as solid as it was originally thought. Furthermore, we increasingly appreciate our incomplete understanding of angiogenesis as a vastly complex biological process. To start with, it is believed that angiogenesis in adults is activated only in response to wound healing, and in the females during the reproductive cycle. Therefore, clinical application of angiogenesis inhibitors are not expected to have serious side effects. However, angiogenesis does occur in skeletal muscle in response to exercise and training. Also in myocardium and other tissues in response to hypoxia and stress. In these situations and others where we have co-existing pathologies, such as peptic ulcers, the use of angiogenesis inhibitors may cause serious problems. We have a special session, where two eminent pathologists (Dr. Haudenschild and Dr. Thompson) will discuss these issues and tell us if Nemesis can be avoided.

Even the concept that tumours are angiogenesis dependent can be challenged. There is considerable variation in the degree of active angiogenesis in different types of tumours and even within the same tumour type. Furthermore, in human tumours angiogenesis is considerable lower than that in animal tumours and much lower that the angiogenesis occurring in the granulation tissue or the reproductive organs. The endothelial cell proliferation index in human prostatic or breast carcinoma is only 0.15%, compared to 6.7% in granulation tissue and 36% in corpus luteum. In other tumours such as the colorectal adenocarcinoma the endothelial cell proliferation index, which is an index of active angiogenesis is 9.9%. In view of these data, the question arises: Are all tumours suitable for effective anti-angiogenic treatment? What will be the criteria for selection?

On the issue of the therapeutic use of promoters of angiogenesis: Can systemic application of such agents provide assurance that latent malignancies will not be activated? Are the new vessels seen in experiments with angiogenic factors truly new vessels or are we seeing remodelling? Can these results be confirmed by endothelial all proliferation index to provide evidence for quantitative angiogenesis?

The third aspect of application of our knowledge on angiogenesis is for diagnostic and prognostic purposes in diseases. This is based on using specific endothelial cell markers, such as VWF, CD31, CD34, PDGF to visualise by histochemical techniques blood vessels in tumour specimens. High vascular densities indicate poor prognosis. However, not all studies are in agreement with this conclusion and the techniques used for identifying «hot spots» in tumours are highly subjective. Furthermore, it is important to keep in mind that high vascular density does not necessarily reflect angiogenesis. Our lungs have the highest vascular density of any tissue, but there is no angiogenesis in a healthy lung.

I am hoping that our clinicians will provide some answers in the near future to these and other important questions such as: Why all angiogenic tumours don't produce metastases? Why in some highly vascularized tumours metastases appear many years later? Why in some cases the primary tumour remains undetectable while the metastases appear first?

In the basic science side, we know that in the adult we have at least three types of angiogenesis: sprouting angiogenesis, intussusceptive microvascular growth, vascular formation from circulating endothelial cells. What are the mechanisms controlling each type? When and where are activated? By what mechanism the physiological angiogenesis returns to quiescent state? Are the new vessels formed in different tissue beds, under physiological and pathological conditions, morphologically, physiologically and functionally similar?

The answers to these questions may help us to understand more key issues such as: Is the derangement of angiogenesis in pathological situations at different sites based on the same mechanism? Can modulation of one factor be sufficient to normalize the derangement in redundant system where many promoters and inhibitors participate? What then is the ideal target for therapeutic intervention?

The discovery of the plethora of endogenous modulators of angiogenesis has lead many scientists to believe that under physiological conditions angiogenesis is controlled by a balance of a redundant inhibitors and promoters of angiogenesis. This balance then can be tipped under pathological conditions by overproduction or deficiency of modulators. Can this be that simple? It is unlikely that such an important physiological process can by controlled only be algebraic addition of the effects of promoters and inhibitors. Strict controls must exist and immediate activation of angiogenesis at a short notice must be possible. This can only be accomplished by intricate interactions of the modulators of angiogenesis. Most likely specific interactions that modulate key angiogenic molecules such as VEGF and its receptors are involved. Some paradigms for such interactions are emerging. We have reported that thrombin promotes angiogenesis by many mechanisms one of which is the upregulation of VEGF receptors. The understanding of all such interactions of modulators of angiogenesis hopefully will lead to a scheme like that of the blood coagulation cascade. When this is accomplished then we can pinpoint the site of derangement in disease states and the most suitable target for intervention. We are moving form the descriptive to the molecular aspects of angiogenesis and we begin to understand the transduction mechanisms involved. Many of the modulators of angiogenesis may have a common path for transmitting the signal from the receptor to the cell nucleus.

I am hoping that many of these issues will be discussed and that in the near future we are going to have a better understanding of angiogenesis in health and disease.

Angiogenic Factors and their Receptors

EXAMINING NEW MODELS FOR THE STUDY OF AUTOCRINE AND PARACRINE MECHANISMS OF ANGIOGENESIS THROUGH FGF2-TRANSFECTED ENDOTHELIAL AND TUMOUR CELLS

Marco Presta, Marco Rusnati, Patrizia Dell'Era, Elena Tanghetti, Chiara Urbinati, Roberta Giuliani, and Daria Leali
Unit of General Pathology and Immunology, Department of Biomedical Sciences and Biotechnology, University of Brescia, 25123 Brescia, Italy

Key words: FGF2, endothelium, angiogenesis, transfection, signal transduction, heparin.

Abstract: Angiogenesis is the process of generating new capillary blood vessels. Uncontrolled endothelial cell proliferation is observed in tumour neovascularization. Several growth factors and cytokines have been shown to stimulate endothelial cell proliferation *in vitro* and *in vivo* and among them FGF2 was one of the first to be characterised. FGF2 is a *Mr* 18,000 heparin-binding cationic polypeptide that induces proliferation, migration, and protease production in endothelial cells in culture and neovascularization *in vivo*. FGF2 interacts with endothelial cells through two distinct classes of receptors, the high affinity tyrosine-kinase receptors (FGFRs) and low affinity heparan sulfate proteoglycans (HSPGs) present on the cell surface and in the extracellular matrix. Besides experimental evidence for paracrine mode of action for FGF2, some observations raise the hypothesis that FGF2 may also play an autocrine role in endothelial cells. FGF2 may therefore represent a target for anti-angiogenic therapies. In order to assess the angiostatic potential of different classes of compounds, novel experimental models have been developed based on the autocrine and/or the paracrine capacity of FGF2.

1. FGF2 AS A PROTOTYPIC ANGIOGENIC GROWTH FACTOR

FGF2 belongs to the family of the heparin-binding growth factors (Basilico and Moscatelli, 1992). The single copy human FGF2 gene encodes multiple FGF2 isoforms with molecular weights ranging from 24 kD to 18 kD. High molecular weight isoforms (HMW-FGF2s) are colinear NH2-terminal extensions of the better characterised 18 kD protein (Florkiewicz and Sommer, 1989) (Fig. 1). Both low and high molecular weight FGF2s exert angiogenic activity *in vivo* and induce cell proliferation, protease production, and chemotaxis in endothelial cells *in vitro* (Gualandris *et al.*, 1994). Also, FGF2 has been shown to stimulate endothelial cells to form capillary-like structures in collagen gels (Montesano *et al.*, 1989) and to invade the amniotic membrane *in vitro* (Mignatti *et al.*, 1989). Moreover, the phenotype induced *in vitro* by FGF2 in endothelial cells includes modulation of integrin expression (Klein *et al.*, 1993), gap-junctional intercellular communication (Pepper and Meda, 1992) and urokinase receptor upregulation (Mignatti *et al.*, 1991a). Studies with neutralising anti-FGF2 antibodies have implicated FGF2 in wound repair (Broadly *et al.*, 1989), vascularization of the chorioallantoic membrane during chick embryo development (Ribatti *et al.*, 1995), and tumour growth under defined experimental conditions (Baird *et al.*, 1986; Gross *et al.*, 1993).

Figure 1. 3D structure of FGF2

As shown in Fig. 2, low and high MW FGF2 isoforms interact with high affinity tyrosine-kinase FGF receptors (FGFRs) (Johnson and Williams, 1993) and low affinity proteoglycans containing heparan sulfate (HS) as polysaccharide (HSPGs) (Gualandris *et al.*, 1994).The physiological effects of the interaction of FGF2 with cell-associated and free HSPGs are manifold. HSPGs protect FGF2 from inactivation in the extracellular environment and modulate the bioavailability of the growth factor (Rusnati *et al.*, 1996a). At the cell surface, free and cell-associated HSPGs may play contrasting roles in modulating the dimerization of FGF2 and its interaction with FGFRs.

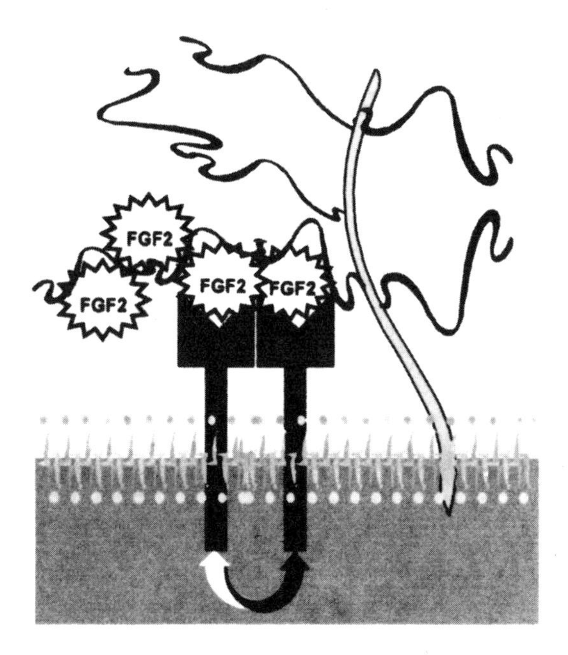

Figure 2. HSPG-mediated FGF2-FGFR interaction

For instance, free heparin induces FGF2-FGFR interaction in HS-deficient cells (Yayon *et al.*, 1991). This relies on the capacity of the glycosaminoglycan (GAG) to form a ternary complex by interacting with both proteins (Guimond *et al.*, 1993; Turnbull and Gallagher, 1993; Rusnati *et al.*, 1994). Indeed, a heparin binding domain in FGFR-1/flg has been identified in the NH2-terminus of IgG-like domain II (Kan *et al.*, 1993). In apparent contrast with these observations, free heparin inhibits the binding of FGF2 to FGFRs when administered to cells bearing surface-associated

HSPGs (Ishihara *et al.*, 1993; Coltrini *et al.*, 1994). This is probably due to the competition of free GAGs with cell-associated HSPGs and FGFRs for the binding to FGF2. The puzzling observation that heparin by itself can activate FGFR in the absence of the growth factor (Gao and Goldfarb, 1995) Finally, HSPGs affect the internalisation and the intracellular fate of FGF2, suggesting that FGF2-HSPG complexes are involved in the intracellular delivery of FGF2 (Gannoun-Zaky *et al.*, 1991; Roghani and Moscatelli, 1993; Rusnati *et al.*, 1993). Thus, the bioavailability and the biological activity of FGF2 on endothelial cells strictly depend on the extracellular GAG milieu, indicating the possibility of modulating the angiogenic activity of FGF2 in vivo by using exogenous GAGs. Recent findings on the capacity of low molecular weight heparin fragments administered systemically to reduce the angiogenic activity of FGF2 support this hypothesis (Norrby and Ostergaard, 1996).

These observations raise the possibility that synthetic molecules able to interfere with HSPG/FGF2/FGFR interaction may act as angiogenesis inhibitors. In particular, heparin-mimicking, polyanionic compounds able to compete with HSPGs for growth factor interaction would hamper the binding of FGF2 to the endothelial cell surface with consequent inhibition of its angiogenic capacity. Among such compounds are suramin (Fig. 3), several suramin analogues (Firsching *et al.*, 1995; Ciomei *et al.*, 1994), pentosan polysulfate (PPS) (Zugmaier *et al.*, 1992), and the polycarboxylated compounds aurin tricarboxylic acid (Gagliardi *et al.*, 1994) and RG-13577 (Miao *et al.*, 1997).

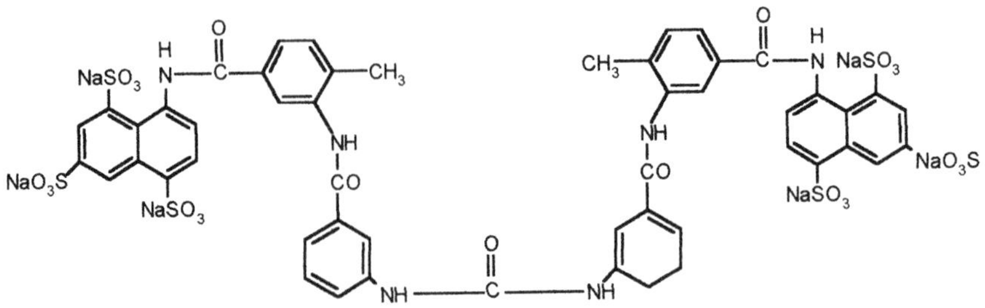

Figure 3. Chemical structure of suramin

The most extensively studied, the polysulfonated naphtylurea suramin, has been shown to block the binding of several growth factors, including FGF2 to its receptors (Rusnati *et al.*, 1996b; Braddock *et al.*, 1994). In addition, suramin exerts a marked inhibitory effect on endothelial cell growth, migration, and urokinase-type plasminogen activator production in vitro (Takano *et al.*, 1994) and angiogenesis and tumour growth in vivo (Gagliardi *et al.*, 1992; Waltz *et al.*, 1991). Suramin has been used in clinical trials on cancer patients with some beneficial effects (Myers *et al.*, 1992).

1.1 Polysulfonates as angiogenesis inhibitors

The sulfonic acid polymers PAMPS [poly(2-acrylamido-2-methyl-1-propanesulfonic acid)], PAS [poly(anetholesulfonic acid)], PSS [poly(4-styrenesulfonic acid)], and PVS [poly(vinylsulfonic acid)] suppress viral replication, including HIV replication (Mohan *et al.*, 1992; Ikeda *et al.*, 1994).

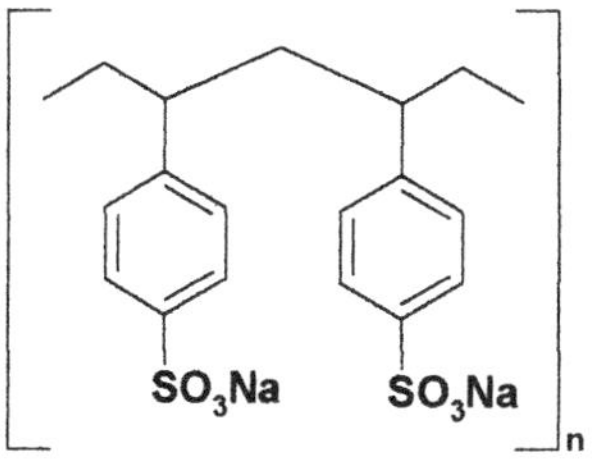

Figure 4. Chemical structure of PSS

We recently reported on the potent inhibitory activity exerted by PAMPS, PAS, and PSS (Fig. 4) on neovascularization that occurs in the chick embryo chorioallantoic membrane (CAM) during development (Liekens *et al.*, 1997). Also, these sulfonic acid polymers exerted an anti-angiogenic effect in the in vitro rat aorta-ring assay and inhibited FGF2-induced human umbilical vein endothelial cell proliferation. Interestingly, a significant correlation was found between the angiostatic activity of these compounds in the CAM assay in vivo and their capacity to inhibit the FGF2-induced mitogenic response in vitro, thus suggesting that FGF2 is a target for sulfonic acid polymers (Liekens *et al.*, 1997).

We have investigated the capacity of sulfonic acid polymers to interact with FGF2 and to affect its biological activity in vitro and in vivo (Liekens *et al.*, 1999). The results indicate that sulfonic acid polymers mimic functional features of heparin/HS by binding to FGF2 and preventing its interaction

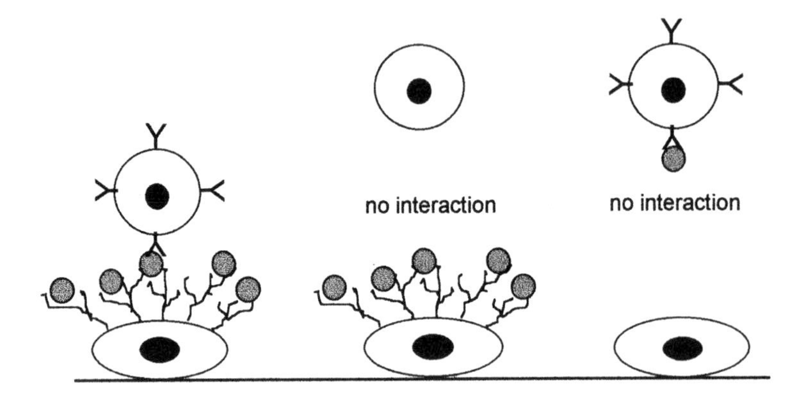

Figure 5. FGF2-mediated cell-cell interaction. FGF2 mediates the interaction of FGFR1-bearing cells with HSPGs of the cell monolayer (left). No interaction occurs in the absence of FGFR1 (center) or of HSPGs (right).

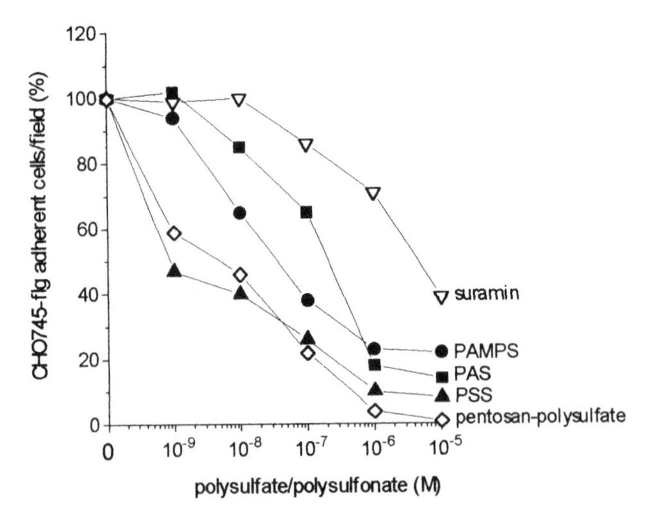

Figure 6. Effect of polysulfated/polysulfonated compounds on FGF2-mediated cell-cell interaction.

with endothelial cell surface HSPGs and FGFRs. Also, the sulfonic acid polymers were evaluated for their capacity to prevent the formation of the HSPG/FGF2/FGFR ternary complex. For this purpose, we utilised an experimental model in which the disruption of the complex abolishes FGF2-mediated cell-cell attachment of HSPG-deficient CHO mutants transfected with FGFR-1 to a monolayer of wild type CHO-K1 cells bearing HSPGs (Richard *et al.*, 1995) (Fig. 5).

In this assay (Fig. 6), all the molecules tested exerted an inhibitory activity, PSS being the most effective (ID_{50} equal to 0.01 μM, 0.06 μM, 0.12 μM, and 0.07 μM for PSS, PAMPS, PAS, and PVS, respectively).

These data demonstrate the ability of sulfonic acid polymers to prevent the interaction of FGF2 with its low and high affinity receptors. Interestingly, PSS appears to be the most potent among the molecules studied, its activity being similar to that exerted by conventional heparin on a molar basis and at least 1,000 more potent than that exerted by suramin when tested under the same experimental conditions. Finally, both PAMPS and PAS inhibit FGF2-mediated angiogenesis in the rabbit cornea (Table 1).

Table 1. Inhibition of FGF2-mediated angiogenesis in the rabbit cornea

Sample	area of neovascularization [a]	p value
FGF2 (650 ng)	16.2 ± 3.3 mm^2	-
FGF2 + PAMPS (100 μg)	8.7 ± 3.8 mm^2	< 0.05
FGF2 + PAS (100 μg)	10.8 ± 2.3 mm^2	< 0.05

a The area of corneal neovascularization was determined 8 days after implantation.

2. AUTOCRINE AND PARACRINE ROLES OF FGF2 IN ANGIOGENESIS

Several cell types, including tumor cells of different origin (Moscatelli *et al.*, 1986; Ohtani *et al.*, 1993; Presta *et al.*, 1986; Schulze-Osthoff *et al.*, 1990; Takahashi *et al.*, 1990; Yamanaka *et al.*, 1993), macrophages (Baird *et al.*, 1985) and T lymphocytes (Blotnik *et al.*, 1994), express FGF2 *in vitro* and *in vivo*. FGF2 lacks a classic signal peptide for secretion (Abraham *et al.*, 1986). However, cell damage may cause the release of FGF2 from producing cells (Gajdusek and Carbon, 1989; McNeil *et al.*, 1989; Witte *et al.*, 1989). Also, an alternative mechanism of exocytosis of FGF2, independent of the endoplasmic reticulum/Golgi pathway, has been proposed (Mignatti *et al.*, 1991b; Mignatti *et al.*, 1992). Accordingly, FGF2 has been found associated

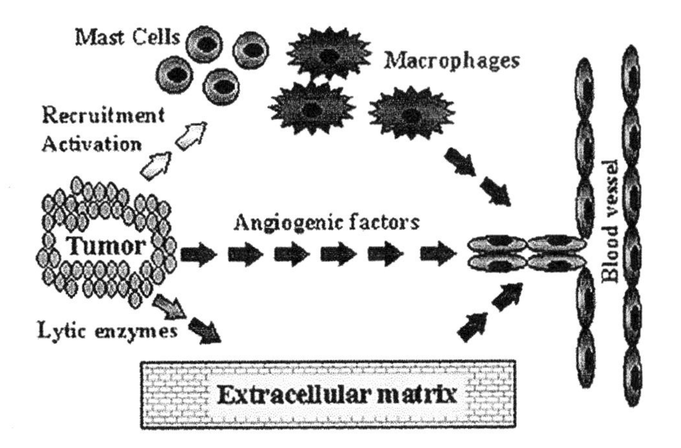

Figure 7. Paracrine activity of FGF2 on endothelium. FGF2 is released by tumor cells and inflammatory cells or is mobilised from ECM. Free FGF2 will act on endothelial cells in a paracrine mode of action.

with the extracellular matrix (ECM) of cell cultures *in vitro* (Rogelj *et al.*, 1989; Vlodavski *et al.*, 1987a) and located in the basement membranes of blood vessels *in vivo* (DiMario *et al.*, 1989; Folkman *et al.*, 1988). On this basis, FGF2 is thought to exert its effects on endothelial cells via a paracrine mode consequent to its release by other cells and/or mobilization from ECM (Fig. 7).

Besides experimental evidence for paracrine mode of action for FGF2, some observations raise the hypothesis that FGF2 may also play an autocrine role in endothelial cells. *In vitro*, it has been shown that different endothelial cells produce FGF2 (Presta *et al.*, 1989; Schweigerer *et al.*, 1987; Vlodavski *et al.*, 1987b) and that endogenous FGF2 modulates cell proliferation and migration, as well as the production of proteinases and their receptors (Itoh *et al.*, 1992; Pepper *et al.*, 1993; Sato and Rifkin, 1988).

In vivo, it has been shown that FGF2 expression occurs in the endothelium adjacent to neoplastic cells in several human tumour types. These neoplasms include neuroblastoma, astrocytoma, glioblastoma and meningioma (Schulze-Osthoff *et al.*, 1990; Takahashi *et al.*, 1990; Zagzag *et al.*, 1990), pheochromocytoma (Statuto *et al.*, 1993), melanoma (Schulze-Osthoff *et al.*, 1990), carcinomas of the stomach and colon (Ohtani *et al.*, 1993; Schulze-Osthoff *et al.*, 1990), and adenocarcinomas of the larynx, endometrium, and cervix. Thus, FGF2 expression is a common feature of vascular endothelium during tumour angiogenesis.

These observations strongly support the hypothesis that neovascularization may be triggered by molecule(s) released by tumour cells and/or infiltrating inflammatory cells that induce FGF2 upregulation in the quiescent endothelium. In keeping with this hypothesis is the observation that tumor cells of different origin release molecule(s) able to interact with endothelium and to upregulate the expression of FGF2 that, in turn, stimulates the fibrinolytic potential of the endothelial cell in an autocrine manner (Peverali *et al.*, 1994). In addition, FGF2 itself, thrombin, and interleukin-2 stimulate FGF2 production in endothelial cells (Cozzolino *et al.*, 1993; Weich *et al.*, 1991).

FGF2 has been detected in cardiac myocytes (Speir *et al.*, 1992) and cells of the coronary vasculature (Hawker and Granger, 1993). Also, cultured coronary endothelium exhibits the FGF receptor on its surface and expresses FGF2 mRNA (Hawker and Granger, 1993), suggesting that FGF2 might induce coronary angiogenesis by an autocrine/paracrine mechanism. Indeed, we have shown (Ziche *et al.*, 1997) that nitric oxide (NO) induces an angiogenic phenotype (including cell proliferation and urokinase-type plasminogen activator upregulation) in coronary venular endothelial cells by inducing endogenous FGF2 and that this pathway mediates the angiogenetic response to the vasoactive neuropeptide substance P (Fig. 8).

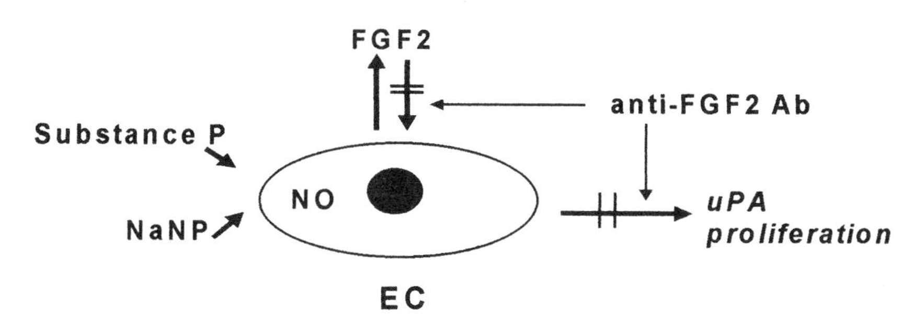

Figure 8. Endogenous FGF2 mediates NO-induced angiogenic phenotype in coronary venular endothelial cells. Substance P and the NO donor nitroprusside (NaNP) induce FGF2 upregulation that stimulates endothelial cell (EC) proliferation and urokinase-type plasminogen activator (uPA) upregulation. Neutralising anti-FGF2 antibody prevents EC response to NO.

3. AUTOCRINE ROLE OF FGF2 IN VASCULAR TUMOURS

Blood vessels may represent the site of origin for neoplasms, hamartomas, and vessel malformations. Neoplasms include benign tumours and tumour-like lesions (hemangioma), tumours of intermediate malignancy (hemangioendothelioma), and malignant tumours (angiosarcoma) (Enzinger, 1995). The pathogenesis of vascular tumours is at present unknown, even though the local, uncontrolled release of growth factors and/or lytic enzymes has been hypothesised to facilitate endothelial cell proliferation and the formation of vascular lacunae (Enzinger, 1995).

A close relationship exists between the formation of vascular tumours and angiogenesis. This relationship is also apparent in Kaposi's sarcoma (KS). Classic KS is a relatively benign, highly vascularized neoplasm. A clinically aggressive form of KS develops in a significant percentage of acquired immune deficiency syndrome (AIDS) patients (Levine, 1993). Histologically, KS is characterised by the presence of spindle-shaped cells, inflammatory cells and newly formed blood vessels (Enzinger, 1995). KS lesions express various markers for vascular endothelial cells, suggesting that KS spindle cells are of endothelial cell lineage (Sturzl *et al.*, 1992).

Several experimental evidences implicate FGF2 in the pathogenesis of vascular lesions, including KS and hemangiomas. *In vitro*, AIDS-KS cells derived from different patients express high levels of FGF2 which is released in the extracellular media (Albini *et al.*, 1994). Antisense oligonucleotides directed against FGF2 mRNA inhibit both the growth of AIDS-KS cells and the angiogenic activity associated with these cells, including the induction of KS-like lesions in nude mice (Ensoli *et al.*, 1994b). FGF2 immunoreactivity is detected both in classic and AIDS-associated KS lesions in humans (Ensoli *et al.*, 1994a) and recombinant FGF2 synergizes with HIV-1-Tat protein in inducing the formation of vascular lesions closely resembling early KS into nude mice (Ensoli *et al.*, 1994). Interestingly, cytokines from activated T cells induce FGF2 upregulation and the acquisition of a AIDS-KS spindle cell-like phenotype in normal endothelial cells (Barillari *et al.*, 1992; Fiorelli *et al.*, 1995). Finally, coexpression of FGF2 and endothelial phenotypic markers CD31 and von Willebrand factor has been found in the proliferating phase of human hemangioma but not in vascular malformations (Takahashi *et al.*, 1994). Taken together, the data suggest that FGF2 produced by cells of the endothelial lineage may play important autocrine and paracrine roles in the pathogenesis of vascular tumours.

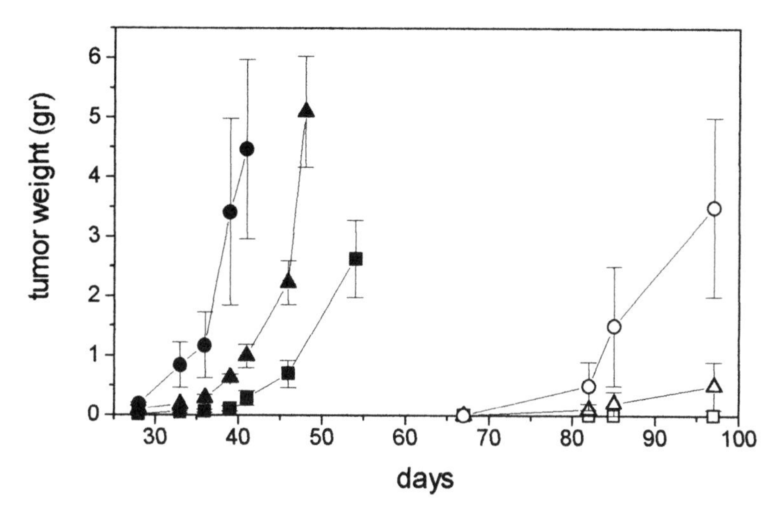

Figure 9. Tumorigenic activity of FGF2-transfected and parental MAE cells in nude mice. Male Swiss nu/nu mice were inoculated s.c. with 0.5×10^6 (squares), 1.0×10^6 (triangles), or 2.0×10^6 (circles) parental MAE cells (open symbols) or pZipbFGF2-MAE cells (closed symbols). The size of the lesion was measured with calipers.

4. FGF2 OVEREXPRESSION IN MOUSE ENDOTHELIAL CELLS

To investigate the biological consequences of endothelial cell activation by endogenous FGF2, immortalised Balb/c mouse aortic endothelial cells (MAE cells) and brain microvascular cells (MBE cells) were transfected with a retroviral expression vector harboring a human FGF2 cDNA (Gualandris *et al.*, 1996a; Gualandris *et al.*, 1996b). FGF2 transfectants express all FGF2 isoforms and are characterized by a transformed morphology and an increased saturation density. FGF2 transfectants show invasive and morphogenetic behaviour in three-dimensional gels which is prevented by anti-FGF2 antibody, revealing the autocrine modality of the process (Gualandris *et al.*, 1996a).

The biological consequences of this autocrine activation were investigated *in vivo*. FGF2-transfected MAE cells induce the growth of highly vascularized tumours (Fig. 9).

Lesions were observed also when cells were injected in x-irradiated syngeneic mice but grew poorly in immunocompetent syngeneic animals, indicating that the growth of these lesions is dependent on the immunological status of the host. Histologically, the tumors have the appearance of hemangioendothelioma with spindled areas resembling KS (Gualandris *et al.*, 1996a) and with numerous CD31-positive blood vessels and lacunae. Southern blot analysis revealed that less than 10% of the cells in the tumor mass were transplanted FGF2-transfected MAE cells. Accordingly, disaggregation of the lesion and *in vitro* cell culture demonstrate that less than 10% of total cell population retain FGF2 overexpression and neomycin-resistance. These data indicate that FGF2-overexpressing endothelial cells cause vascular lesions in the immunocompromised host that are sustained to a large extent by recruitment of host cells, including endothelial cells.

In agreement with these observations, FGF2-transfected MAE cells induce an angiogenic response when implanted in the avascular rabbit cornea (Gualandris *et al.*, 1996a). Also, they cause an increase in vascular density (Gualandris *et al.*, 1996a) and formation of hemangiomas in the chorioallantoic membrane when injected into the allantoic sac of the chick embryo.

Thus, the data demonstrate that pZipbFGF2-MAE cells induce highly vascularized spindle-cell hemangioendotheliomas in immunodeficient mice that are sustained by recruitment of host elements, including endothelial cells (Fig. 10).

4.1 Suramin derivatives as angiogenesis inhibitors

The sulfonated distamycin derivative PNU 153429 is a non cytotoxic compound endowed with antiangiogenic and antimetastatic activity and of FGF2-binding capacity (Fig. 11). Since vascular tumours may represent an interesting model for the evaluation of angiostatic therapies, we evaluated the effect of PNU 153429 on tumour growth induced by s.c. injection of 0.5×10^6 pZipbFGF2-MAE cells into nude mice. PNU 153429 administered i.p. once a week at the dose of 100 mg/kg causes a significant delay in tumour take, resulting in inhibition of tumour growth (Fig. 12A).

To assess whether the inhibitory effect exerted by PNU 153429 on vascular lesions was due, at least in part, to its angiostatic activity, microvessel density was measured in tissue sections immunostained with anti-CD31 antibody to highlight newly formed blood vessels.

For each tumor section the most vascularized area was selected and CD31-positive microvessels in a $\times160$-field were counted. The results demonstrate that PNU 153429 causes ~50% decrease in CD31-positive microvessel density of pZipbFGF2-MAE cell-induced lesions (Fig. 12B).

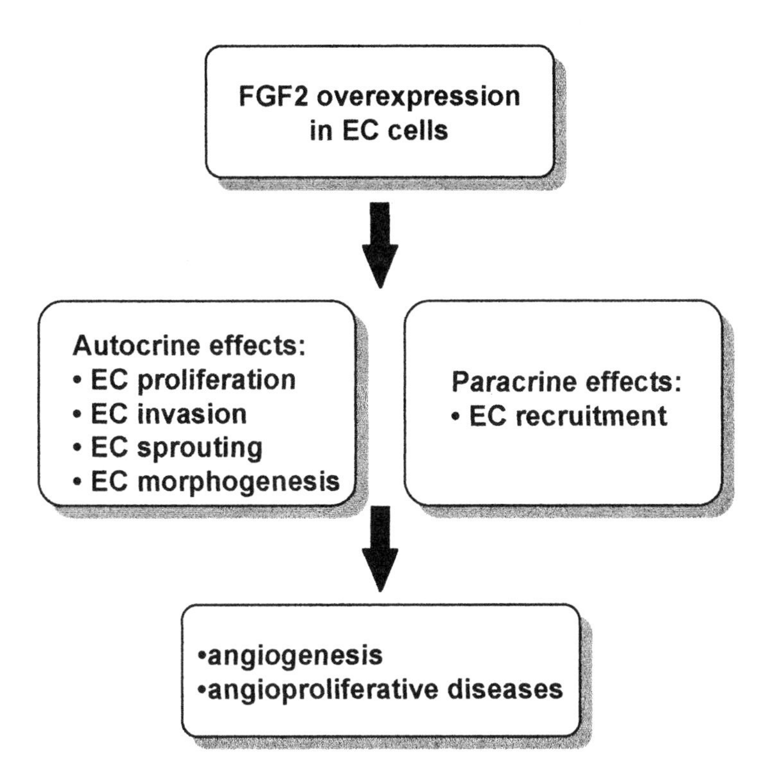

Figure 10. Biological consequences of FGF2 overexpression in endothelial cells

Figure 11. Chemical structure of PNU 153429

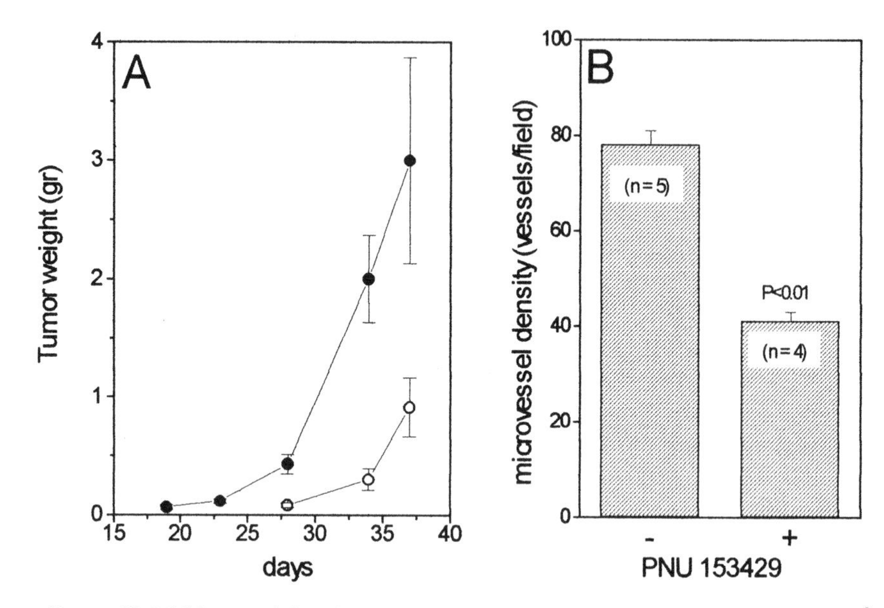

Figure 12. Inhibitory activity of PNU 153429. A) Mice were inoculated s.c. with 0.5×10^6 pZipbFGF2-MAE cells and treated i.p. with PNU 153429 at 100 mg/kg (closed circles) or with vehicle (open circles) once a week. B) Density of CD31-positive microvessels was evaluated on tumor biopsies obtained from control and treated animals.

In conclusion, the sulfonated distamycin A derivative PNU 153429 inhibits the growth and vascularization of pZipbFGF2-MAE cell-induced lesions. Previous observations had shown that the growth of vascular tumours induced by PmT-transformed endothelial cells (eEnd.1 cells) is susceptible to the inhibitory activity of the matrix metalloprotease inhibitor batimastat (Taraboletti *et al.*, 1995) and that batimastat is able to inhibit *in vitro* endothelial cell chemoinvasion and *in vivo* neovascularization induced by eEnd.1 cell supernatant (Taraboletti *et al.*, 1995). Interferon α-2a, that inhibits angiogenesis in mice, induces regression of hemangiomas of infancy in humans (Ezekowitz *et al.*, 1992). Taken together, the data indicate that vascular tumours are suitable targets for angiostatic therapy.

The combination of angiogenesis inhibitors have been hypothesized to be useful in tumour treatment (Brem *et al.*, 1993). Also, anti-angiogenic agents interact in a positive manner with cytotoxic therapies (Teicher *et al.*, 1996). Relevant to this point, preliminary observations have shown that doxorubicin (four i.v. injections at 6 mg/kg given at 4-day interval starting at day 14) exerts an inhibitory activity on the growth of pZipbFGF2-MAE cell-induced lesions similar to that exerted by PNU 153429 causing however only a limited decrease in microvessel density (F. Sola, M.P. Molinari Tosatti, and M. Presta, unpublished observations). This suggests that the two molecules may exert their inhibitory activity on pZipbFGF2-MAE cell-induced lesions *via* different mechanisms of action. The opportunistic vascular tumours caused by pZipbFGF2-MAE cells may therefore represent an useful model for the study of angiostatic compounds in single agent or combination therapy of vascular tumours, including KS.

4.2 Purine analogues as angiogenesis inhibitors

The use of purine analogue antimetabolites in combination chemotherapy of solid tumours has been proposed. To assess the possibility that selected purine analogues may affect tumour neovascularization, 6-methylmercaptopurine riboside (6-MMPR), 6-methylmercaptopurine, 2-aminopurine, and adenosine were evaluated for the capacity to inhibit angiogenesis *in vitro* and *in vivo* (Fig. 13).

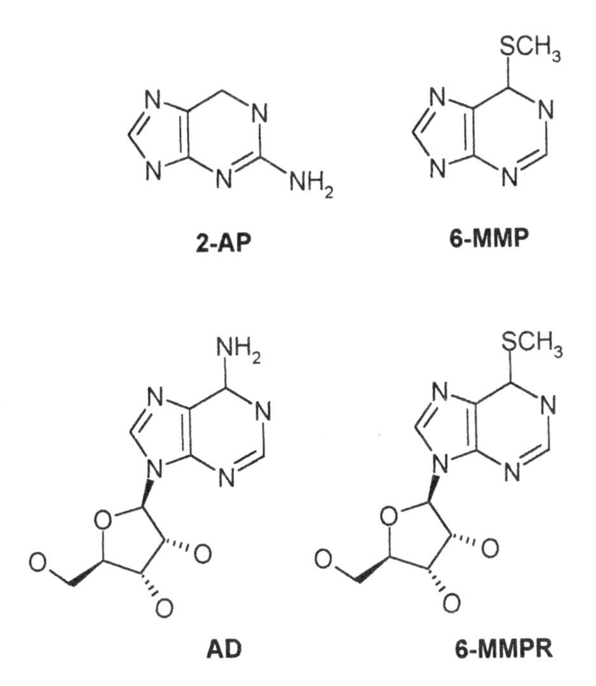

Figure 13. Structure of the purine analogues

6-MMPR inhibited FGF2-induced proliferation and delayed the repair of mechanically wounded monolayer in endothelial GM 7373 cell cultures. 6-MMPR also inhibited the formation of solid sprouts within fibrin gel by FGF2-treated murine brain microvascular endothelial cells and the formation of capillary-like structures on Matrigel by murine aortic endothelial cells transfected with FGF2 cDNA. 6-MMPR affected FGF2-induced intracellular signaling in murine aortic endothelial cells by inhibiting the phosphorylation of extracellular signal-regulated kinase-2. The other molecules were ineffective in all the assays. *In vivo*, 6-MMPR inhibited vascularization in the chick embryo chorioallantoic membrane (CAM) and prevented blood vessel formation induced by human endometrial adenocarcinoma specimens grafted onto the CAM. Also, topical administration of 6-MMPR caused the regression of newly formed blood vessels in the rabbit cornea (Presta et al., 1999). Thus, 6-MMPR specifically inhibits both the early and the late phases of the angiogenesis process *in vitro* and exerts a potent anti-angiogenic activity *in vivo*. These results suggest a new rationale for the use of selected purine analogues in combination therapy of solid cancer.

5. PARACRINE MECHANISM OF ACTION OF FGF2

Various tumour cell lines express FGF2 in vitro (Presta *et al.*, 1986; Moscatelli *et al.*, 1986; Halaban *et al.*, 1993, Okumura *et al.*, 1989; Nakano *et al.*, 1992). In situ hybridization and immunolocalization experiments have shown the presence of FGF2 mRNA and/or protein in neoplastic cells, endothelial cells, and infiltrating cells within human tumours of different origin (Schulze-Osthoff *et al.*, 1990; Zagzag *et al.*, 1990; Ohtani *et al.*, 1993; Takahashi *et al.*, 1990; Statuto *et al.*, 1993). Antisense-FGF2 and FGF receptor-1 cDNAs inhibit neovascularization and growth of human melanomas in nude mice (Wang and Becker, 1997). Also, a significant correlation between the presence of FGF2 in cancer cells and advanced tumor stage has been reported (Yamanaka *et al.*, 1993).

In the last few years, various angiogenic factors other than FGF2 have been identified. Among them, VEGF appears to play an important role in tumor neovascularization (Martiny and Marmè, 1995). Indeed, VEGF antagonists, including neutralizing antibodies (Kim *et al.*, 1993), antisense-VEGF cDNA (Saleh *et al.*, 1996), and dominant-negative VEGF receptor mutant (Millauer *et al.*, 1994), can inhibit tumour growth in different experimental models. Also, VEGF levels in tumour biopsies correlate with blood vessel density of the neoplastic tissue and may be of prognostic significance (Samoto *et al.*, 1994; Takahashi *et al.*, 1996).

At variance with VEGF, FGF2 lacks a hydrophobic signal sequence needed to enter the secretory pathway and is usually poorly secreted by producing cells. On the other hand FGF2 may be released by an alternative secretion pathway Mignatti *et al.*, 1991b; Mignatti *et al.*, 1992) and accumulates in the extracellular matrix (ECM), from where it is mobilised by ECM-degrading enzymes (Yeoman 1993) FGF2 is detectable in urine of patients with a wide spectrum of cancers (Chodak *et al.*, 1988; Nguyen *et al.*, 1994) and in cerebrospinal fluid of children with brain tumours (Li *et al.*, 1994). Interestingly, the appearance of an angiogenic phenotype correlates with the export of FGF2 during the development of fibrosarcoma in a transgenic mouse model (Kandell *et al.*, 1991). These data suggest that FGF2 release may occur in vivo and may influence solid tumour growth and neovascularization by autocrine and paracrine modes of action. Accordingly, neutralizing anti-FGF2 antibodies affect tumour growth under defined experimental conditions (Czubayko *et al.*, 1997; Rak and Kerbel, 1997).

Relevant to this point is the recent observation that a secreted FGF-binding protein that mobilises stored extracellular FGF2 can serve as an angiogenic switch for different tumour cell lines, including squamous cell carcinoma and colon cancer cells (Czubayko *et al.*, 1997). Interestingly, targeting of FGF-binding protein with specific ribozymes reduces significantly the growth and vascularization of xenografted tumours in mice (Czubayko *et al.*, 1997) despite the high levels of VEGF produced by these cells (see Rak and Kerbel, 1997 for a further discussion). These data suggest that modulation of FGF2 expression, release, and mobilisation may allow a fine tuning of the angiogenesis process even in the presence of significant levels of VEGF. This hypothesis is supported by the capacity of the two factors to act synergistically in stimulating angiogenesis in vitro and in vivo (Goto *et al.*, 1993; Asahara *et al.*, 1995).

Recently we have described an export-dependent mechanism of action for FGF2 in the human endometrial adenocarcinoma HEC-1B cell line (Coltrini *et al.*, 1995). After transfection with an expression vector harbouring a human FGF2 cDNA, different FGF2-expressing clones were obtained and one of them (FGF2-B9 clone) showed the capacity to secrete significant amounts of the growth factor. This clone stimulates angiogenesis in the avascular rabbit cornea and forms highly vascularized tumours growing faster than the non-FGF2 releasing clones (FGF2-A8 and FGF2-B8 clones) when transplanted s.c. into nude mice line (Coltrini *et al.*, 1995). FGF2-transfected HEC-1-B clones may therefore represent an useful experimental model to study the effects on the microvascular architecture of the neoplastic tissue consequent to modifications of tumour microenvironment due to FGF2 expression and release (Fig. 14).

We also studied the microvascular pattern of tumours grown in nude mice originated from HEC-1B cell clones expressing and secreting different levels

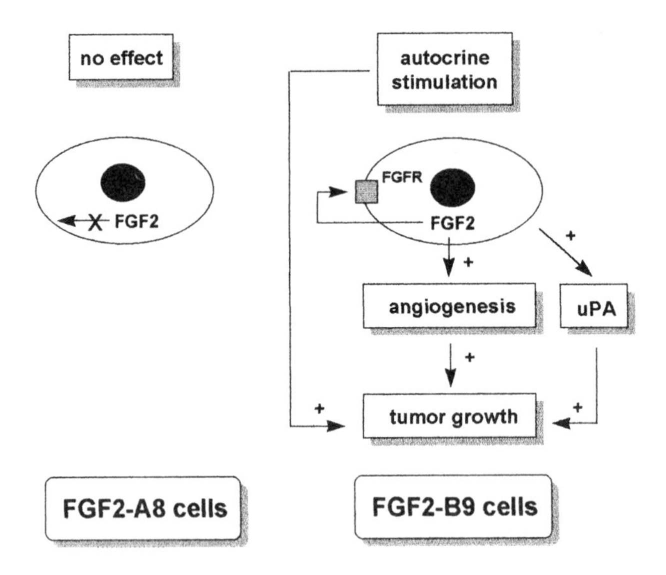

Figure 14. FGF2 export decides the biological behaviour of FGF2-transfected HEC-1-B cells

of FGF2 by means of the microvascular corrosion casting technique. Microvascular corrosion casting allows for qualitative as well as quantitative insights into the tumour vascular system. All relevant parameters defining the microvascular network, such as interbranching distances, intervascular distances, branching angles, and vessel diameters can be determined using 3D stereo pairs of the casted tumour vascularity (Malkusch *et al.*, 1995). We used this powerful method to assess whether differences in FGF2 expression and secretion may differently affect the microvascular architecture of the tumor in our experimental model (Konerding et al., 1998). FGF2-B9 cells grew faster in nude mice compared to FGF2-A8 and FGF2-B8 clones. The total amount of new vessel formation was higher in FGF2-B9 tumours than in FGF2-B8 or FGF2-A8 tumours. Also, vessel courses were more irregular and blind ending vessels and evasates were more frequent in FGF2-B9 tumours (Fig. 15).

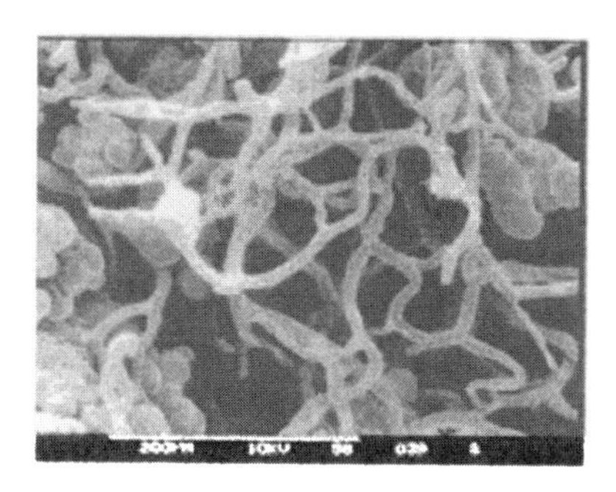

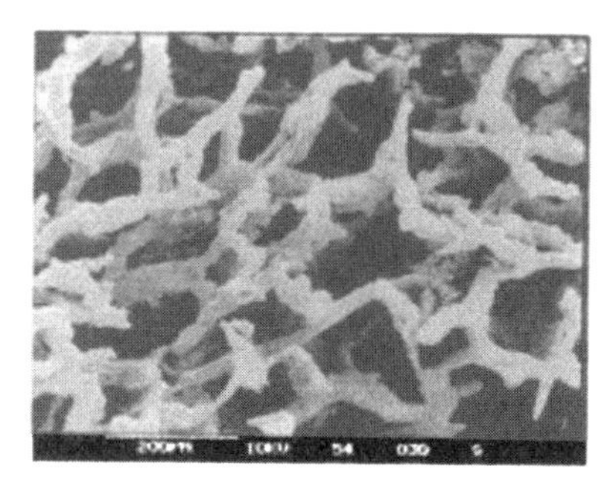

Figure 15. Corrosion casts of FGF2-B8 (top) and FGF2-B9 (bottom) tumours

Moreover, FGF2-B9 tumour microvasculature was characterized by a wider average vascular diameter and by an extreme variability of the diameter of each individual vessel along its course between two ramifications. No statistical differences were instead observed when the distribution curves of the values of intervascular distances, interbranching distances, and branching angles of the microvessel network were compared among the different experimental groups. The distinctive features of the microvasculature of FGF2-B9 tumours were retained, at least in part, in the smaller lesions produced by injection of a limited number of cells.

The data indicate that FGF2 production and release confer to FGF2-B9 cells the ability to stimulate the formation of new blood vessels with distinctive morphological features. Neovascularization of FGF2-B9 lesions parallels the faster rate of growth of the neoplastic parenchyma. This does not affect the overall architecture of the microvessel network that appears to be primed by characteristics of the HEC-1-B tumour cell line and/or by the microenvironment of the host.

6. CONCLUDING REMARKS

FGF2 exerts angiogenic activity *in vivo* and induce a pro-angiogenic phenotype in cultured endothelial cells. In vivo, FGF2 exerts paracrine effects on endothelial cells when released by tumour and/or inflammatory cells. FGF2 may also play an autocrine role in endothelial cells in vitro and in vivo (Fig. 16).

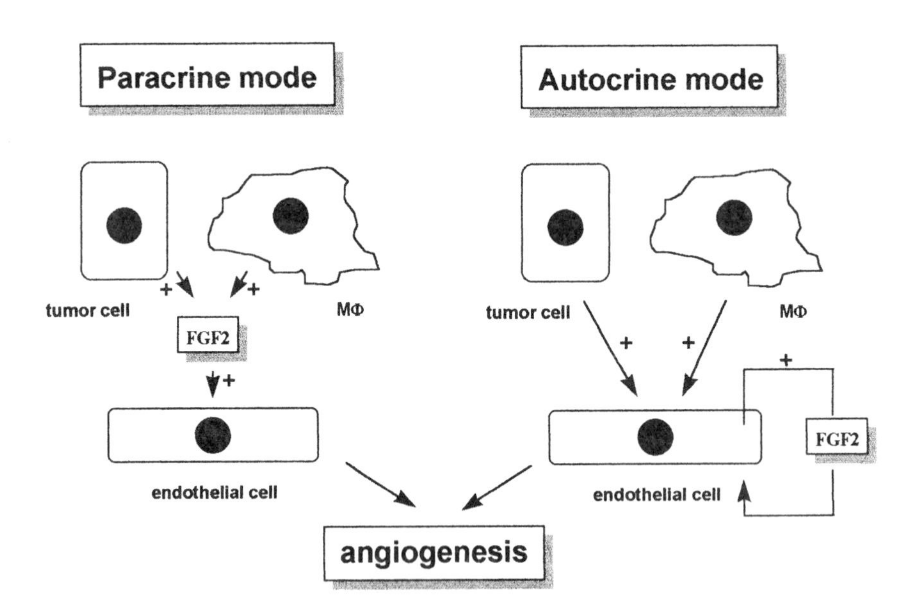

Figure 16. Paracrine and autocrine activity of FGF2 on endothelium. Tumour cells and inflammatory cells (Mϕ) release FGF2 which acts on endothelial cells in a paracrine mode of action. Alternatively, endogenous FGF2 is upregulated in endothelial cells causing an autocrine loop of stimulation.

FGF2 may therefore represent a target for anti-angiogenic therapies. In order to assess the angiostatic potential of different classes of compounds, novel experimental models have been developed based on the autocrine and/or the paracrine capacity of FGF2.

ACKNOWLEDGMENTS

The work from our laboratory described in the present paper was supported by grants from Associazione Italiana per la Ricerca sul Cancro, Istituto Superiore di Sanità (AIDS Project), European Communities (Human Capital Mobility Project "Mechanisms for the Regulation of Angiogenesis"), C.N.R. (Target Project Biotechnology), and M.U.R.S.T (Project "Inflammation: Biology and Clinics") to M.P.

REFERENCES

Abraham, J.A., Mergia, A., Whang, J.L., Tumolo, A., Friedman, J., Hjerrild, J., Gospodarowicz, D., and Fiddes, J., 1986, Nucleotide sequence of a bovine clone encoding the angiogenic protein, basic fibroblast growth factor, *Science* **233**:545-548.

Albini, A., Fontanini, G., Masiello, L., Tacchetti, C., Bigini, D., Luzzi, P., Noonan, D.M., and Stetler-Stevenson, W.G., 1994, Angiogenic potential in vivo by Kaposi sarcoma cell-free supernatants and HIV1-tat product: inhibition of KS-like lesions by TIMP-2, *AIDS* **8**:1237-1244.

Asahara, T., Bauters, C., Zheng. L.P., Takeshita, S., Bunting. S., Ferrara, N., Symes, J.F., and Isner, J.M., 1985, Synergistic effect of vascular endothelial growth factor and basic fibroblast growth factor on angiogenesis in vivo, *Circulation* **92**:365-371.

Baird, A., Mormède, P., and Bohlen, P., 1985, Immunoreactive fibroblast growth factor in cells of peritoneal exudate suggests its identity with macrophage-derived growth factor, *Biochem. Biophys. Res. Commun.* **126**:358-364.

Baird, A., Mormède, P., and Bohlen, P., 1986, Immunoreactive fibroblast growth factor (FGF) in a transplantable chondrosarcoma: inhibition of tumor growth by antibodies to FGF, *J. Cell. Biochem.* **30**:79-85.

Barillari, G., Buonaguro, L., Fiorelli, V., Hoffman, J., Michaels, F., Gallo, R.C., and Ensoli, B., 1992, Effects of cytokines from activated immune cells on vascular cell growth and HIV-1 gene expression, *J. Immunol.* **149**:3727-3734.

Basilico, C., and Moscatelli, D., 1992, The FGF family of growth factors and oncogenes, *Adv. Cancer Res.* **59**:115-165.

Blotnick, S., Peoples, G.E., Freeman, M.R., Eberlein, T.J., and Klagsbrun, M., 1994,. T lymphocytes synthesize and export heparin-binding epidermal growth factor-like growth factor and basic fibroblast growth factor, mitogens for vascular cells and fibroblasts:

differential production and release by CD4+ and CD8+ T cells, *Proc. Natl. Acad. Sci. U.S.A.* **91**:2890-2894.

Braddock, P.S., Hu, D.-E., Fan, T.-P.D., Stratford, I.J., Harris, A.L., and Bicknell, R.A., 1994, A structure-activity analysis of antagonism of the growth factor and angiogenic activity of basic fibroblast growth factor by suramin and related polyanions, *Br. J. Cancer* **69**:890-898.

Brem, H., Gresser, I., Grosfeld, J., and Folkman, J., 1993, The combination of antiangiogenic agents to inhibit primary tumor growth and metastasis, *J. Pediatr. Surg.* **28**: 1253-1257.

Broadly, K.N., Aquino, A.M., Woodward, S.C., Buckley-Sturrock, A., Sato, Y., Rifkin, D.B., and Davidson, J.M., 1989, Monospecific antibodies implicate basic fibroblast growth factor in normal wound repair, *Lab. Invest.* **61**:571-575.

Chodak, G.W., Hospelhorn, V., Judge, S.M., Mayforth, R., Koeppen, H., and Sasse, J., 1988, Increased levels of fibroblast growth factor-like activity in urine from patients with bladder or kidney cancer, *Cancer. Res.* **48**:2083-2088.

Ciomei, M., Pastori, W., Mariani, M., Sola, F., Grandi, M., and Mongelli, N., 1994, New sulfonated distamycin A derivatives with bFGF complexing activity, *Biochem. Pharmacol.* **47**:296-302.

Coltrini, D., Rusnati, M., Zoppetti, G., Oreste, P., Grazioli, G., Naggi, A., and Presta, M., 1994, Different effects of mucosal, bovine lung and chemically modified heparin on selected biological properties of basic fibroblast growth factor, *Biochem. J.* **303**:583-590.

Coltrini, D., Gualandris, A., Nelli, E.E., Parolini, S., Molinari-Tosatti, M.P., Quarto, N., Ziche, M., Giavazzi, R., Presta, M., 1995, Growth advantage and vascularization induced by basic fibroblast growth factor overexpression in endometrial HEC-1-B cells: an export-dependent mechanism of action, *Cancer Res.* **55**:4729-4738.

Cozzolino, F., Torcia, M., Lucibello, M., Morbidelli, L., Ziche, M., Platt, J., Fabiani, S., Brett, J., and Stern, D., 1993, Cytokine-mediated control of endothelial cell growth: interferon-α and interleukin-2 synergistically enhance basic fibroblast growth factor synthesis and induce release promoting cell growth in vitro and in vivo, *J. Clin. Invest.* **91**:2504-2512.

Czubayko, F., Liaudet-Coopman, E.D.E., Aigner, A., Tuveson, A.T., Berchem, G.J., and Wellstein, A., 1997, A secreted FGF-binding protein can serve as the angiogenic switch in human cancer, *Nature Medicine* **3**:1137-1140.

DiMario, J., Buffinger, N., Yamada, S., and Strohman, R.C., 1989, Fibroblast growth factor in the extracellular matrix of dystrophic (mdx) mouse muscle, *Science* **244**:688-690.

Ensoli, B., Gendelman, R., Markham, P., Fiorelli, V., Colombini, S., Raffeld, M., Cafaro, A., Chang, H.K., Brady, J.N., and Gallo, R.C., 1994a, Synergy between basic fibroblast growth factor and HIV-1 Tat protein in induction of Kaposi's sarcoma, *Nature* **371**:674-680.

Ensoli, B., Markham, P., Kao, V., Barillari, G., Fiorelli, V., Gendelman, R., Raffeld, M., Zon, G., and Gallo, R.C., 1994b, Block of AIDS-Kaposi's sarcoma (KS) cell growth, angiogenesis, and lesion formation in nude mice by antisense oligonucleotide targeting basic fibroblast growth factor. A novel strategy for the therapy of KS, *J. Clin. Invest.* **94**:1736-1746.

Enzinger, F.M., and Weiss S.W., 1995, *Soft tissue tumors*, pp. 579-677, Mosby-Year Book Inc., St. Louis.

Ezekowitz, R.A.B., Mulliken, J.B., and Folkman, J., 1992, Interferon alfa-2a therapy for life-threatening hemangiomas of infancy, *New Engl. J. Med.* **326**: 1456-1463.

Fiorelli, V., Gendelman, R., Samaniego, F., Markham, P.D., and Ensoli, B., 1995, Cytokines from activated T cells induce normal endothelial cells to acquire the phenotypic and functional features of AIDS-Kaposi's sarcoma spindle cells, *J. Clin. Invest.* **95**:1723-1734.

Firsching, A., Nickel, P., Mora, P., and Allolio, B., 1995, Antiproliferative and angiostatic activity of suramin analogues, *Cancer Res.* **55**:4975-4961.

Florkiewicz, R.Z., and Sommer, A., 1989, Human basic fibroblast growth factor gene encodes four polypeptides: three initiate translation from non-AUG codons, *Proc. Natl. Acad. Sci. U.S.A.* **86**:3978-3981.

Folkman, J., Klagsbrun, M., Sasse, J., Wadzinski, M., Ingber, D., and Vlodavski, I., 1988. A heparin-binding angiogenic protein -basic fibroblast growth factor- is stored within basement membran,. *Am. J. Pathol.* **130**:393-400.

Gagliardi, A., Hadd, H., and Collins, D.C., 1992, Inhibition of angiogenesis by suramin, *Cancer Res.* **52**:5073-5075.

Gagliardi, A.R., and Collins, D.C., 1994, Inhibition of angiogenesis by aurintricarboxylic acid, *Anticancer Res.* **14**:475-479.

Gajdusek, C.M., and Carbon, S., 1989, Injury-induced release of basic fibroblast growth factor from bovine aortic endothelium, *J. Cell. Physiol.* **139**:570-579.

Gannoun-Zaky, L., Pieri, I., Badet, J., Moenner, M., Barritault, D., 1991, Internalization of basic fibroblast growth factor by chinese hamster lung fibroblast cells: involvement of several pathways, *Exp. Cell Res.* **197**:272-279.

Gao, G., and Goldfarb, M., 1995 Heparin can activate a receptor tyrosine kinase, *EMBO J.* G:2183-2190.

Goto, F., Goto, K., Weindel, K., and Folkman, J., 1993, Synergistic effects of vascular endothelial growth factor and basic fibroblast growth factor on the proliferation and cord formation of bovine capillary endothelial cells within collagen gels, *Lab. Invest.* **69**:508-517.

Gross, J.L., Herblin, W.F., Dusak, B.A., Czerniak, P., Diamond, M.D., Sun, T., Eidsvoog, K., Dexter, D.L., and Yayon, A., 1993, Effects of modulation of basic fibroblast growth factor on tumor growth in vivo, *J. Natl. Cancer Inst.* **85**:121-131.

Gualandris, A., Urbinati, C., Rusnati, M., Ziche, M., and Presta, M., 1994, Interaction of high molecular weight basic fibroblast growth factor (bFGF) with endothelium: biological activity and intracellular fate of human recombinant Mr 24,000 bFGF, *J. Cell. Physiol.* **161**:149-159.

Gualandris, A., Rusnati, M., Belleri, M., Nelli, E.E., Bastaki, M., Molinari-Tosatti, M.P., Bonardi, F., Parolini, S., Albini, A., Morbidelli, L., Ziche, M., Corallini, A., Possati, L., Vacca, A., Ribatti, D., and Presta, M., 1996a, Basic fibroblast growth factor overexpression in endothelial cells: an autocrine mechanism for angiogenesis and angioproliferative diseases, *Cell Growth & Differ.* 7:147-160.

Gualandris, A., Rusnati, M., Belleri, M., Molinari-Tosatti, M.P., Bonardi, F., Parolini, S., Albini, A., Ziche, M., and Presta, M., 1966b, Angiogenic phenotype induced by basic fibroblast growth factor transfection in brain microvascular endothelial cells: an in vitro autocrine model of angiogenesis in brain tumors, *Int. J. Oncol.* **8**:567-573.

Guimond, S., Maccarana, M., Olwin, B.B., Lindahl, U., and Rapraeger, A.C., 1993, Activating and inhibitory heparin sequences for FGF-2 (basic FGF), *J. Biol. Chem.* **268**:23906-23914.

Halaban, R., Kwon, B.S., Ghosh, S., Delli-Bovi, P., and Baird, A., 1993, bFGF as an autocrine growth factor for human melanomas, *Oncogene Res.* **3**:177-186.

Hawker, J.R.H., and Granger, J., 1993, Tyrosine kinase inhibitors impair fibroblast growth factor signaling in coronary endothelial cells, *Am. J. Physiol.* **266**:H107-H120.

Ikeda, S., Neyts, J., Verma, S., Wickramasinghe, A., Mohan, P., and De Clercq, E., 1994, *In vitro* and *in vivo* inhibition of ortho- and paramyxovirus infections by a new class of sulfonic acid polymers interacting with virus-cell binding and/or fusion, *Antimicrob. Agents Chemother.* **38**:256-259.

Ishihara, M., Tyrrell, D.J., Stauber, G.B., Brown, S., Cousens, L.S., and Stack, R.J., 1993, Preparation of affinity-fractionated, heparin-derived oligosaccharides and their effects on selected biological activities mediated by basic fibroblast growth factor, *J. Biol. Chem.* **268**:4675-4683.

Itoh, H., Mukoyama, M., Pratt, R.E., and Dzau, V.J., 1992, Specific blockade of basic fibroblast growth factor gene expression in endothelial cells by antisense oligonucleotide, *Biochem. Biophys. Res. Commun.* **188**:1205-1213.

Johnson, D.E., and Williams, L.T. 1993, Structural and functional diversity in the FGF receptor multigene family, *Adv. Cancer Res.* **60**:1-41.

Kandell, J., Bossy-Wetzei, E., Radvanyi, F., Klagsbrun, M., Folkman, J., and Hanahan, D., 1991, Neovascularization is associated with a switch to the export of bFGF in the multistep development of fibrosarcoma, *Cell* **66**:1095-1104.

Kim, K.J., Li, B., Winer, J., Armanini, M., Gillett, N., Phillips, H.S., and Ferrara, N., 1993, Inhibition of vascular endothelial growth factor-induced angiogenesis suppresses tumor growth in vivo, *Nature* **362**:841-844.

Kan, M., Wang, F., Xu, J., Crabb, J.W., Hou, J., and McKeehan, L.W., 1993, An essential heparin-binding domain in the fibroblast growth factor receptor kinase, *Science* **259**:1918-1921.

Klein, S., Giancotti, F.G., Presta, M., Albelda, S.M., Buck, C.A., and Rifkin, D.B., 1993, Basic fibroblast growth factor modulates integrin expression in microvascular endothelial cells, *Mol. Biol. Cell.* **4**:973-982.

Konerding, M.A., Miodonski, A.J., and Lametschwandtner, A., 1995, Microvascular corrosion casting in the study of tumor vascularity: a review, *Scanning Microsc.* **9**:1233-1244.

Konerding, M.A., Fait, E., Dimitropoulou, C., Malkusch, W., Ferri, C., Giavazzi, R., Coltrini, D., and Presta, M., 1998, Impact of fibroblast growth factor-2 on tumor microvascular architecture. A tridimensional morphometric study, Am. J. Pathol. **152**:1607-1616.

Levine, A.M., 1993, AIDS-related malignancies: the emerging epidemic, *J. Natl. Cancer Inst.* **85**:1382-1387.

Li, V.W., Folkerth, R.D., Watanabe, H., Yu, C., Rupnick, M., Barnes, P., Scott, R.M., Black, P.M., Sallan, S.E., and Folkman, J., 1994, Microvessel count and cerebrospinal fluid basic fibroblast growth factor in children with brain tumors, *Lancet* **344**:82-86.

Liekens, S., Neyts, J., Degrève, B., and De Clercq, E., 1997, The sulfonic acid polymers PAMPS [poly(acrylamido-2-methyl-1-propanesulfonic acid)] and related analogues are highly potent inhibitors of angiogenesis, *Oncol. Res.* **9**:173-181.

Liekens, S., Leali, D., Neyts, J., Esnouf, R., Rusnati, M., Dell'Era, P., Maudgal, P.C., De Clercq, E., and Presta, M., 1999, Modulation of fibroblast growth factor-2 receptor binding, signaling, and mitogenic activity by heparin-mimicking polysulfonated compounds, *Mol. Pharmacol.* in press.

Malkusch, W., Konerding, M.A., Klapthor, B., and Bruch, J., 1995, A simple and accurate method for 3-D measurements in microcorrosion casts illustrated with tumor vascularization, *Anal. Cell. Pathol.* **9**:69-81.

Martiny-Baron, G., and Marmé, D., 1995, VEGF-mediated tumour angiogenesis: a new target for cancer therapy, *Curr. Opinion. Biotech.* **6**:675-680.

McNeil, P.L., Muthukrishnan, L., Warder, E., and D'Amore, P., 1989, Growth factors are released by mechanically wounded endothelial cells, *J. Cell Biol.* **109**:811-822.

Miao, H.-Q., Ornitz, D.M., Aingorn, E., Ben-Sasson, S.A., and Vlodavsky, I., 1997, Modulation of fibroblast growth factor-2 receptor binding, dimerization, signaling and angiogenic activity by a synthetic heparin-mimicking polyanionic compound, *J. Clin. Invest.* **99**:1565-1575.

Mignatti, P., Tauboi, R., Robbins, E., and Rifkin, D.B., 1989, In vitro angiogenesis on the human amniotic membrane: requirement for basic fibroblast growth factor, *J. Cell Biol.* **108**:671-682.

Mignatti, P., Mazzieri, R., and Rifkin, D.B., 1991a, Expression of the urokinase receptor in vascular endothelial cells is stimulated by basic fibroblast growth factor, *J. Cell Biol.* **113**:1193-1201.

Mignatti, P., Morimoto, T., and Rifkin, D.B., 1991b, Basic fibroblast growth factor released by single, isolated cells stimulates their migration in an autocrine manner, *Proc. Natl. Acad. Sci. U.S.A.* **88**:11007-11011.

Mignatti, P., Morimoto, T., and Rifkin, D.B., 1992, Basic fibroblast growth factor, a protein devoid of secretory signal sequence, is released by cells via a pathway independent of the endoplasmic reticulum-Golgi complex, *J. Cell. Physiol.* **151**:81-93.

Millauer, B., Shawver, K.L., Plate, K.H., Risau, W., and Ullrich, A., 1994, Glioblastoma growth inhibited in vivo by a dominant-negative Flk-1 mutant, *Nature* **367**:576-579.

Montesano, R., Vassalli, J.D., Baird, A., Guillemin, R., and Orci, L., 1986, Basic fibroblast growth factor induces angiogenesis in vitro, *Proc. Natl. Acad. Sci. U.S.A.* **83**:7297-7301.

Moscatelli, D., Presta, M., Joseph-Silverstein, J., and Rifkin, D.B., 1986, Both normal and tumor cells produce basic fibroblast growth factor, *J. Cell. Physiol.* **129**:273-276.

Mohan, P., Schols, D., Baba, M., and De Clercq, E., 1992, Sulphonic acid polymers as a new class of human immunodeficency virus inhibitors, *Antiviral Res.* **18**:139-150.

Myers, C., Cooper, M., Stein, C., LaRocca, R., McClellan, M.W., Weiss, G., Choyke, P., Dawson, N., Steinberg, S., Uhrich, M.M., Cassisy, J., Kohler, D.R., Trepel, J., and Linehan, M., 1992, Suramin: a novel growth factor antagonist with activity in hormone-refractory metastatic prostate cancer, *J. Clin. Oncol.* **10**:881-889.

Nakamoto, T., Chang, C., Li, A., and Chodak, G.W., 1992, Basic fibroblast growth factor in human prostate cancer cells, *Cancer Res.* **52**:571-577

Nguyen, M., Watanabe, H., Budson, A.E., Richie, J.P., Hayes, D.F., and Folkman, J., 1994, Elevated levels of an angiogenic peptide, basic fibroblast growth factor, in the urine of patients with a wide spectrum of cancers, *J. Natl. Cancer. Inst.* **86**:356-361.

Norbby, K., and Ostergaard, P., 1996, Basic-fibroblast-growth-factor-mediated de novo angiogenesis is more effetively suppressed by low-molecular weight than by high-molecular-weight heparin. *Int. J. Microcirc. Clin. Exp.* **16**:8-15.

Ohtani, H., Nakamura, S., Watanabe, Y., Mizoi, T., Saku, T., and Nagura, H., 1993, Immunocytochemical localization of basic fibroblast growth factor in carcinomas and inflammatory lesions of the human digestive tract, *Lab. Invest.* **68**:520-527.

Okumura, N., Takimoto, K., Okada, M., and Nakagawa, H., 1989, C6 glioma cells produce basic fibroblast growth factor that can stimulate their own proliferation, *J. Biochem* **106**:904-909.

Pepper, M.S., and Meda, P., 1992, Basic fibroblast growth factor increases junctional communication and connexin 43 expression in microvascular endothelial cells, *J. Cell. Physiol.* **153**:196-205.

Pepper, M.S., Sappino, A.P., Stocklin, R., Montesano, R., Orci, L., and Vassalli, J.D., 1993, Upregulation of urokinase receptor expression on migrating endothelial cells, *J. Cell Biol.* **122**:673-684.

Peverali, F.A., Mandriota, S.J., Ciana, P., Marelli, R., Quax, P., Rifkin,D.B., Della Valle, G., and Mignatti, P., 1994, Tumor cells secrete an angiogenic factor that stimulates basic fibroblast growth factor and urokinase expression in vascular endothelial cells, *J. Cell. Physiol.* **161**:1-14.

Presta, M., Moscatelli, D., Joseph-Silverstein, J., and Rifkin, D.B., 1986, Purification from a human hepatoma cell line of a basic fibroblast growth factor-like molecule that stimulates capillary endothelial cell plasminogen activator production, DNA synthesis, and migration, *Mol. Cell. Biol.* **6**:4060-4066.

Presta, M., Maier, J.A.M., Rusnati, M., and Ragnotti, G., 1989, Basic fibroblast growth factor: production, mitogenic response, and post-receptor signal transduction in cultured normal and transformed fetal bovine aortic endothelial cells, *J. Cell. Physiol.* **141**:517-526.

Presta, M., Rusnati, M., Belleri, M., Morbidelli, L., Ziche, M., and Ribatti, D., 1999, Purine analogue 6-methylmercaptopurine riboside inhibits early and late phases of the angiogenesis process, Cancer Res. **59**:2417-2424.

Rak, J., and Kerbel, R.S., 1997, bFGF and tumor angiogenesis - Back in the limelight?, *Nature Medicine* **3**:1083-1084.

Ribatti, D., Urbinati, C., Nico, B., Rusnati, M., Roncali, L., and Presta, M., 1995, Endogenous basic fibroblast growth factor in the vascularization of the chick embryo chorioallantoic membrane, *Dev. Biol.* **170**:39-49.

Richard, C., Liuzzo, J.P., and Moscatelli, D., 1995, Fibroblast growth factor-2 can mdiate cell attachment by linking receptors and heparan sulfate proteoglycans on neighboring cells, *J. Biol. Chem.* **270**:24188-24196.

Rogelj, S., Klagsbrun, M., Atzmon, R., Kurokawa, M., Haimovitz, A., Fuks, Z., and Vlodavski, I., 1989, Basic fibroblast growth factor is an extracellular matrix component required for supporting the proliferation of vascular endothelial cells and the differentiation of PC12 cells, *J. Cell Biol.* **109**:823-831.

Roghani, M., and Moscatelli, D., 1993, Basic fibroblast growth factor is internalized through both receptor-mediated and heparan sulfate-mediated mechanisms, *J. Biol. Chem.* **267**:22156-22162.

Rusnati, M., Urbinati, C., Presta, M., 1993, Internalization of basic fibroblast growth factor (bFGF) in cultured endothelial cells: role of the low affinity heparin-like bFGF receptors, *J. Cell. Physiol.* **154**:152-161.

Rusnati, M., Coltrini, D., Caccia, P., Dell'Era, P., Zoppetti, G., Oreste, P., Valsasina, B., Presta, M., 1994, Distinct role of 2-O-, N-, and 6-O-sulfate groups of heparin in the formation of the ternary complex with basic fibroblast growth factor and soluble FGF receptor-1, *Biochem. Biophys. Res. Commun.* **203**:450-458.

Rusnati, M., and Presta, M., 1996a, Interaction of angiogenic basic fibroblast growth factor with endothelial cell heparan sulfate proteoglycans, *Int. J. Clin. Lab. Res.* **26**:15-23.

Rusnati, M., Dell'Era, P., Urbinati, C., Tanghetti, E., Massardi, M.L., nagamine, Y., Monti, E., and Presta, M., 1996b, A distinct basic fibroblast growth factor (FGF-2/FGF receptor interaction distinguishes urokinase-type plasminogen activator induction from mitogenicity in endothelial cells, *Mol. Biol. Cell* **7**:369-381.

Saleh, M., Stacker, S.A., and Wilks, A.F., 1996, Inhibition of growth of C6 glioma cells in vivo by expression of antisense vascular endothelial growth factor sequence, *Cancer Res.* **56**:393-401.

Samoto, K., Ikezaki, K., Ono, M., Shono, T., Kohno, K., Kuwano, M., and Fukui, M., 1995, Expression of vascular endothelial growth factor and its possible relation with neovascularization in human brain tumors, *Cancer Res.* **55**:1189-1193.

Sato, Y., and Rifkin, D.B., 1988, Autocrine activities of basic fibroblast growth factor: regulation of endothelial cell movement, plasminogen activator synthesis, and DNA synthesis, *J. Cell Biol.* **107**:1199-1205.

Schulze-Osthoff, K., Risau, W., Vollmer, E., and Sorg, C., 1990, In situ detection of basic fibroblast growth factor by highly specific antibodies, *Am. J. Pathol.* **137**:85-92.

Schweigerer, L. Neufeld, G., Friedman, J., Abraham, J.A., Fiddes, J.C., and Gospodarowicz, D., 1987, Capillary endothelial cells express basic fibroblast growth factor, a mitogen that promotes their own growth, *Nature* **325**:257-259.

Speir, E., Tanner, V., Gonzales, A.M., Farris, J., Baird, A., and Casscells, W., 1992, Acid and basic fibroblast growth factors in adult rat heart myocytes: localization, regulatio in culture, and effects on DNA synthesis, *Circ. Res.* **71**:251-259.

Statuto, M., Ennas, M.G., Zamboni, G., Bonetti, F., Pea, M., Bernardello, F., Pozzi, A., Rusnati, M., Gualandris, A., and Presta, M., 1993, Basic fibroblast growth factor in human pheochromocytoma: a biochemical and immunohistochemical study, *Int. J. Cancer.* **53**:5-10.

Sturzl, M., Brandstetter, H., and Roth, W.K., 1992, Kaposi's sarcoma: a review of gene expression and ultrastructure of KS spindle cells in vivo, *AIDS Res. Human Retrov.* **8**:1753-1763.

Takahashi, J.A., Mori, H., Fukumoto, M., Igarashi, K., Jaye, M., Oda, Y., Kikuchi, H., and Hatanaka, M., 1990, Gene expression of fibroblast growth factors in human gliomas and meningiomas: demonstration of cellular source of basic fibroblast growth factor mRNA and peptide in tumor tissues, *Proc. Natl. Acad. Sci. U.S.A.* **87**:5710-5714.

Takahashi, K., Mulliken, J.B., Kozakewich, H.P.W., Rogers, R.A., Folkman, J., and Ezekowitz, R.A.B., 1994, Cellular markers that distinguish the phases of hemangioma during infancy and childhood, *J. Clin. Invest.* **93**:2357-2364.

Takahashi, Y., Cleary, K.R., Mai, M., Kitadai, Y., Bucana, C.D., and Ellis, L.M., 1996, Significance of vessel count and vascular endothelial growth factor ant its receptor (KDR) in intestinal-type gastric cancer, *Clin, Cancer Res,* **2**:1679-1684.

Takano, S., Gately, S., Neville, M.E., Herblin, W.F., Gross, J.L., Engelhard, H., Perricone, M., Eidsvoog, K., and Brem, S., 1994, Suramin, an anticancer and angiosuppressive agent, inhibits endothelial cell binding of basic fibroblast growth factor, migration, proliferation and induction of urokinase-type plasminogen activator, *Cancer Res.* **54**:2654-2660.

Taraboletti, G., Garofalo, A., Belotti, D., Drudis, T., Borsotti, P., Scanziani, E., Brown, P.D., and Giavazzi, G. 1995, Inhibition of angiogenesis and murine hemangioma growth by batimastat, a synthetic inhibitor of matrix metalloproteinases, *J. Natl. Cancer Inst.* **87**: 293-298.

Teicher, B.A., Holden, S.A., Ara, G., Korbut, T., and Menon, K., 1996, Comparison of several antiangiogenic regimens alone and with cytotoxic therapies in the Lewis lung carcinoma, *Cancer Chemother. Pharmacol.* **38**: 169-177.

Turnbull, J.E., and Gallagher, J.T., 1993, Heparan sulfate: functional role as modulator of fibroblast growth factor activity, *Biochem. Soc. T.* **21**:477-482.

Vlodavski, I., Folkman, J., Sullivan, R., Friedman, R., Ishai-Michaell, R., Sasse, J., and Klagsbrun, M., 1987a, Endothelial cell-derived basic fibroblast growth factor: synthesis and deposition into subendothelial extracellular matrix, *Proc. Natl. Acad. Sci. U.S.A.* **84**:2292-2296.

Vlodavski, I., Friedman, R., Sullivan, R., Sasse, J., and Klagsbrun, M., 1987b, Aortic endothelial cells synthesize basic fibroblast growth factor which remains cell associated and platelet-derived growth factor-like protein which is secreted, *J. Cell. Physiol.* **131**:402-408.

Waltz, T.M., Abdiu, A., Wingren, S., Smeds, S., Larsson, S.E., and Wasteson, A., 1991, Suramin inhibits growth of human osteosarcoma xenografts in nude mice, *Cancer Res.* **51**:3585-3589.

Wang, Y., and Becker, D., 1997, Antisense targeting of basic fibroblast growth factor and fibroblast growth factor receptor-1 in human melanomas blocks intratumoral angiogenesis and tumor growth, *Nature Medicine* **3**:887-893.

Weich, H., Iberg, N., Klagsbrun, M., and Folkman, J., 1991, Transcriptional regulation of basic fibroblast growth factor gene expression in capillary endothelial cells, *J. Cell. Biochem.* **47**:158-194.

Witte, L., Fuka, Z., Haimovitz, F.A., Vlodavski, I., Goodman, D.S., and Eldor, A., 1989, Effects of irradiation on the release of growth factors from cultured bovine, porcine, and human endothelial cells, *Cancer Res.* **49**:5066-5072.

Zagzag, D., Miller, D.C., Sato, Y., Rifkin, D.B., and Burstein, D.E., 1990, Immunohistochemical localization of basic fibroblast growth factor in astrocytomas, *Cancer Res.* **50**:7393-7398.

Ziche, M., Parenti, A., Ledda, F., Dell'Era, P., Granger, H.J., Maggi, C.A., and Presta, M., 1997, Nitric oxide promotes proliferation and plasminogen activator production by coronary venular endothelium through endogenous bFGF, *Circ. Res.* in press.

Zugmaier, G., Lippman, M.E., and Wellstein, A., 1992, Inhibition by pentosan polysulfate (PPS) of heparin-binding growth factors released from tumor cells and blockage by PPS of tumor growth in animals, 1992, *J. Natl. Cancer Inst.* **84**:1716-1724.

Yamanaka, Y., Friess, H., Buchler, M., Beger, H.G., Uchida, E., Onda, M., and Kobrin, M.S., 1993, Overexpression of acidic and basic fibroblast growth factors in human pancreatic cancer correlates with advanced tumor stage, *Cancer. Res.* **53**:5289-5296.

Yayon, A., Klagsbrun, M., Esko, J.D., Leder, P., and Ornitz, D.M., 1991, Cell surface, heparin-like molecules are required for binding of basic fibroblast growth factor to its high affinity receptor, *Cell* 64:841-848.

Yeoman, L.C., 1993, An autocrine model for cell-associated and matrix-associated fibroblast growth factor, *Oncol. Res.* **5**:489-499.

TIE-1 RECEPTOR TYROSINE KINASE ENDODOMAIN INTERACTION WITH SHP2: POTENTIAL SIGNALLING MECHANISMS AND ROLES IN ANGIOGENESIS

Marrie B Marron, David P Hughes, Mark J McCarthy, Eleanor R Beaumont* and Nicholas PJ Brindle
*University of Leicester Cardiovascular Research Institute and Department of Surgery and *Department of Biochemistry, University of Leicester, Leicester LE1 7RH UK*

Key words: endothelial, receptor, tyrosine kinase, angiogenesis

Abstract: The endothelial receptor tyrosine kinase plays an essential role in vascular development where it is thought to be required for vessel maturation and stabilization. The ligands responsible for activating Tie-1, its signalling pathways and specific cellular functions are however not known. As with some other receptor tyrosine kinases, Tie-1 is subject to extracellular proteolytic cleavage generating a membrane bound receptor fragment comprising the intracellular and transmembrane domains. Here we examine the signalling potential of this Tie-1 endodomain. We show that the Tie-1 endodomain has poor ability to induce tyrosine phosphorylation. However, on formation the endodomain physically associates with a number of tyrosine phosphorylated signalling intermediates including the tyrosine phosphatase and adaptor protein SHP2. The assembly of this multimolecular complex is consistent with the endodomain having a ligand-independent signalling role in the endothelial cell. The potential roles of ectodomain cleavage and cleavage activated signalling in regulating microvessel stability in angiogenesis, vessel remodelling and regression are considered.

1. INTRODUCTION

The Tie family of receptor tyrosine kinases (RTKs) consists of two members, Tie-1 and Tie-2/Tek. These receptors are expressed predominantly in endothelial and haematopoietic cells [1-3] and are essential for vascular development [4,5]. Tie-1 and Tie-2 share similar structural features. The extracellular domains of both receptors have two immunoglobulin-like repeats separated by three EGF-homology domains and followed by three fibronectin III-like repeats. The intracellular portion contains a tyrosine kinase domain interrupted by a kinase insert sequence. At the amino acid level, they share an over-all identity of 44%, which increases to 76% in the intracellular domain. Four ligands, angiopoietins 1 to 4, have been described for Tie-2 [6-8]. Some of these ligands have opposing effects on Tie-2. Angiopoietin-1 activates Tie-2 and in endothelial cells this activation is antagonised by angiopoietin-2, which can bind Tie-2 with equal affinity [7]. Therefore, the activity of the Tie-2 receptor appears to be tightly regulated. Neither of these ligands binds Tie-1. The cellular functions controlled by Tie-2 have not been extensively characterised, although indirect evidence suggests stimulation of the receptor by angiopoietin-1 promotes stabilisation and maturation of immature vessels, possibly by enhancing recruitment of peri-endothelial cells [9]. A number of downstream signalling molecules have been identified which interact with Tie-2 *in vivo* . Tie-2 may activate PI3-kinase and Akt and has also been shown to interact with the signalling intermediates GRB2 and SH-PTP2 [10,11]. Furthermore, a novel Dok-related docking protein, Dok-R, has been cloned recently and represents the first downstream substrate of the activated Tie-2 receptor [12].

The ligands for Tie-1 have yet to be described and much less is known about this member of the Tie family. Targeted inactivation of Tie-1 gene in mice results in a lethal phenotype with mice dying between late gestation and birth. These animals exhibited normal patterning of early vessels but died as a result of extensive haemorrhage and oedema due to the lack of integrity of the microvasculature [4,5]. Close examination of microvessel $Tie1^{-/-}$ endothelial cells reveal them to have highly dynamic cytoskeletal features, including lamellae and filopodial extensions into the vessel lumen and extensive stretching [13]. These data would suggest that Tie-1 signals to suppress endothelial activation and promote vascular stability.

2. TIE-1 HOLORECEPTOR SIGNALLING

In order to examine the signalling and downstream effects of Tie-1 activation we have constructed chimeric receptors comprising the extracellular domain of the nerve growth factor receptor TrkA and the transmembrane and intracellular domains of Tie-1. However, when expressed in endothelial cells these receptors are poorly phosphorylated in the absence or presence of NGF and fail to activate tyrosine phosphorylation of cellular proteins (MBM & NB submitted). This is in contrast to TrkA/Tie-2 chimeras which we find highly active in inducing tyrosine phosphorylation. It is possible the poor ability of the Tie-1 chimeras to induce tyrosine phosphorylation may be due to low intrinsic kinase activity of Tie-1. Alternatively the phosphotyrosine signalling capacity of these receptors could be suppressed, for example by specific tyrosine phosphatases, in the endothelial cell. We have also examined, therefore, the ability of Tie-1 to induce tyrosine phosphorylation in a different cell background. The intracellular domain of Tie-1 was expressed in the yeast *Saccharomyces cerevisiae* and its phosphorylation state examined (Fig. 1).

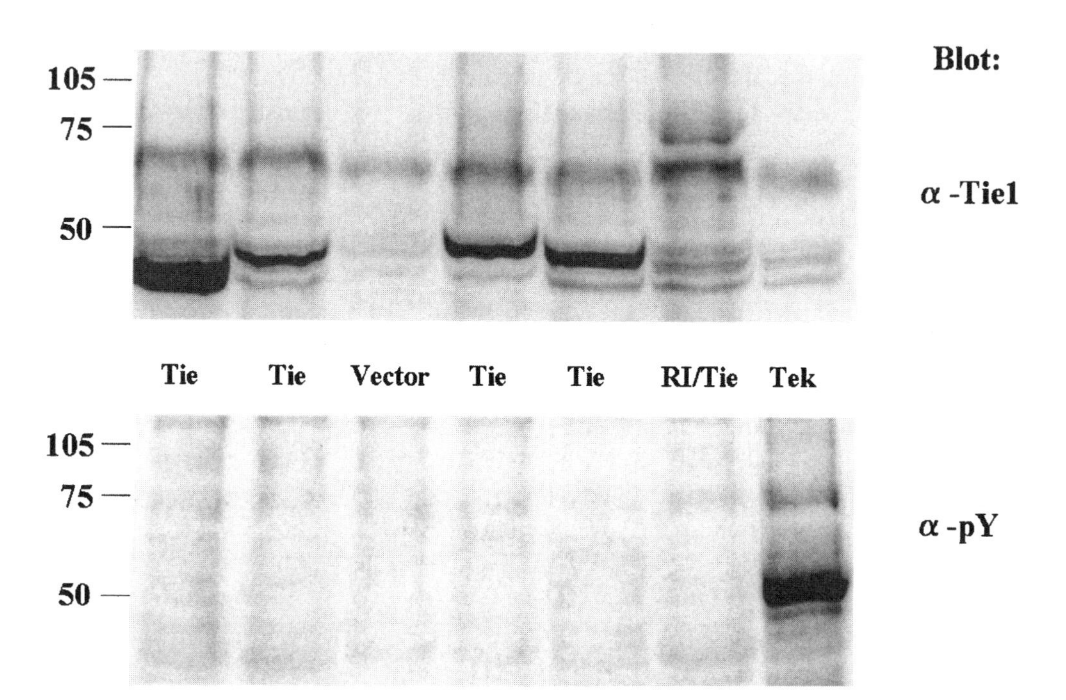

Figure 1. Expression of Tie-1 endodomain in yeast. Western blots of whole cell lysates from yeast expressing various Tie-1 (Tie) or Tie-2 (Tek) constructs including a fusion protein incorporating the RI dimerization (RI/Tie)

As found in the endothelial cell background, expression of the Tie-1 tyrosine kinase fails to induce tyrosine phosphorylation of either cellular proteins or itself. In contrast the intracellular domain of Tie-2 induces marked tyrosine phosphorylation in this system (Fig 1). Receptor tyrosine kinases are activated by ligand-induced dimerization or oligomerization. This allows transphosphorylation of individual receptors on activation domain tyrosine residues, leading to elevated kinase activity, and at other sites providing docking sites for phosphotyrosine binding domains of signalling intermediates [14]. In an attempt to promote maximal Tie-1 kinase activity in the yeast system the intracellular domains were also expressed as fusion proteins incorporating a dimerization motif. This was accomplished by inserting the RI dimerization domain of the type I alpha regulatory subunit of cyclic AMP-dependent protein kinase, at the N-terminal region of the intracellular domain. Again, even in this form the Tie-1 exhibits a poor ability to induce tyrosine phosphorylation (Fig 1). Taken together these data suggest in the cellular environment Tie-1, in contrast to Tie-2, has little phosphotyrosine signalling capacity, at least when activated by homodimerization.

3. ECTODOMAIN RELEASE ACTIVATES ASSOCIATION OF TIE-1 WITH SIGNALLING INTERMEDIATES

Recently it has been reported that Tie-1 is subject to regulated proteolytic cleavage that results in release of its extracellular ligand-binding domain [15]. Stimulation of endothelial cells with phorbol ester causes loss of the 135kDa form of Tie-1 and appearance of a 46kDa fragment. The 135kDa form of Tie-1 is the fully glycosylated surface expressed receptor and the 46kDa fragment corresponds to the intracellular and transmembrane domains [16]. This cleavage is mediated by a metalloprotease, most likely of the metalloprotease/disintegrin/cysteine-rich (MDC) family [17,16]. Surprisingly, the endodomain generated as a result of cleavage is not rapidly degraded but persists in the cell for several hours. Furthermore subcellular fractionation reveals the endodomain to be localised in a membrane fraction [17].

In the absence of ligand the ectodomains of some receptor tyrosine kinases suppress signalling by their respective intracellular domains. A truncated form of the *sevenless* receptor tyrosine kinase lacking ectodomain activates both signalling and downstream events in Drosophila [18]. Expression of ectodomain negative mutants of several growth factor receptors, including epidermal growth factor receptor and platelet-derived growth factor receptor B, results in activation of their signalling activity in mammalian cells [19]. Except in situations such as mutation leading to oncogenic transformation [20], it is unlikely that truncation of these receptors would occur *in vivo*. However, in the case of receptors which can undergo ectodomain cleavage it is possible that, provided the endodomain is not rapidly degraded, it may have signalling potential. Thus in addition to inhibiting the ability of Tie-1 to respond to its ligands, it is possible that ectodomain release may activate endodomain signalling. We have previously shown that endodomain generated following phorbol ester activation is not itself detectably tyrosine phosphorylated but is physically associated with phosphotyrosine containing proteins [17]. Among the tyrosine-phosphorylated proteins that Tie-1 endodomain co-immunoprecipitated with was one of approximately 70kDa. A possible candidate for this protein is the tyrosine phosphatase SHP2 [21]. To investigate whether SHP2 associates with endodomain we expressed cDNA constructs corresponding to truncated Tie-1, consisting of the transmembrane and intracellular portions of Tie-1 but

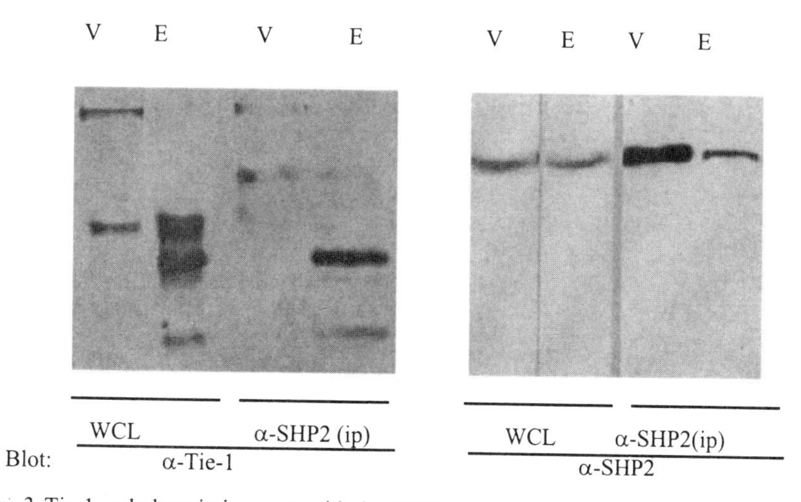

Figure 2. Tie-1 endodomain interacts with the SHP2 phosphatase. Transiently expressed Tie-1 holoreceptor- (H) and endodomain- (E) in BAEC immunoprecipitated with anti-SHP2 were probed with anti-Tie-1 antibody or anti-SHP2 antibody. WCL- whole cell lysates

lacking the extracellular domain. Cell lysates were prepared from these cells and immunoprecipitated with an antibody recognizing full length and truncated Tie-1 (Fig 2). Truncated Tie-1 was clearly detected in the SHP2 immunoprecipitates whereas little full length Tie-1 was observed. In similar experiments endothelial cells were activated with phorbol ester to induce ectodomain cleavage and lysates immunoprecipitated with anti-SHP2. Again the Tie-1 endodomain but not full length receptor was recovered in immunoprecipitates (data not shown). These data demonstrate that on ectodomain release Tie-1 becomes physically associated with SHP2. This tyrosine phosphatase has been shown to be a positive effector in signal transduction by a number of receptor tyrosine kinases, including platelet-derived growth factor receptor B [22]. In addition to its phosphatase activity SHP2 has been suggested to act as an adaptor protein [23].

The mechanism by which SHP2 binds to Tie-1 endodomain is not clear. SHP2 possesses two SH2 domains that allow it to bind phosphotyrosine residues in activated receptors [21]. However, the phosphatase binds to endodomain generated following phorbol ester treatment of endothelial cells and this has no detectable tyrosine phosphorylation. This suggests the

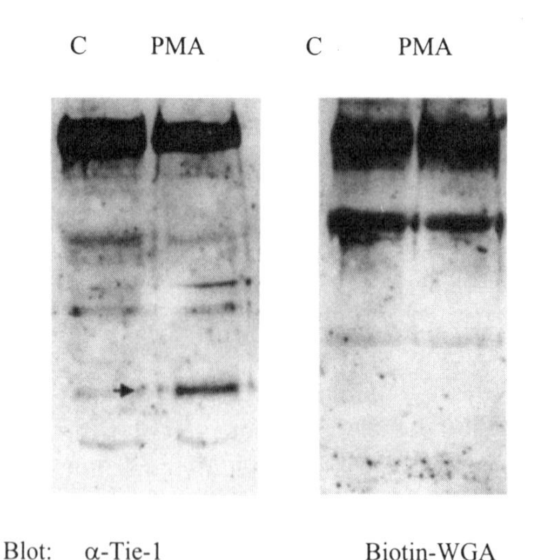

Figure 3. PMA activated Tie-1 endodomain binds to a glycoprotein. Western blot of wheatgerm agarose (WGA) precipitations from cells untreated (C) or treated with 10 ng/ml PMA (PMA) probed with anti-Tie-1 antibodies or biotin-WGA. Endodomain position indicated

interaction may be indirect involving a tyrosine-phosphorylated intermediate that binds SHP2 but binds to endodomain in a phosphotyrosine-independent manner. We were interested to determine whether the endodomain associates with other proteins. We hypothesized that the endodomain may bind a cell surface glycoprotein. To investigate this we activated endothelial cells with phorbol ester to generate endodomain and recovered N-acetyl-glucosamine containing glycoproteins from cell lysates by their ability to bind lectin. Recovered proteins were analysed for the presence of Tie-1 by Western blotting. As shown in figure 3, Tie-1 endodomain is present in the fraction binding lectin, as is full-length receptor. While Tie-1 holoreceptor would be

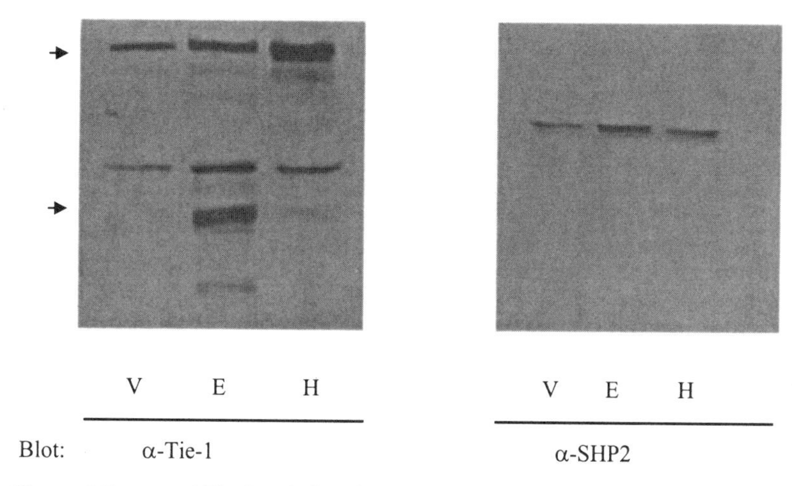

Figure 4. Expressed Tie-1 endodomain binds a glycoprotein. Western blot of WGA precipitations from endothelial cells transfected with vector (V), endodomain (E) or holoreceptor (H) probed with antibodies to Tie-1 or SHP2. Endodomain and holoreceptor indicated

expected to bind lectin by virtue of its glycosylation, Tie-1 endodomain would not. Reprobing of blots with biotinylated lectin confirms the lectin does not bind directly to endodomain. This indicates that the endodomain physically associates with one or more glycoproteins.

To verify that this association was independent of any effects of phorbol ester, similar experiments were performed in cells expressing full length and truncated Tie-1 (Fig 4). Both Tie-1 holoreceptor and truncated receptor were recovered in the lectin-binding fraction from cell lysates. In addition, reprobing blots with an antibody against SHP2 demonstrated this phosphatase was also recovered in the lectin-binding fraction, indicating SHP2. The transmembrane glycoprotein SHPS-1 has been shown to recruit SHP2 and to bind receptor tyrosine kinases [24]. We are currently examining whether SHPS-1 participates in SHP2/Tie-1 endodomain interaction.

Taken together these data demonstrate that ectodomain release results in assembly of a complex involving Tie-1 endodomain, the signalling intermediate and phosphatase SHP2 and at least one glycoprotein. The ability of Tie-1 endodomain to participate in formation of a multimolecular complex provides a possible mechanism for transduction of an endodomain signal. Thus, modulation of the activity of a signalling intermediate by direct protein:protein interaction is a potential route by which an endodomain signal may be initiated. Tie-1 endodomain is membrane associated, raising the possibility that it may act by recruitment of adapter molecules such as SHP2 or other effectors to the membrane.

4. REGULATORS OF TIE-1 TRUNCATION

We have previously reported that Tie-1 endodomain is found in tissues *in vivo* [17] and others have detected Tie-1 ectodomain in blood [25]. We were

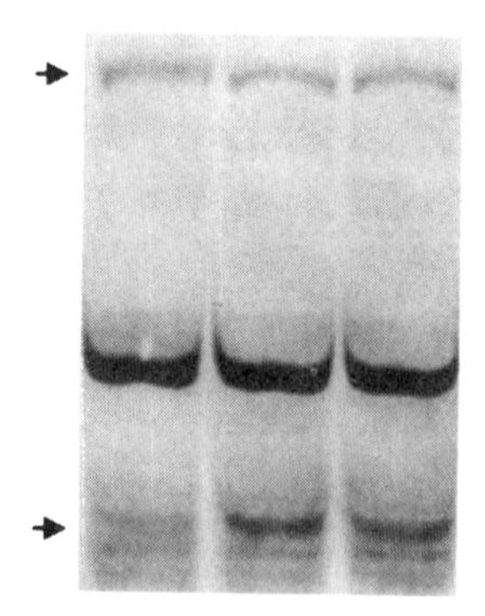

Figure 5. Tie-1 endodomain generation is activated by VEGF. Western blot of whole cell lysates from HUVECs untreated (C), treated with 100 ng/ml VEGF165 or treated with 10 ng/ml PMA probed with Tie-1 antibodies. Positions of holoreceptor and endodomain indicated

interested therefore in identifying agonists that could induce Tie-1 cleavage *in vivo*. We examined the effect of a number of agonists on Tie-1 cleavage in endothelial cells and found vascular endothelial growth factor (VEGF) to be a potent activator of cleavage. Treatment of human endothelial cells with VEGF165 led to a rapid cleavage of the ectodomain and generation of 46kDa endodomain (fig 5). Others have very recently also reported cleavage to be activated by VEGF and tumour necrosis factor α (TNFα) [16].

5. ROLE OF TIE-1 ECTODOMAIN CLEAVAGE IN ANGIOGENESIS

Data from transgenic animals deficient in Tie-1 indicate that this receptor has roles in the latter stages of angiogenesis during neovessel maturation and stabilization. It is likely that these effects are mediated via Tie-1 holoreceptor. In the initial stages of neovessel growth quiescent endothelial cells in stabilized microvessels become activated and partially released from the suppressive influence of surrounding basement membrane and supporting pericytes and smooth muscle cells [26]. This initial destabilization

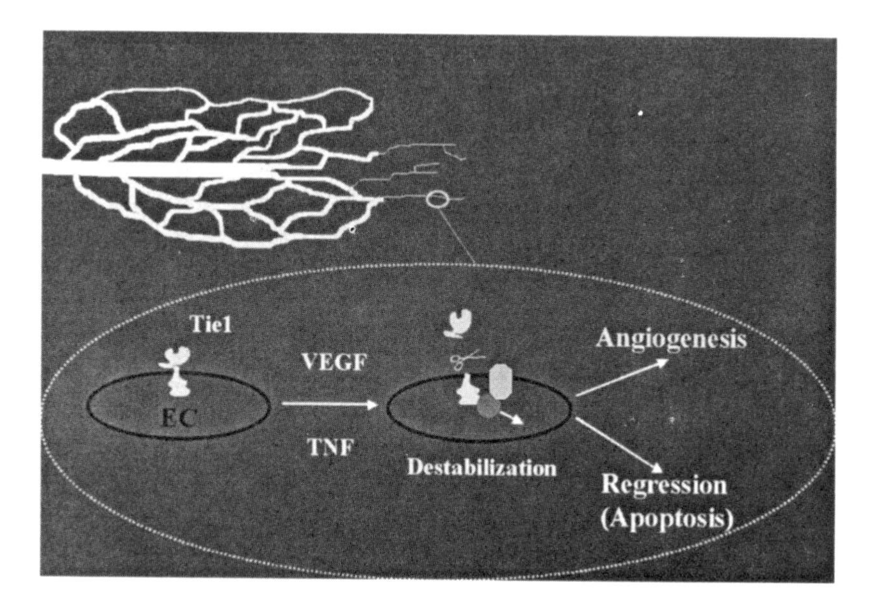

Figure 6. Model of the role of Tie-1 ectodomain cleavage in angiogenesis.

is required for angiogenesis to progress and is initiated by angiogenic growth factors such as VEGF. In the absence of these factors destabilized vessels undergo regression resulting from endothelial apoptosis [27]. The ability of VEGF to induce Tie-1 cleavage therefore may have a role in antagonizing any stabilizing effect of Tie-1 holoreceptor on the microvasculature and thus aiding in preparation of the vessel for growth. Signalling activity of Tie-1 endodomain in complex with SHP2 and other molecules could contribute to the VEGF stimulated phenotype. Interestingly, TNFα is known to induce endothelial apoptosis [28] as well as Tie-1 cleavage. Thus if Tie-1 cleavage is involved in vessel destabilization TNFα would be expected to induce vessel regression in the absence of anti-apoptotic factors [29]. This model is outlined in figure 6.

6. CONCLUSIONS

Tie-1 ectodomain release induced by VEGF and TNFα will inhibit ligand dependent signalling of full length receptor. Data presented here suggests that, in addition, cleavage generates a truncated form of Tie-1 which recruits a number of proteins including the signalling intermediate SHP2. Such a complex has the potential to initiate ligand independent signalling. Thus ectodomain cleavage of a Tie-1 receptor tyrosine kinase can result in more than simply loss of ligand binding but could also be a mechanism for initiating signalling and downstream effects. These effects may have a role in vessel destabilization prior to vessel remodelling, growth or regression.

ACKNOWLEDGMENTS

We thank the British Heart Foundation (PG/97104) and Wellcome Trust (048255/055) for supporting this work

REFERENCES

1. Dumont D. J., Yamaguchi T. P., Conlon R. A., Rossant J. and Breitman M. L. ,1992, tek, a novel tyrosine kinase gene located on mouse chromosome 4, is expressed in endothelial cells and their presumptive precursors. *Oncogene* 7: 71-80.

2. Partanen J., Armstrong E., Makela T. P., Korhonen J., Sandberg M., Renkonen R., et al. ,1992, A novel endothelial cell surface receptor tyrosine kinase with extracellular epidermal growth factor homology domains. *Mol. Cell. Biol.* 12: 1698-1707.
3. Yano M., Iwama A., Nishio H., Suda J., Takada G. and Suda T. ,1997, Expression and function of murine receptor tyrosine kinases, TIE and TEK, in hematopoietic stem cells. *Blood* 89: 4317-4326.
4. Puri M., Rossant J., Alitalo K., Bernstein A. and Partanen J. ,1995, The receptor tyrosine kinase TIE is required for integrity and survival of vascular endothelial cells. *EMBO J.* 14: 5884-5891.
5. Sato T., Tozawa Y., Deutsch U., Wolburg-Bucholz K., Fujiwara Y., Gendron-Maguire M., et al. ,1995, Distinct roles of the receptor tyrosine kinases Tie-1 and Tie-2 in blood vessel formation. *Nature* 376: 70-74.
6. Davis S., Aldrich T., Jones P., Acheson A., Compton D., Jain V., et al. ,1996, Isolation of angiopoietin-1, a ligand for the TIE2 receptor, by secretion-trap expression cloning. *Cell* 87: 1161-1169.
7. Maisonpierre P. C., Suri C., Jones P. F., Bartunkova S., Weigand S. J., Radziejewski C., et al. ,1997, Angiopoietin-2, a natural antagonist for Tie-2 that disrupts in-vivo angiogenesis. *Science* 277: 55-60.
8. Valenzuela D. M., Griffiths J. A., Rojas J., Aldrich T. H., Jones P. F., Zhou H., et al. ,1999, Angiopoietins 3 and 4: Diverging gene counterparts in mice and humans. *Proc. Natl. Acad. Sci. (USA)* 96: 1904-1909.
9. Hanahan D. ,1997, Signaling vascular morphogenesis and maintenance. *Science* 277: 48-50.
10. Huang L., Turck C., Rao P. and Peters K. ,1995, GRB2 and SH-PTP2:potentially important endothelial signaling molecules downstream of the TEK/TIE2 receptor tyrosine kinase. *Oncogene* 11: 2097-2103.
11. Kontos C. D., Stauffer T. P., Yang W.-P., York J. D., Huang L., Blanar M. A., et al. ,1998, Tyrosine 1101 of tie-2 is the major site of association of p85 and is required for activation of phosphatidylinositol 3-kinase and Akt. *Mol. Cell. Biol.* 18: 4131-4140.
12. Jones N. and Dumont D. J. ,1998, The Tek/Tie2 receptor signals through a novel Dok-related docking protein, Dok-R. *Oncogene* 17: 1097-1108.
13. Patan S. ,1998, TIE1 and TIE2 receptor tyrosine kinases inversely regulate embryonic angiogenesis by the mechanism of intussusceptive microvascular growth. *Microvasc Res* 56: 1-21.
14. Heldin C.-H. ,1996, Protein tyrosine kinase receptors. *Cancer Surv.* 27: 7-23.
15. Yabkowitz R., Myer S., Yanagihara D., Brankow D., Staley T., Elliot G., et al. ,1997, Regulation of tie receptor expression on human endothelial cells by protein kinase C-mediated release of soluble tie. *Blood* 90: 706-715.
16. Yabkowitz R., Meyer S., Black T., Elliott G., Merewether L. A. and Yamane H. K. ,1999, Inflammatory cytokines and vascular endothelial growth factor stimulate the release of soluble tie receptor from human endothelial cells via metalloprotease activation. *Blood* 93: 1969-79.
17. McCarthy M. J., Burrows R., Bell S. C., Christie G., Bell P. R. F. and Brindle N. P. J. ,1999, Potential roles of metalloprotease mediated ectodomain cleavage in signaling by the endothelial receptor tyrosine kinase Tie-1. *Lab. Invest.* 79: *in press.*

18. Basler K., Christen B. and Hafen E. ,1991, Ligand-independent activation of the sevenless receptor tyrosine kinase changes the fate of cells in the developing Drosophila eye. *Cell* 64: 1069-1081.
19. Chantry A. ,1995, The kinase domain and membrane localization determine intracellular interactions between epidermal growth factor receptors. *J. Biol. Chem.* 270: 3068-3073.
20. Rodrigues G. A. and Park M. ,1994, Oncogenic activation of tyrosine kinases. *Current Opinion in Genetics and Development* 4: 15-24.
21. Freeman R. M. J., Plutzky J. and Neel B. G. ,1992, Identification of a human src homology 2-containing protein-tyrosine-phosphatase: a putative homolog of Drosophila corkscrew. *Proc. Natl. Acad. Sci. (USA)* 89: 11239-11243.
22. Roche S., McGlade J., Jones M., Gish G. D., Pawson T. and Courtneidge S. A. ,1996, Requirement of phospholipase Cgamma, the tyrosine phosphatase Syp and the adaptor proteins Shc and Nck for PDGF-induced DNA synthesis: Evidence for the existence of Ras-dependent and Ras-independent pathways. *EMBO Journal.* 15: 4940-4948.
23. Li W., Nishimura R., Kashishian A., Batzer A. G., Kim W. J. H., Cooper J. A., et al. ,1994, A new function for a phosphotyrosine phosphatase: Linking GRB2-Sos to a receptor tyrosine kinase. *Molecular & Cellular Biology.* 14: 509-517.
24. Fujioka Y., Matozaki T., Noguchi T., Iwamatsu A., Yamao T., Takahashi N., et al. ,1996, A novel membrane glycoprotein, SHPS-1, that binds the SH2-domain-containing protein tyrosine phosphatase SHP-2 in response to mitogens and cell adhesion. *Mol. Cell. Biol.* 16: 6887-6899.
25. Vuorela P., Matikainen M. T., Kuusela P., Ylikorkala O., Alitalo K. and Halmesmäki E. ,1998, Endothelial tie receptor antigen in maternal and cord blood of healthy and preeclamptic subjects. *Obstetrics and Gynecology* 92: 179-83.
26. Beck L. and D'Amore P. A. ,1997, Vascular development: cellular and molecular regulation. *FASEB J.* 11: 365-373.
27. Benjamin L. E., Hemo I. and Keshet E. ,1998, A plasticity window for blood vessel remodelling is defined by pericyte coverage of the preformed endothelial network and is regulated by PDGF-B and VEGF. *Development* 125: 1591-1598.
28. Karsan A., Yee E. and Harlan J. M. ,1996, Endothelial cell death induced by tumor necrosis factor-alpha is inhibited by the Bcl-2 family member, A1. *J. Biol. Chem.* 271: 27201-4.
29. Spyridopoulos I., Brogi E., Kearney M., Sullivan A. B., Cetrulo C., Isner J. M., et al. ,1997, Vascular endothelial growth factor inhibits endothelial cell apoptosis induced by tumor necrosis factor-alpha: balance between growth and death signals [published erratum appears in J Mol Cell Cardiol 1998 Apr;30(4):897]. *J. Mol. Cell. Cardiol.* 29: 1321-30.

ON THE MECHANISM(S) OF THROMBIN INDUCED ANGIOGENESIS

Michael E. Maragoudakis and Nikos E. Tsopanoglou
University of Patras Medical School, Department of Pharmacology, 261 10 Rio, Patras, GREECE

Key words: angiogenesis, cancer, flt-1, KDR, thrombin, thrombin receptor, thrombosis, vascular endothelial growth factor (VEGF)

Abstract: Promotion of tumour progression by thrombin is suggested by several clinical and laboratory observations. A plausible explanation for this effect of thrombin may be related to our previous findings that thrombin is a potent promoter of angiogenesis in the chick chorioallantoic membrane system (CAM) and in the Matrigel system *in vivo*. In this report we summarise the cellular and molecular actions of thrombin that could be contributing to the activation of angiogenic cascade. Treatment of endothelial cells with thrombin leads to activation of gelatinase A, which may allow for local dissolution of basement membrane, an essential first step of angiogenesis. Similarly thrombin-treated endothelial cells have diminished ability to adhere to collagen type IV and laminin. This new phenotype of endothelial cells can migrate and survive without attachment to extracellular matrix. Thrombin-treatment of endothelial cells increases the vectorial secretion of extracellular matrix proteins, a process essential at the final steps of angiogenesis. In addition, thrombin potentiates the VEGF-induced mitogenesis of endothelial cells. This can be explained by the upregulation of the VEGF receptors (KDR & flt-1) by thrombin treatment. All the aforementioned effects of thrombin are receptor mediated, dose-dependent and require only brief exposure of endothelial cells to thrombin for these actions of thrombin. The transduction mechanisms involved are via protein kinase C (PKC) and MAP-kinase pathways.

Angiogenesis: From the Molecular to Integrative Pharmacology
Edited by Maragoudakis, Kluwer Academic / Plenum Publishers, New York, 2000

1. INTRODUCTION

The frequency of blood coagulation in cancer patients, first reported by Trousseau in 1872, was subsequently confirmed by many investigators, who provided laboratory, clinical and pharmacological evidence (Rickles & Edwards, 1983). It has been established that many tumour cell types elicit pro-coagulant activity and have detectable thrombin present (Zacharsky et al., 1995). These results can explain the hyper-coagulability observed in cancer, but does not answer the question if thrombin and thromboembolism can contribute to tumour progression. This is suggested by epidemiological studies (Sorensen et al., 1998) and animal experiments where thrombin treatment of B16 melanoma cells increases dramatically the number of lung metastasis in mice (Nierodzik et al., 1992).

The mechanism by which thrombin promotes tumour growth and metastasis may be related to our previous findings that thrombin is a potent promoter of angiogenesis (Tsopanoglou et al., 1993 & Haralabopoulos et al., 1997).

Activation of the angiogenic cascade by thrombin is likely not only in tumours but in other conditions such as diabetic retinopathy, wound healing, in the endometrium during ovulation etc. In all these situations we have bleeding, therefore, blood coagulation and thrombin generation. In support of this view that blood coagulation supports neovascularization is the very common clinical observation that when blood clots in large veins, the thrombi are often recanalized by blood capillaries, which grow into thrombi as evidenced by angiography. In this report we summarise the cellular and molecular mechanisms involved in the activation of angiogenesis by thrombin.

2. MATERIALS AND METHODS

The methods employed for studying angiogenesis *in vitro* and *in vivo* have been described previously (Tsopanoglou et al., 1993 and Haralabopoulos et al., 1997). The methodology used for culturing endothelial cells and for the cellular actions of thrombin are described in detail in previous publications (Tsopanoglou and Maragoudakis, 1998 & 1999).

3. RESULTS

Angiogenic action of thrombin in the chick chorioallantoic membrane system (CAM) and in the Matrigel system *in vivo.*

Using these systems we have shown previously that thrombin is a potent promoter of angiogenesis. Both α and γ-thrombin promote angiogenesis to the same extent. Because γ-thrombin can not form fibrin, therefore, blood clotting, we conclude that the action of thrombin on angiogenesis is independent of fibrin formation. This effect of thrombin is specific (hirudin or heparin abolish this effect) and dose dependent. The catalytic site of thrombin is essential since PPACK thrombin , which has the catalytic site chemically inactivated, is without effect and competitively inhibits the angiogenic action of thrombin.

Thrombin proteolytically cleaves the receptor thus generating a new NH_2 terminal peptide, which acts as a tethered receptor agonist by binding to an as yet unidentified site of the receptor to affect cell activation. A synthetic peptide TRAP consisting of 14 amino-acids representing the NH_2-terminus of the activated receptor, mimics many of the cellular effects of thrombin (Grant et al., 1996) including the promotion of angiogenesis (Maragoudakis et al., 1995).

Therefore, the promotion of angiogenesis by thrombin is a receptor-mediated event, which is distinct from the blood clotting mechanism.

Angiogenesis *in vivo* using the Matrigel plug system.

Matrigel plugs containing 0.3-3.0 IU thrombin/ml appeared pink and blood vessels could be seen leading into the Matrigel plug, while the controls (no thrombin) were clear. Histological evaluations of the sections by image analysis showed 15-20-fold increase in the area of cells that have infiltrated in the Matrigel plug containing thrombin (Haralabopoulos et al., 1997).

Cellular effects of thrombin related to the angiogenic action.

a) *Inhibition of endothelial cell attachment to extracellular matrix:* Endothelial cells (HUVECs) when applied to plastic wells coated with either collagen type IV or laminin adhere to an extend of 32% and 39% respectively within 60 min. However, when these cells were exposed to thrombin even briefly (5-10 min) the ability for attachment is diminished by 50%. This effect of thrombin on endothelial cell attachment to extracellular matrix proteins is dose dependent (IC_{50} at 1 IU/ml) and fully reversible. Cells washed free of thrombin after thrombin-treatment and subsequently incubated for further 15 min in fresh endothelial cell growth medium had the same adhesion characteristics as the non-thrombin treated cells. This is a thrombin receptor mediated phenomenon since TRAP has the same effect as thrombin on cell adhesion. The proteolytic activation of thrombin receptor is required since P-PACK thrombin is without effect and hirudin abolishes the effect of thrombin (Tsopanoglou & Maragoudakis, 1998).

b) *Activation of progelatinase A by thrombin in vascular endothelial cells:* It has been shown by Zucker et al. (1995) that when endothelial cells were exposed to thrombin, activation of the 72 KD gelatinase was evident. We find that in addition to activation there is an increase in gelatinase secretion into the medium and that this effect of thrombin is mimicked by TRAP (unpublished observations).

c) Increase in vectorial secretion of extracellular matrix proteins of endothelial cells by thrombin: Papadimitriou et al (1997) have shown that brief exposure of endothelial cells to thrombin causes within 3 hrs an increase in basolateral deposition of the extracellular matrix proteins. Maximal amounts of deposited proteins increased between 2.5 fold for fibronectin and 4-fold for collagen type I over baseline values. Similar results were obtained with TRAP while P-PACK thrombin, DIP-thrombin and other serin proteases as well as collagenase type IV were without effect.

d) *Thrombin potentiates VEGF-induced mitogenesis of endothelial cells:* Mitogenic activity of endothelial cells can be stimulated by a variety of growth factors including VEGF and thrombin. We have reported recently (Tsopanoglou & Maragoudakis, 1999) that when endothelial cell were exposed to thrombin for a brief period (10-15 min) they become activated. When these cells were subsequently exposed to VEGF, the increase in DNA synthesis was far greater than the additive effect obtained by either thrombin or VEGF alone. This synergistic effect of thrombin with VEGF on endothelial cell DNA synthesis and cell proliferation was evident at least 8 hrs after a brief exposure to thrombin. Continuous presence of thrombin for the 8 hrs was not necessary for this synergistic effect. This effect is specific to thrombin, since hirudin completely cancels out this effect and P-PACK thrombin is without effect. TRAP has the same effect as thrombin, thus establishing that the activated receptor of thrombin is involved in this phenomenon also.

Molecular events involved in the synergistic effect of thrombin with VEGF.

It was shown that the potentiating effect of thrombin on VEGF-induced endothelial cell mitogenesis can be explained by the up-regulation of VEGF receptors (Tsopanoglou & Maragoudakis, 1999). With a sensitive RT-PCR technique we have shown that both KDR & flt-1 gene expression was increased after exposure of endothelial cells to thrombin. The increase in KDR messenger RNA was about 170% as compared with untreated cells. The up-regulation of KDR mRNA was evident 8-12 hours after thrombin stimulation. This increase is transient. At earlier times 2-4 hours or after 16 hours of thrombin treatment the mRNA levels of KDR in the thrombin treated cells were comparable to controls. A similar increase in mRNA levels for flt-1 was also evident 8 hours after thrombin treatment. This effect is dose-dependent and is maximal at 1.5 IU/ml of thrombin while at 5.0 IU/ml the effect of thrombin declined to control levels. This bell-shaped effect of thrombin is observed in many of the actions of thrombin including angiogenesis (Tsopanoglou et al., 1993).

The possibility that thrombin may cause a generalised increase of mRNA synthesis was investigated using primers for fibroblast growth factor receptor 1 (FGFR1). Under the same conditions as above thrombin did not cause any change in mRNA for FGFR1.

We also have excluded the possibility that thrombin may be causing stabilisation of mRNA rather than new synthesis. For that we measured KDR m-RNA in the presence of activomycin D to inhibit transcription. Under these conditions the rate of decay of KDR mRNA was the same in the presence and absence of thrombin. The half-life of KDR m-RNA was about 2.8 hours and identical for control and thrombin treated cells. Furthermore, with nuclear run-on experiments it was established that the rate of transcription for KDR and GAPDH in HUVEC nuclei is linear for at least one hour. Thrombin (1.5 IU/ml) increases the rate of KDR gene transcription, which reaches approximately 80% over that in controls at one hour after the beginning of the in vitro transcription.

The transduction mechanisms involved in the up-regulation of VEGF receptors are through PKC and MAP-kinase-dependent pathways. While PMA, the PKC activator, has the same effect as thrombin, calphostin C a selective PKC inhibitor abolishes the effects of PMA and thrombin. In contrast to PMA, forskolin the selective activator of c-AMP dependent protein kinase (PKA) was without effect. MAP kinase is also involved since the selective inhibitor of MAP kinase PD98059 abolished the thrombin-induced increase of KDR and flt-1 m-RNA. All these results point to up regulation of VEGF receptors by thrombin via activation of PKC and MAP kinase signalling pathways.

We have also shown that thrombin treatment of endothelial cells increases not only mRNA of VEGF receptors but also increases the synthesis of new functional protein of KDR. Cell lysates were immuno-precipitated using an affinity-purified rabbit anti-KDR polyclonal antibody. The immunoprecipitates were electrophoresed and transferred into nitro-cellulose membranes and immuno blotted with the same anti-KDR antibody. A band at 210 KDa corresponding to KDR was obtained. The functionality of KDR receptor was established by identification of the phosphorylated KDR receptor after VEGF treatment.

4. DISCUSSION

Many laboratory and clinical studies support the notion that thrombin has a tumour promoting effect. A plausible explanation may be our previous finding that thrombin is a potent angiogenic factor (Tsopanoglou et al., 1993).

In this report we summarise the experimental evidence for this new action of thrombin in various models of angiogenesis. In addition the cellular actions of thrombin on endothelial cells, which may contribute to the angiogenic action of thrombin are reviewed. For example: Thrombin-treatment of endothelial cells diminish their ability to attach to the extracellular matrix proteins such as laminin and collagen type IV (Tsopanoglou and Maragoudakis, 1998). This action of thrombin on endothelial cells may contribute to migration and survival of endothelial cells in the early stages of angiogenesis.

Another action of thrombin related to activation of angiogenesis is the activation of gelatinase A. This may result in local dissolution of basement membrane, the first essential step in the angiogenic cascade (Zucker et al., 1995).

One of the final steps in the angiogenic process is the synthesis and deposition of extracellular matrix proteins to complete the structure of the new blood vessel. Papadimitriou et al (1997) have shown that thrombin modulates the vectorial secretion of extracellular matrix proteins in cultured endothelial cells.

Thrombin has many actions on platelets, smooth muscle cells (Kanthou et al., 1998), which may promote and support the angiogenic process in the various pathophysiological conditions.

All these effects of thrombin on endothelial cells as well as in other cells involved in angiogenesis may have synergistic effects in the activation of angiogenesis. The relative importance of the aforementioned cellular effects of thrombin in the promotion of angiogenesis is likely to depend on the particular site and pathology involved. Thrombin thus may orchestrate these events both temporally and spatially in order to activate, amplify and maintain the angiogenic cascade.

The discovery of the plethora of endogenous angiogenic and anti-angiogenic factors led many to believe that activation of angiogenesis may be the result of an imbalance of angiogenic and anti-angiogenic factors (Folkman J., 1985). However, angiogenesis being such an important physiological process is unlikely to be controlled only by algebraic additions of the actions of the redundant modulators of angiogenesis. Most probably strict controls must exist and immediate activation of angiogenesis must be possible through specific interactions of the modulators of angiogenesis like the factors involved in blood coagulation. The role of thrombin in the blood coagulation cascade may play the VEGF, the specific endothelial cell mitogen. Modulation of VEGF by other angiogenic factors may be involved in the activation of angiogenesis under the various physiological and pathological conditions. However, little is known about such interactions of VEGF and its receptor with other angiogenic factors. Only recently b-FGF has been shown to increase the VEGF receptor KDR (Pepper and Mandriota, 1998). This may be explained by the increase of endogenous VEGF expression by b-FGF, which in turn may increase the level of its receptors (Seghetti et al., 1998; Shen et al., 1998).

We have reported recently (Tsopanoglou & Maragoudakis, 1999) that thrombin causes an up-regulation of the VEGF receptors. This leads to a synergism between thrombin and VEGF on endothelial cell proliferation.

In addition thrombin has been reported to increase the release of VEGF from platelets (Möhle et al., 1997).

Through all the aforementioned cellular actions, thrombin may activate angiogenesis and promote tumour growth and metastasis. Since many of these effects of thrombin as well as angiogenesis - are thrombin-receptor - mediated and can be mimicked by TRAP, (the agonist peptide for the activated thrombin receptor), this opens up the possibility for using thrombin-peptide mimetics to promote angiogenesis. Non-thrombogenic analogs of the activated thrombin receptor have potential therapeutic applications in conditions where promotion of angiogenesis is desirable e.g. wound healing, ischemic conditions etc.

On the contrary inhibitors of thrombin or peptide antagonists to the activated receptor, which do not interfere with blood coagulation may be valuable for anti-angiogenic therapy in cancer and other angiogenic disease.

ACKNOWLEDGMENTS

This work was supported in part by grants from the Greek Ministry of Research and Technology, The European Community (Biomed 2 BMH-4-CT960069) and a NATO collaborative grant SA.5-2-05(CRG940677) 138/98/AHJ-514.

REFERENCES

1. Rickles, F.R., and Edwards, R.L.,1983, *Blood* 64, 14-31.
2. Zacharsky, L.R., Memoli, V.A., Morain, W.D., Schlaeppi, J.M., and Rousseau, S.M., 1995, *Thromb. Haemost.* 73, 793-797.
3. Sörensen, H.T., Mellem, K.L., Steffensen, F.H., Olsen, J.H., and Nielsen, G.L., 1998, *N. Engl. J. Med.* 338, 1169-1173.
4. Nierodzik, M.L., Kajumo, F. and Karpatikin, S., 1992, *Cancer Res.* 52, 3267-3272.
5. Tsopanoglou, N.E., Pipili-Synetos, E. and Maragoudakis, M.E., 1993, *Am. J. Physiol.* 264, C1302-C1307.
6. Haralabopoulos, G.C., Grant, D.S., Kleinman, H.K. and Maragoudakis, M.E., 1997, *Am. J. Physiol.* 273, C239-C242.
7. Grand, R.J.A., Turnell, A.S., and Grabham, R.W.,1996, *Biochem J.* 313, 353-368
8. Maragoudakis, M.E., Tsopanoglou, N.E., Sakoula, E., and Pipili-Synetos, E., 1995, *FASEB J.*, 9: A587.

9. Tsopanoglou, N.E., and Maragoudakis, M.E. ,1998, *Angiogenesis* 1, 192-200.
10. Zucker, S., Mirza, H., Conner, C.E., Lorenz, A.F., Drews, M.H., and Bahou, W.F. , 1998, *Int. J. Cancer* 75, 780-786.
11. Papadimitriou E., Manolopoulos, V. G., Hayman, G. T., Maragoudakis, M.E., Unsworth, B.R., Fenton II, J.W. & P.I. Lelkes,1997, *Am. J. Physiol.* 272 (Cell Physiol. 41): C1112-C1122.
12. Tsopanoglou, N.E. & Maragoudakis, M.E., 1999, *J. Biol. Chem.* 274 (34), 23969-23976.
13. Kanthou, C., Kakkar, V.V., and Benzakour, O.,1998, In: Angiogenesis: Models, Modulators, and Clinical Applications, M.E. Maragoudakis, editor. *Plenum Press* 298, 263-282.
14. Folkman, J., 1985, *Adv. Cancer Res.* 43, 172-203.
15. Pepper, M.S. and Mandriota, S.J. , 1998, *Exp. Cell Res.* 241, 414-425.
16. Seghetti, G., Patel, S., Ren, C-J., Gualandris, A., Pintucci, G., Robbins, E.S., Shapiro, R.L., Galloway, A.C., Rifkin, D.B., & Mignatti, P. , 1998, *J. Cell Biol.* 141, 1659-1673.
17. Shen, B-Q., Lee, D.Y., Gerber, H-P., Keyt, B.A., Ferrara, N., and Zioncheck, J.F., 1998, *J. Biol. Chem.* 273, 29979-29985.
18. Möhle, R., Green, D., Moore, M.A., Nachman, R.L., and Rafii, S. ,1997, *Proc. Natl. Acad. Sci. (USA)* 94, 663-668.

ENDOTHELIAL RECEPTOR TYROSINE KINASES INVOLVED IN BLOOD VESSEL DEVELOPMENT AND TUMOR ANGIOGENESIS

Georg Breier
Department of Molecular Cell Biology, Max-Planck-Institut für physiologische und klinische Forschung, Bad Nauheim, Germany

In memoriam Werner Risau (1953-1998)

1. INTRODUCTION

The cardiovascular system is the first functional organ of the vertebrate embryo. It forms by two distint but related processes, vasculogenesis and angiogenesis (Risau and Flamme, 1995). During vasculogenesis, endothelial progenitor cells (angioblasts) differentiate in the mesodermal compartments of embryonic and extra-embryonic tissues to form a primary capillary plexus. Subsequently, the primitive vasculature is refined by sprouting and non-sprouting angiogenesis, and undergoes extensive remodeling into small and large vessels (Risau, 1997). The functional maturation of blood vessels is accompanied by the recruitment of perivascular cells, such as pericytes and smooth cells. The embryonic vascular system develops in anticipation of the demands of the growing embryo for oxygen and nutrients. In contrast, angiogenesis in the adult organism occurs in response to the metabolical requirements of tissues, and is triggered by hypoxia. Deregulated angiogenesis plays an important role in the pathogenesis of a variety of diseases, including retinopathy and solid tumor growth (Folkman, 1995).

Angiogenesis: From the Molecular to Integrative Pharmacology
Edited by Maragoudakis, Kluwer Academic / Plenum Publishers, New York, 2000

The formation of new blood vessels is orchestrated by a plenitude of different proteins, including cell adhesion molecules, extracellular matrix components, angiogenic growth factors and their receptors (Risau, 1997). A body of evidence suggests that endothelium-specific receptor tyrosine kinases and their ligands govern this complex biological process. Two central endothelial signal transduction systems were identified, consisting of (i) vascular endothelial growth factor (VEGF) and the high affinity VEGF receptors (Flt-1 and Flk-1/KDR), and (ii) the Angiopoietins (Ang-1, Ang-2, Ang-3) and the Tie2 receptor (Figure 1).

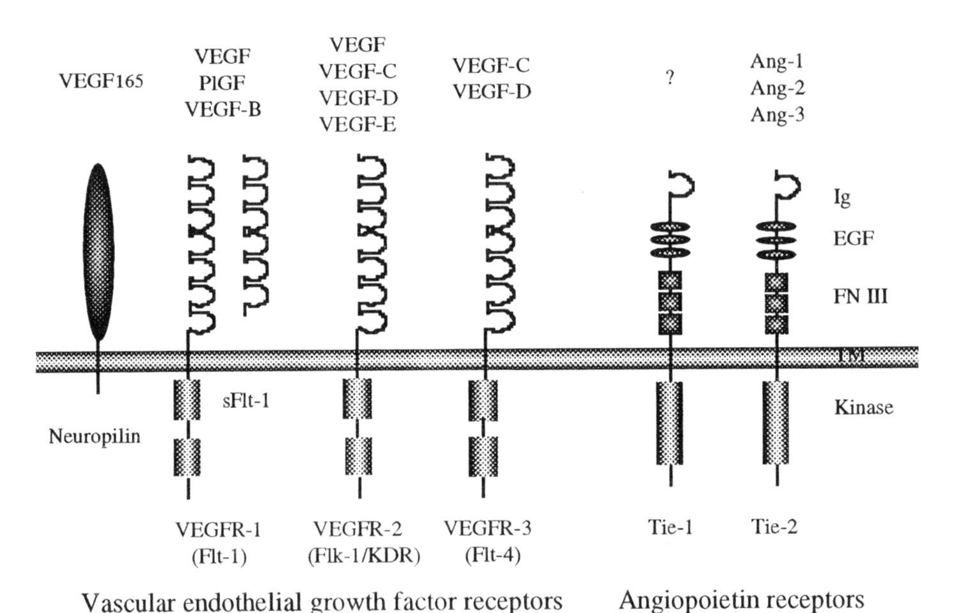

Figure 1. Endothelial signaling systems. Vascular endothelial growth factor (VEGF) receptors are receptor tyrosine kinases that bind different members of the VEGF family of growth factors. VEGF receptors are characterized by a split tyrosine kinase domain. Neuropilin is a non tyrosine kinase receptor specific for the $VEGF_{165}$ isoform. Angiopoietins (Ang-1, -2, -3) are the ligands of the Tie2 receptor tyrosine kinase. The Tie1 ligand has not been identified. EGF, epidermal growth factor-like repeat, FN, fibronectin-like domain, Ig, immunoglobulin-like domain, sFlt-1, soluble Flt-1, TM, transmembrane domain. Modified from Breier et al., 1997.

This article summarizes the functions of the VEGF/VEGF receptor system and of the Angiopoietin-1/Tie2 system, and focuses on the gene regulation of the VEGF receptor Flk-1 which is critically involved in the differentiation of the endothelial lineage and in tumor angiogenesis.

2. VEGF AND ANGIOPOIETIN IN VASCULAR DEVELOPMENT

VEGF was initially identified as endothelium-specific mitogen and potent vascular permeability factor (Ferrara, 1999). In the meanwhile, a whole VEGF family is known (Veikkola et al., 1999; Figure 1). Gene targeting experiments have provided insights into the function of the VEGF/VEGF receptor system during embryonic vascular development. The founding member of the VEGF family, VEGF (also designated VEGF-A), is critically required for early stage vascular development. The formation of blood vessels is severely impaired in mice lacking a single VEGF allele, leading to embryonic death around day 10 (Carmeliet et al., 1996; Ferrara et al., 1996). Various processes of early stage vascular development are affected: the formation of large vessels, such as the dorsal aorta; the remodeling of the yolk sac vasculature; and the sprouting of capillaries from the dorsal aorta. The severe haploid insufficient phenotype revealed that VEGF acts in a strictly dose-dependent manner to maintain the differentiation of the early embryonic vasculature. The inactivation of each individual VEGF receptor also caused embryonic lethality at mid-gestation. Interestingly, the phenotypes of the VEGF receptor deficient mouse embryos are distinct, indicating that they exert different functions. Mice lacking functional Flk-1 receptors show the most dramatic phenotype: a complete failure of vasculogenesis, endothelial cell differentiation and hematopoiesis (Shalaby et al., 1995). In contrast, Flt-1 deficient mouse embryos have abnormally enlarged blood vessels, resulting from an increased commitment of angioblasts (Fong et al., 1995). Several lines of evidence suggest that Flk-1 is the main signaling VEGF receptor in endothelial cells. In contrast, the main function of Flt-1 in endothelial cells appears to be ligand binding, because gene targeting experiments have shown that the intracellular portion of the receptor is dispensable for embryonic development (Hiratsuka et al., 1998).

The targeted inactivation of Ang-1 or Tie2 resulted in defective vascular development and subsequent embryonic death (Sato et al., 1995; Suri et al., 1996). The phenotypes of Ang-1 and Tie2 knockout mice are very similar. Vasculogenesis proceeds normally in Ang-1 and Tie2 deficient mouse embryos, indicating that this signaling system functions downstream of the VEGF/VEGF receptor system. The following processes are affected in Ang-1 or Tie2 deficient mouse embryos: the trabeculation in the heart, the remodeling of the yolk sac vasculature, and the sprouting of capillaries in the neural tube. The heart trabeculation phenotype indicates that Ang-1 and Tie2 are involved in the recruitment of perivascular cells. Ang-1 also stimulates the formation of endothelial cell sprouts *in vitro* (Koblizek et al., 1997), and it acts synergistically with VEGF. Thus, Ang-1 and VEGF appear to cooperate during angiogenesis.

2.1 Signal transduction by Flk-1

Ligand binding to receptor tyrosine kinases leads to the activation of the intrinsic kinase activity, and the dimerization of receptor molecules. Subsequently, the dimeric partners trans-phosphorylate each other in a process known as receptor autophosphorylation. Multiple tyrosine residues in the intracellular portion become phosphorylated and serve as docking sites for cytoplasmic signaling molecules. We found that Flk-1 receptor autophosphorylation leads to the binding and tyrosine phophorylation of the signaling molecule SHC. To determine the critical residues involved in Flk-1 autophosphorylation and SHC phosphorylation, we mutated individual tyrosine residues in the intracellular portion of the Flk-1 receptor and analysed the mutants for their ability to become phosphorylated. Among various mutations, the double mutation of tyrosine residues 1052 and 1057 leads to an almost complete loss of Flk-1 receptor autophosphorylation. This receptor mutant is capable of binding, but not of phophorylating SHC. Thus, residues 1052 and 1057 of Flk-1 are critical for receptor autophosphorylation and SHC phosphorylation.

2.2 Transcriptional regulation of the *Flk-1*gene

The activation of the Flk-1 receptor appears to be the principal event involved in the differentiation of the endothelial lineage. Upstream regulators of Flk-1 can therefore be considered as master regulators of vascular development. Moreover, as discussed below (chapter 3), the upregulation of Flk-1 in the vasculature of various tumors is a critical event during the switch to the angiogenic state. It is therefore of great importance to study the transcriptional regulation of the Flk-1 gene. To identify the regulatory elements involved, we have generated reporter constructs containing the LacZ gene under the transcriptional control of various regions of the mouse Flk-1 gene, and have used them to generate transgenic mice. Upstream sequences alone were not sufficient to reproducibly target reporter gene expression to the vasculature. However, a 939 bp promoter fragment, in combination with a 2.3 kb fragment of the first intron, mediated a uniform and reproducible reporter gene expression in the vasculature of transgenic mouse embryos. The intronic sequences were also capable of conferring endothelium-specificity to the heterologous thymidine kinase promoter and acted in a position-independent manner. Thus, they fulfilled all criteria of an autonomous tissue-specific enhancer. The LacZ reporter gene, like the endogenous Flk-1 gene, was expressed throughout vascular development and was down-regulated in most adult vascular beds of transgenic mice. Thus, we have identified the regulatory elements sufficient for endothelium-specific expression of Flk-1 during vascular development. Finer mapping localized the intron enhancer in a 430 bp fragment that contains several potential binding sites for transcription factors of the Ets-, GATA and Scl/Tal-1 families. Gene targeting experiments have shown that these transcription factors exert important functions in hematopoiesis. In addition, they have been implicated in angiogenesis, however, direct evidence for a role in mouse vascular development is missing. The individual mutation of any of two Scl/Tal-1 motifs in the Flk-1 enhancer led to a reduced LacZ reporter gene expression in transgenic mice. In contrast, the individual mutation of a single GATA site or of a single Ets site resulted in ectopic reporter gene expression. Thus, transcription factors of the GATA and Ets families are involved in specifying endothelium-specific expression by interacting with the Flk-1 intron enhancer.

Results of transfection experiments (Rönicke et al., 1996) and the analysis of transgenic reporter mice (Kappel et al., 1999) indicate that the Flk-1 promoter, rather than specifying endothelium-specificity, contributes to a strong expression in endothelial cells. To identify transcription factors involved, we co-transfected Flk-1 promoter/LacZ reporter gene constructs with expression vectors encoding candidate regulatory transcription factors in A293 cells. The Flk-1 promoter was strongly stimulated by Ets-1 and Ets-2. The mutation of a single Ets binding site in the Flk-1 promoter in transgenic reporter mouse embryos resulted in a reduced lacZ expression, indicating that an Ets factor stimulates Flk-1 promoter activity during vascular development. *In vitro*, the Flk-1 promoter was also strongly activated by hypoxia-inducible factor-2α (HIF-2α), a basic helix-loop-helix transcription factor that is expressed during embryonic vascular development (Flamme et al., 1997). HIF-2α, like the related HIF-1α, upregulates VEGF expression in response to hypoxia, and may thus represent both an endogenous and exogenous regulator of blood vessel formation, by stimulating the expression of both receptor and ligand. It should be noted, however, that Flk-1/KDR mRNA, in contrast to Flt-1 mRNA, is not directly upregulated

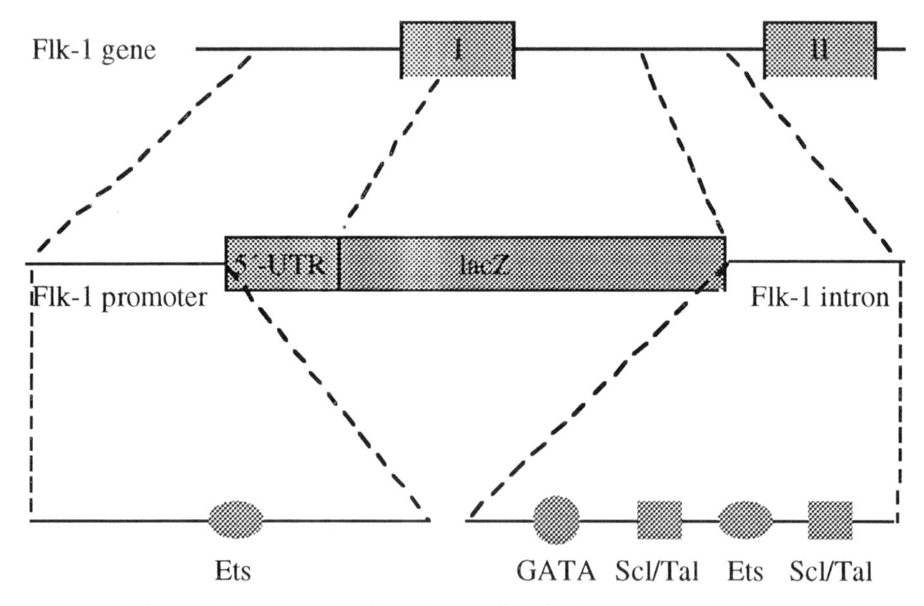

Figure 2. Transcription factor binding sites in the Flk-1 promoter and enhancer regions. The structure of the lacZ reporter gene used to generate transgenic mice is shown.

in endothelial cells in response to hypoxia (Kremer et al., 1997). Despite this fact, we observed that Flk-1 is upregulated in response to hypoxia by an indirect mechanism in cerebral slice cultures of postnatal mouse brain. This upregulation is most likely mediated by the ligand itself because VEGF, which is upregulated in response to hypoxia, is able to induce endothelial Flk-1 expression in this model.

3. VEGF AND VEGF RECEPTORS IN TUMOR ANGIOGENESIS

The upregulation of VEGF in tumor cells and the induction of VEGF receptors in the tumor vasculature is a hallmark of the switch to the angiogenic state in various tumors. VEGF is considered as a major regulator of tumor angiogenesis because the inhibition of VEGF activity strongly reduced tumor growth and angiogenesis in a variety of experimental tumors (Kim et al., 1993; Millauer et al., 1994; 1996). It is well established that hypoxia is a major stimulator of VEGF expression in tumors. *In vitro* studies have shown that the 5´-flanking sequences of the VEGF gene, which harbor a consensus HIF-1 binding site, mediate the transcriptional stimulation of the VEGF gene whereas sequences derived from the 3´-UTR are involved in regulating VEGF mRNA stability. By using reporter gene studies in experimental glioma, we demonstrate that these mechanisms of VEGF upregulation in response to hypoxia are functional *in vivo* (Damert et al., 1997). A strong and reproducible reporter gene expression in perinecrotic palisading cells is observed only when both the 5´-flanking sequences and the 3´-UTR were included in the construct. Although the expression of angiogenic growth factors is modulated by the microenvironment, genetic alterations in tumor cells, such as the activation of the Ras oncogene, set the stage for tumor angiogenesis. The mechanisms of VEGF receptor induction are under investigation. As discussed above (see chapter 2.2), Flt-1 is directly upregulated in response to hypoxia, and Flk-1 expression can be induced by VEGF. It seems likely however, that additional, as yet unidentified factors are also relevant for this process.

Experiments employing tyrosine kinase deficient dominant-negative VEGF receptor mutants have previously shown that angiogenesis in various experimental tumors is dependent on Flk-1 signaling (Millauer et al., 1994; 1996). These results have encouraged the development of low molecular weight inhibitors of Flk-1 signaling that efficiently inhibit angiogenesis in experimental tumors (Fong et al., 1999). In contrast, the function of Flt-1 in tumor angiogenesis is unknown. We observed that Flt-1 is capable of forming heterodimers with Flk-1, indicating that the two VEGF receptors may interact functionally in endothelial cells. To address the role of Flt-1, we have generated truncated kinase-deficient Flt-1 mutants. The retrovirus-mediated gene transfer of these mutants strongly inhibits the neovascularization and growth of C6 glioma in nude mice. These experiments identify Flt-1 as a potential target for glioma therapy.

4. CONCLUSION

Our studies underline the central role of the VEGF signaling system in tumor angiogenesis. We provide evidence that transcription factors of the GATA-, Ets- and Scl/Tal-1 families control the expression of Flk-1, a principal regulator of embryonic blood vessel formation and tumor angiogenesis. Thus, these transcription factors are not only involved in hematopoiesis, but have also central functions in vascular development. An important problem to be addressed in future research is the transcriptional regulation of the Flk-1 gene in tumor endothelium. Knowledge of the regulatory sequences of the Flk-1 gene should allow to study and eventually target the signaling pathways leading to the upregulation of Flk-1 in tumors. And finally, these regulatory sequences should be extremely useful for targeting therapeutic genes specifically to the vasculature of tumors.

ACKNOWLEDGMENTS

I wish to thank Sabine Blum, Dr. Matthias Clauss, Dr. Annette Damert, Dr. Urban Deutsch, Regina Heidenreich, Andreas Kappel, Dr. Marcia Machein, Dr. Karl H. Plate, and Cornelia Weiss for valuable contributions. The work from our laboratory is supported by the Max Planck Society, Deutsche Krebshilfe, BMBF and Sugen.

REFERENCES

Breier, G., Damert, A., Plate, K. H., and Risau, W., 1997, Angiogenesis in embryos and ischemic diseases. *Thromb Haemost* **78**: 678-683.

Carmeliet, P., Ferreira, V., Breier, G., Pollefeyt, S., Kieckens, L., Gertsenstein, M., Fahrig, M., Vandenhoeck, A., Harpal, K., Eberhardt, C., Declercq, C., Pawling, J., Moons, L., Collen, D., Risau, W., and Nagy, A., 1996, Abnormal blood vessel development and lethality in embryos lacking a single VEGF allele. *Nature* **380**: 435-439.

Damert, A., Machein, M., Breier, G., Fujita, M. Q., Hanahan, D., Risau, W., and Plate, K. H., 1997, Up-regulation of vascular endothelial growth factor expression in a rat glioma is conferred by two distinct hypoxia-driven mechanisms. *Cancer Res* **57**: 3860-3864.

Ferrara, N., Carver-Moore, K., Chen, H., Dowd, M., Lu, L., O'Shea, K. S., Powell-Braxton, L., Hillan, K. J., and Moore, M. W., 1996, Heterozygous embryonic lethality induced by targeted inactivation of the VEGF gene. *Nature* **380**: 439-442.

Ferrara, N., 1999, Role of vascular endothelial growth factor in the regulation of angiogenesis. *Kidney Int* **56**: 794-814.

Flamme, I., Fröhlich, T., von Reutern, M., Kappel, A., Damert, A., and Risau, W., 1997, HRF, a putative basic helix-loop-helix-PAS-domain transcription factor is closely related to hypoxia-inducible factor-1 alpha and developmentally expressed in blood vessels. *Mech Dev* **63**: 51-60.

Folkman, J., 1995, Angiogenesis in cancer, vascular, rheumatoid and other disease. *Nat. Med.* **1**: 27-31.

Fong, G. H., Rossant, J., Gertsenstein, M., and Breitman, M. L., 1995, Role of the Flt-1 receptor tyrosine kinase in regulating the assembly of vascular endothelium. *Nature* **376**: 66-70.

Fong, T. A., Shawver, L. K., Sun, L., Tang, C., App, H., Powell, T. J., Kim, Y. H., Schreck, R., Wang, X., Risau, W., Ullrich, A., Hirth, K. P., and McMahon, G., 1999, SU5416 is a potent and selective inhibitor of the vascular endothelial growth factor receptor (Flk-1/KDR) that inhibits tyrosine kinase catalysis, tumor vascularization, and growth of multiple tumor types. *Cancer Res* **59**: 99-106.

Hiratsuka, S., Minowa, O., Kuno, J., Noda, T., and Shibuya, M., 1998, Flt-1 lacking the tyrosine kinase domain is sufficient for normal development and angiogenesis in mice. *Proc Natl Acad Sci U S A* **95**: 9349-9354.

Kappel, A., Ronicke, V., Damert, A., Flamme, I., Risau, W., and Breier, G., 1999, Identification of vascular endothelial growth factor (VEGF) receptor-2 (Flk-1) promoter/enhancer sequences sufficient for angioblast and endothelial cell-specific transcription in transgenic mice. *Blood* **93:** 4284-4292.

Kim, K. J., Li, B., Winer, J., Armanini, M., Gillett, N., Phillips, H. S., and Ferrara, N., 1993, Inhibition of vascular endothelial growth factor-induced angiogenesis suppresses tumour growth in vivo. *Nature* **362**: 841-844.

Koblizek, T. I., Weiss, C., Yancopoulos, G. D., Deutsch, U., and Risau, W., 1998, Angiopoietin-1 induces sprouting angiogenesis in vitro. *Curr Biol* **8**: 529-532.

Kremer, C., Breier, G., Risau, W., and Plate, K. H., 1997, Up-regulation of flk-1/vascular endothelial growth factor receptor 2 by its ligand in a cerebral slice culture system. *Cancer Res* **57**: 3852-3859.

Millauer, B., Shawver, L. K., Plate, K. H., Risau, W., and Ullrich, A., 1994, Glioblastoma growth inhibited in vivo by a dominant-negative Flk-1 mutant. *Nature* **367:** 576-579.

Millauer, B., Longhi, M. P., Plate, K. H., Shawver, L. K., Risau, W., Ullrich, A., and Strawn, L. M., 1996, Dominant-negative inhibition of Flk-1 suppresses the growth of many tumor types in vivo. *Cancer Res* **56**: 1615-1620.

Risau, W. and Flamme, I., 1995, Vasculogenesis. *Annu. Rev. Cell Dev. Biol.* **11**: 73-91.

Risau, W., 1997, Mechanisms of angiogenesis. *Nature* **386**: 671-674.

Rönicke, V., Risau, W., and Breier, G., 1996, Characterization of the endothelium-specific murine vascular endothelial growth factor receptor-2 (Flk-1) promoter. *Circulation Res.* **79**: 277-285.

Sato, T. N., Tozawa, Y., Deutsch, U., Wolburg-Buchholz, K., Fujiwara, Y., Gendron-Maguire, M., Gridley, T., Wolburg, H., Risau, W., and Qin, Y., 1995, Distinct roles of the receptor tyrosine kinases Tie-1 and Tie-2 in blood vessel formation. *Nature* **376**: 70-74.

Shalaby, F., Rossant, J., Yamaguchi, T. P., Gertsenstein, M., Wu, X. F., Breitman, M. L., and Schuh, A. C., 1995, Failure of blood-island formation and vasculogenesis in Flk-1-deficient mice. *Nature* **376**: 62-66.

Suri, C., Jones, P. F., Patan, S., Bartunkova, S., Maisonpierre, P. C., Davis, S., Sato, T. N., and Yancopoulos, G. D., 1996, Requisite role of angiopoietin-1, a ligand for the TIE2 receptor, during embryonic angiogenesis. *Cell* **87**: 1171-1180.

Veikkola, T., and Alitalo, K., 1999, VEGFs, receptors and angiogenesis. *Semin Cancer Biol* **9**: 211-220.

ONTOGENY OF THE ENDOTHELIAL SYSTEM IN THE AVIAN MODEL

LUC PARDANAUD AND FRANÇOISE DIETERLEN-LIÈVRE
Institut d'Embryologie Cellulaire et Moléculaire du CNRS, 49bis, Avenue de la Belle Gabrielle, 94736 Nogent-sur-Marne Cedex, France. E-mail: pardanau@infobiogen.fr

Summary The avian model provides an experimental approach for dissecting the origin, migrations and differentiation of cell lineages in early embryos. In this model, the endothelial network was shown to take place through two processes depending on the origin of endothelial precursors: vasculogenesis when angioblasts emerge *in situ*, angiogenesis when angioblasts are extrinsic. Two different mesodermal territories produce angioblasts, the somite which only gives rise to endothelial cells and the splanchnopleural mesoderm which also produces hemopoietic stem cells. Potentialities of the mesoderm are determined by a positive influence from the endoderm and a negative control from the ectoderm. The presence of circulating endothelial precursors in the embryonic blood stream is also detected.

1. INTRODUCTION

In vertebrates, the development of the endothelial network is an early event occurring in the mesodermal layer. Two different mechanisms of vascularization are at work depending on the origin of endothelial precursors[1,2]. When these precursors emerge *in situ* the process is designated as vasculogenesis. When the precursors are extrinsic the vascularization occurs *via* the angiogenesis process. The development of arteries, veins and capillaries from primitive endothelial network involves specific molecules[3,4,5] then the differentiation of vessel wall[6,7].

The development of endothelial cells is tightly linked to the emergence of hemopoietic cells in the yolk sac and the embryo proper. This may indicate the existence of a common ancestor now referred to as the hemangioblast, according to a term originally coined by Murray. In avian, murine, pig and human embryo, hemopoietic clusters bud in the aortic floor from the endothelium[9,10,11,12,13,14]. In birds, these clusters do not express the endothelial-specific molecule VEGFR2[15] and acquire hemopoietic-specific markers such as the c-*myb* transcriptional factor[16,17] and the CD45 molecule[12,13,15]. The floor of the aorta has been identified as the first intraembryonic location that produces hemopoietic stem cells. The analysis of cellular and molecular controls underlying these cytological aspects is in progress and the identification of these factors will be important to understand how the embryonic development and the further stability of the differentiation state are dependent on the vascularization. Furthermore, the molecules at work in embryonic life are probably later involved in the physiology of the vessels.

2. EMERGENCE OF ANGIOBLASTS IN THE EMBRYO

Early studies on the development of the vascular tree relied on microscopic observations of embryos, alive[18], fixed and stained with classical dyes, injected with india ink[19,20] or prepared for scanning and transmission electronic microscopy[21,22]. The identification of endothelial cells was based on their integration in tubular structures; the first endothelial cells were identified around the 4-5 somite stage. The advent of the QH1 monoclonal antibody, specific for the endothelial and hemopoietic lineages of the quail species, permitted to study earlier events of vascular development.

During the first two days of development, the avian embryo grows in a plane. Thus, we initiated an *in toto* immunostaining approach in which whole quail blastodiscs were treated with QH1 revealed by a fluorescent marker[23]. We found the first angioblasts at the head-process stage in the yolk sac and at the 1-somite stage in the embryonic area. As soon as they emerge, $QH1^{+}$ cells express two earlier markers of the endothelial lineage, the c-*ets*1 transcriptional factor[16,24,25] and the receptor 2 of the vascular endothelial cell growth factor (VEGF)[26]. These cells progressively interconnect, thus building up the profiles of the heart and dorsal aortae. The emergence and connectivity of new angioblasts then rapidly progress in both rostral and caudal directions. The aortic rudiments begin to form cords and, at the 7-somite stage, the heart becomes interconnected with the extraembryonic network. The blood islands are visible caudally in the yolk sac, then the anterior horns of the vascular tree move up in front of the brain vesicles. At this stage, the extraembryonic and intraembryonic endothelial plexus mingle and no limit can be observed between these two regions. At the 13-somite stage, the vascular tree covers the whole embryonic and extraembryonic territories with the exception of the proamnion and Hensen's node region where gastrulation is still underway. The edge of the vascular area becomes entirely limited by the QH1$^+$ *sinus marginalis*.

Further descriptive studies followed our initial observations, using QH1 immunostained quail wholemounts[27,28,29,30]. Angioblast emergence and the morphogenesis of the endothelial network have also been studied in mouse wholemounts immunostained for von Willebrand factor (VWF)[31]. In this species, the aortic endothelium assembles from single VWF$^+$ cells while VWF$^-$ precursors give rise to other vessels as the intersegmental arteries.

3. THE SPLANCHNOPLEURAL MESODERM HAS A HEMANGIOPOIETIC POTENTIAL

The intermediate germ layer emerges during the gastrulation as a homogeneous monolayer which differentiates in distinct compartments in which some potentialities are mapped. However, the distribution of the endothelial potential of the mesoderm is more recent. When cultured, the epiblast (the upper layer of the blastodisc) has a hemangiopoietic potential[32,33], i.e., it can produce both endothelial and hemopoietic cells. This potential is located in the caudal region as gastrulation occurs. To determine the further distribution of this capacity in paraxial (somite) and lateral mesoderm (dorsal somatopleural mesoderm and ventral splanchnopleural mesoderm), chick somites were replaced *in ovo* by quail somites, somatopleural mesoderm or splanchnopleural mesoderm[17]. The splanchnopleural mesoderm produces angioblasts able to colonize all the chick host embryo. These endothelial progenitors which can be identified by

QH1 affinity migrate around the neural tube (Fig 1), in the body wall and the limb buds, in the mesonephros and visceral organs. Concerning the aorta, quail angioblasts reach the dorsal and ventral endothelium (Fig 1) of the vessel. In the aortic floor these cells participate to the hemopoietic clusters. Thus, the splanchnopleural mesoderm gives rise to both endothelial and hemopoietic lineages and the hemopoietic capacity only occurs when the precursors integrate a specific site, the aortic floor.

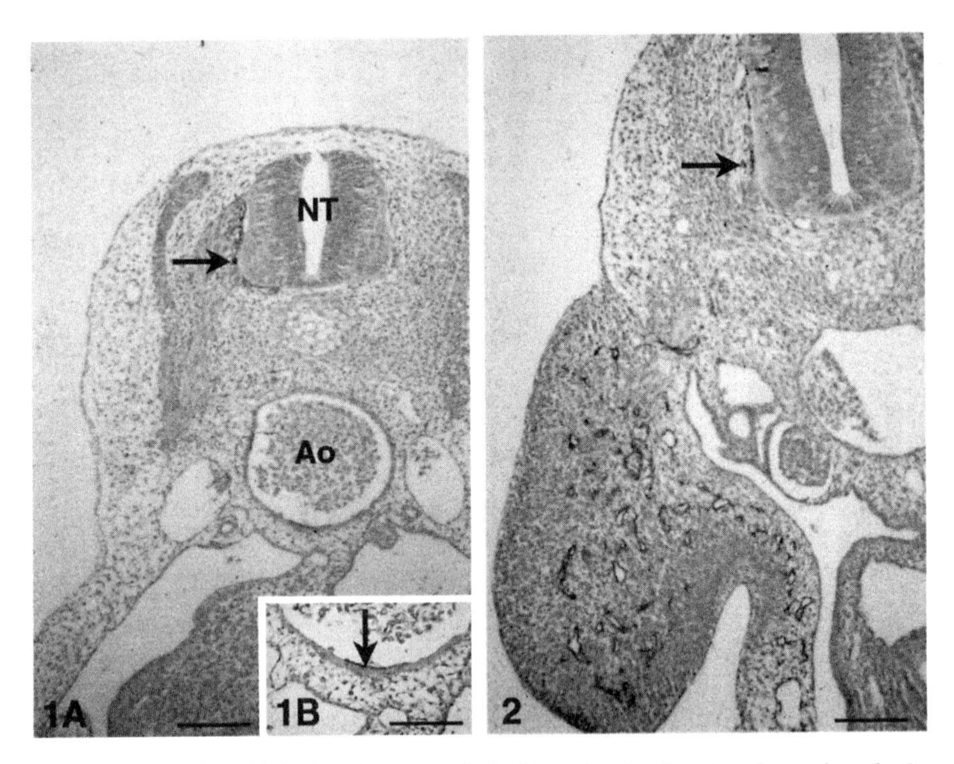

Figure 1: Three-day-old chick embryo engrafted with quail splanchnopleural mesoderm for 2 days. QH1 staining. A) $QH1^+$ endothelial cells, restricted to the side of the graft, migrated around the neural tube (NT) and reached the aortic floor (Ao). Bar: 100μm. B) Enlargement of the ventral endothelium of the aorta showing the positive endothelial cells (arrow). Bar: 55μm.

Figure 2: Three-day-old chick embryo engrafted with quail somites for 2 days. QH1 staining. $QH1^+$ endothelial cells are distributed in the host dorsal region, in particular the neural tube (NT), the wing bud and the body wall. $Qh1^+$ endothelial cells also reached the dorsolateral endothelium of the aorta (Ao) in other sections of the same embryo. Bar: 100 μm.

The double capacity in the splanchnopleural mesoderm soon becomes restricted: grafted buds of visceral organs obtain their vascularization from intrinsic precursors, but are colonized by host extrinsic hemopoietic stem cells (erythropoiesis and granulopoiesis in the spleen, macrophagic differentiation in the lung, *etc*...)[2]. The rule according to which intraembryonic organs are colonized by extrinsic hemopoietic stem cells has been known for a long time[34].

4. THE SOMATOPLEURAL MESODERM HAS NO ANGIOPOIETIC POTENTIAL

When limb buds are grafted from quail to chick or *vice versa*, the endothelial network and the hemopoietic stem cells colonizing the bone marrow differentiate from host extrinsic precursors. Thus the somatopleural mesoderm has no angiopoietic capacity, i.e., it does not produce angioblasts. In this case, the endothelial and hemopoietic lineages are not dissociated by the transplantation. What is the origin of these lineages? Is there a double colonization by independent precursors or is there one colonization by a common precursor?

These results put forward a paradox: when the limb buds are explanted, a primitive endothelial network is already present but its derivatives are lost at the end of the graft. This observation can be explained by recent experiments[35]: when the somatopleural mesoderm is separated from the ectoderm before grafting, intrinsic endothelial cells differentiate. Thus, the ectoderm seems to play a negative role (see paragraph 6).

5. THE ANGIOPOIETIC POTENTIAL OF THE SOMITES

The angiopoietic capacity of the somite was previously defined[36, 37,38,39]. Orthotopic transplantations of quail somites in place of chick somites demonstrate that the angioblasts colonizing the somatopleural mesoderm come from the axial mesoderm[17]. However, these endothelial precursors never invade the visceral organs. Their behaviour is different when compared to the splanchnic angioblasts which are able to colonize all parts of the host embryo. The distribution of somitic angioblasts is restricted to the body wall, neural tube, limb buds and mesonephros (Fig 2). In the aorta, the integration of somitic angioblasts is also restricted; they reach the roof and sides of the vessel but never integrate the aortic floor nor participate to the ventral aortic hemopoietic clusters. Thus, the potential of the somitic angioblasts is purely angiopoietic. Furthermore, these experiments demonstrate the double origin of the aorta.

In the same way, when quail somatopleural mesoderm, already colonized by somitic angioblasts, is grafted, endothelial cells vascularize the body wall and the aortic roof of the chick host but are barred from the aortic floor and visceral organs.

6. THE ANGIOPOIETIC/HEMANGIOPOIETIC POTENTIAL OF THE MESODERM CAN BE COMMUTED AND/OR RESTRICTED BY THE OTHER EMBRYONIC GERM LAYERS OR BY GROWTH FACTORS

In normal development, the splanchnopleural mesoderm is closely associated with the endoderm and the somatopleural mesoderm with the ectoderm. According to previous findings, the endoderm would induce hemopoiesis in the yolk sac mesoderm[40] and the ectoderm is known to create an avascular zone in the underlying mesoderm of the limb bud[41,42]. We dissected the role of these different embryonic layers in angiopoiesis and hemangiopoiesis. We devised an *in vivo* assay based on the polarized homing of precursors to the floor of the aorta in the quail/chick model. We chose to graft the quail rudiments on top of the chick host splanchnopleura in the early coelom (Fig 3)[35]. This lateral grafting site gave the same results as the more dorsal site and allowed to transplant larger pieces of tissues. In these experiments, the homing patterns and types of grafted-derived $QH1^+$ cells were analyzed two days later and the diagnosis took into account only whether these cells became integrated in the roof and sides or also in the floor of the aorta and whether they displayed a typical endothelial phenotype or budded into the aortic lumen in hemopoietic cell-fashion.

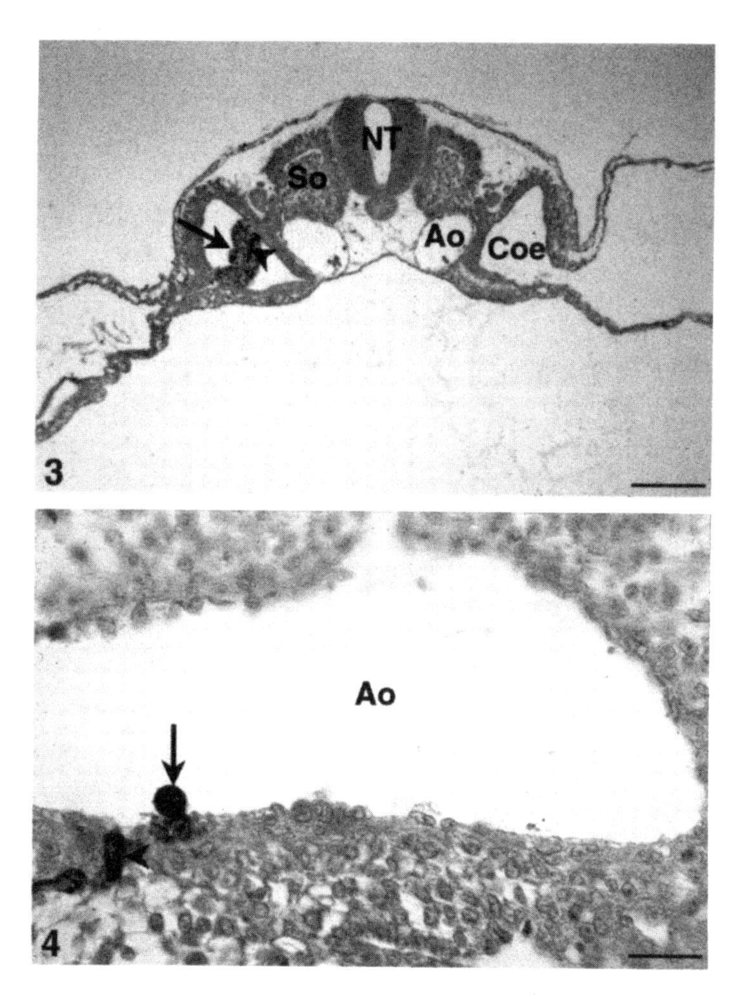

Figure 3: 22-somite stage chick embryo, 3 hours after the graft of quail mesoderm onto the splanchnopleura. QH1 staining. The graft (arrow) is inserted in the coelomic cavity (Coe). The dorsal tissues have healed. A $QH1^+$ endothelial cell (arrowhead) is visible inside the graft. Ao: aorta; NT: neural tube; So: somite. Bar: 100 μm.

Figure 4: Colonization of the aortic floor (Ao) by quail cells from somatopleural mesoderm treated by VEGF. The host was a 3-day-old chick embryo. One $QH1^+$ hemopoietic cell (arrow) buds into the lumen of the vessel. $QH1^+$ endothelial cells were migrating in the dorsal mesentery (arrowhead). Bar: 20 μm.

Before the transplantation in the chick host, quail mesodermal explants (somitic or somatopleural) were transiently cultured in a semisolid medium in association with chick endoderm or in presence of different growth factors (VEGF, bFGF, TGFβ1). In these conditions, somitic and somatopleural mesoderms acquired the hemangiopoietic potential in a great percentage of explants: precursors from the quail graft integrated into the floor of the chick aorta and gave rise to hemopoietic clusters (Fig 4). Thus, our experiments demonstrate that the endoderm does play an instructive role by imparting novel potentials to somitic and somatopleural mesoderm. Conversely, the splanchnopleural mesoderm losts its hemangiopoietic potential when treated with ectoderm, EGF, TGFα or HGF: no precursor reach the ventral endothelium of the aorta and the number of migrating angioblasts decreased[35].

We postulate that two gradients, one positive from the endoderm and one negative from the ectoderm, modulate the potential of the mesoderm, thus achieving the different grades of angiopoiesis/ hemangiopoiesis (Fig 5).

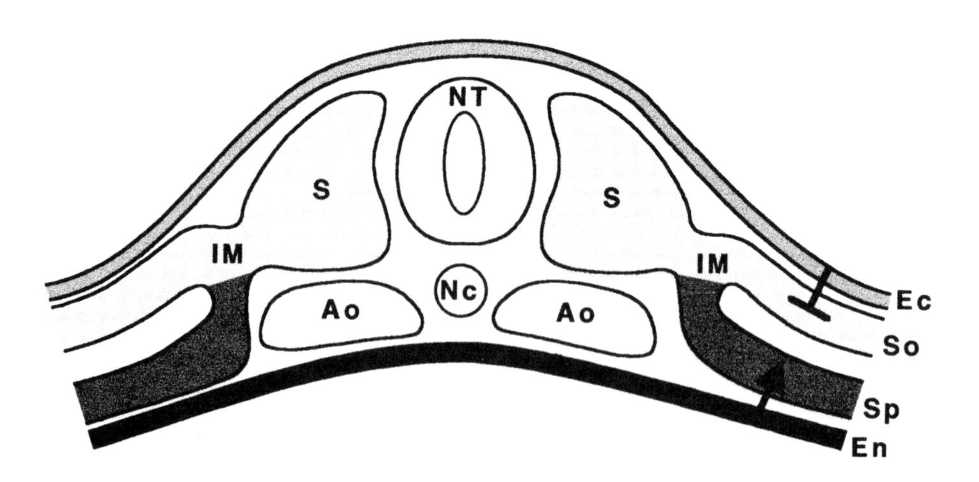

Figure 5: Scheme of the putative gradient controlling the angiopoietic/hemangiopoietic potential. Dorsally, the ectoderm inhibits the hemangiopoietic capacity in the somatopleural and somitic mesoderms while, ventrally, the endoderm promotes this potential in the splanchnopleural mesoderm. Ao: aorta; Ec: ectoderm; En: endoderm; IM: intermediate mesoderm; Nc: notochord; NT: neural tube; S: somite; So: somatopleural mesoderm; Sp: splanchnopleural mesoderm.

7. ENDOTHELIAL PRECURSORS ARE PRESENT IN THE EMBRYONIC BLOOD STREAM

Recent studies demonstrated that endothelial precursors are present in the adult blood[43,44]. Isner's group isolated putative angioblasts from human peripheral blood by magnetic bead selection on the basis of cell surface antigen expression. *In vitro*, these cells differentiated into endothelial cells. *In vivo*, these precursors can be incorporated into sites of active vascularization. Furthermore, these cells are mobilized endogenously in response to tissue ischemia or exogenously by cytokines and thereby increase neovascularization of ischemic tissues. Therapeutic applications from these findings would be very interesting: these endothelial precursors may be useful for delivering anti- or pro-angiogenic molecules, respectively, to sites of pathological or physiological angiogenesis.

However, what about the embryo? Do endothelial precursors circulate in the embryonic blood stream? Recently, we described that the avian allantois produces progenitors belonging to the blood system[45]. The quail allantoic bud grafted in a chick host produced hemopoietic progenitors and also angioblasts, which colonized the bone marrow. We postulated that this colonization occurs *via* the circulation. However, we had no solid evidence for this hypothesis.

Always using the quail/chick system, we firstly ascertained that angioblasts or endothelial cells exhibit homing and differentiation capacities when injected in the circulation. Cell suspensions of quail somites or splanchnopleural mesoderm were injected in the heart of a chick host. The next day, the host was killed and we sought $QH1^+$ endothelial cells in chick tissues. Indeed, quail endothelial cells participated to the vascularization of the host demonstrating that angioblasts or endothelial cells survived and preserved their features when they were introduced in the embryonic blood stream.

In a second approach, we isolated blood from quail embryos at different stages of development and separated the white cells on a Ficoll gradient. These cells are injected in the circulation of chick hosts through the heart. According to preliminary unpublished results, $QH1^+$ endothelial cells from 3, 4 and 5 day quail embryos are found in host chimeric vessels demonstrating the presence of angioblasts in the embryonic blood stream. The next step is to seek whether these precursors are present throughout the ontogeny and what their roles are.

8. CONCLUSION

Experimental approaches in the avian model showed that the emergence of the endothelial network and hemopoietic precursors are associated. Then, angiopoiesis and hemangiopoiesis exhibit striking developmental translocations and the early connection between endothelial and hemopoietic lineages appears to decrease during development. The experimental effects of various growth factors on the potential of distinct mesodermal subsets lead the way for identifying those operating within the embryo. It will be critical to determine at the protein level the *in situ* expression patterns of the various growth factors shown here to promote or downgrade hemangiopoiesis. Other approaches, for instance the expression of dominant negative receptors, will be required to determine which growth factors are actually involved. Unraveling these processes in their detail should help to understand the controls that ensure appropriate developmental and steady state vascularization. Aspects of these processes to which our findings may bear relevance are the elaboration of organ specific vascular patterns and the non-occurrence of hemopoietic stem cell determination in the rudiments of definitive hemopoietic organs.

ACKNOWLEDGMENTS

We thank Francis Beaujean for the illustrations and Colette Spire for the preparation of the text.

REFERENCES

1. Risau, W. and Lemmon, V., 1988, *Dev. Biol.* 125: 441-450.
2. Pardanaud, L., Yassine, F. and Dieterlen-Lièvre, F., 1989, *Development* 105: 473-485.
3. Mcbride, J.L. and Ruiz, J.C., 1998, *Mech.Dev.* 77: 201-204.
4. Wang, H.U., Chen, Z.F. and Anderson, D.J., 1998, *Cell* 93: 741-753.
5. Adams R.H., Wilkinson, G.A., Weiss, C., Diella, F., Gale, N.W., Deutsch, U., Risau, W. and Klein, R., 1999, *Genes And Development* 13: 295-306.
6. Davis, S., Aldrich, T.H., Jones, P.F., Acheson, A., Compton, D.L., Jain, V., Ryan, T.E., Bruno, J., Radziejewski, C., Maisonpierre, P.C. and Yankopoulos, G.D., 1996, *Cell* 87: 1161-1169.

7. Maisonpierre, P.C., Suri, C., Jones, P.F., Bartunkova, S., Wiegand, S.J., Radziejewski, C., Compton, D., Mcclain, J., Aldrich, T.H., Papadopoulous, N., Daly, T.J., Davis, S., Sato, T.N. and Yankopoulos, G.D., 1997, *Science* 277: 55-60.
8. Murray, P. D. F.,1932, *Strangeways Res. Labor. Cambridge* 497-521.
9. Emmel, V.E., 1916, *Am. J.Anat.* 19: 401-421.
10. Dieterlen-Lièvre, F., 1984, *Dev. Comp.Immunol.* 3: 75-80.
11. Garcia-Porrero, J., Godin, I. and Dieterlen-Lièvre, F., 1995, *Anat. Embryol.* 192: 301-308.
12. Tavian, M., Coulombel, L., Luton, D., San Clemente, H., Dieterlen-Lièvre, F. and Péault, B., 1996, *Blood* 87: 67-72.
13. Tavian, M., Hallais, M.F. and Péault, B., 1999, *Development*, 126: 793-803.
14. Wood, H. B., May, G., Healy, L., Enver, T. and Morrisskay, G. M., 1997, *Blood* 90: 2300-2311.
15. Jaffredo, T., Gautier, R., Eichmann A. and Dieterlen-Lièvre, F., 1998, *Development* 125: 4575-4583.
16. Vandenbunder, B., Pardanaud, L., Jaffredo, T., Mirabel, M.A. and Stéhelin, D., 1989, *Development* 106: 265-274.
17. Pardanaud, L., Luton, D., Prigent, M., Bourcheix, L.M., Catala, M. and Dieterlen-Lièvre, F., 1996, *Development*, 1996: 122: 1363-1371.
18. Sabin, F.R., 1920, *Carnegie Inst. Wash. Pub. Contrb. Embryol.* 9: 214-262.
19. Evans Hm., 1909, *Anat. Rec.* 3: 498-518
20. Flamme, I., 1989, *Anat. Embryol.* 180: 259-272.
21. Hirakow, R. and Hiruma, T., 1981, *Anat. Embryol.* 163: 299-306.
22. Hirakow, R. and Hiruma, T., 1983, *Anat. Embryol.* 166: 307-315
23. Pardanaud, L., Altmann, C., Kitos, P., Dieterlen-Lièvre, F. and Buck, C.A., 1987, *Development* 100: 339-49.
24. Pardanaud, L. and Dieterlen-Lièvre, F., 1993, *Cell Adh. Comm.* 1: 151-160.
25. Quéva, C., Leprince, D., Stéhelin, D. and Vandenbunder, B., 1993, *Oncogene* 8: 2511-2520.
26. Eichmann, A., Marcelle, C., Bréant, C. and Le Douarin, N.M., 1993, *Mech. Dev.* 42: 33-48.
27. Coffin, J.D. and Poole T.J., 1988, *Development* 102: 735-748.
28. Poole, T.J. and Coffin, J.D., 1989, *J. Exp. Zool.* 251: 224-231.
29. Drake, C.J., Davis, L.A., Walters, L. and Little, C.D., 1990, *J. Exp. Zool.* 255: 309-322.
30. Drake, C. J., Brandt, S. J., Trusk, T. C. and Little, C. D., 1997, *Dev. Biol.* 192: 17-30.
31. Coffin, J.D., Harrison, J., Schwartz, S. and Heimark, R., 1991, *Dev. Biol.* 148: 51-62.
32. Flamme, I. and Risau, W., 1992, *Development* 116: 435-439.
33. Eichmann, A., Corbel, C., Nataf, V., Vaigot, P., Bréant, C. and Le Douarin, N.M., 1997, *Proc. Natl. Acad. Sci. Usa.* 94: 5141-5146.
34. Metcalf, D. and Moore, M.A.S., 1971, In *Haemopoietic Cells*, (Neuberger A, Tatum El, Eds), Amsterdam, North Holland Publ, Pp550.
35. Pardanaud, L. and Dieterlen-Lièvre, F., 1999, *Development* 126: 617-627.
36. Noden, D. M.,1989, *An. Rev. Pulmon. Dis.* 140: 1097-1103.
37. Schramm, C. and Solursh, M., 1990, *Anat. Embryol.* 182: 235-247.
38. Pardanaud, L. and Dieterlen-Lièvre, F., 1995, *Anat. Embryol.* 192: 425-435.

39. Wilting, J., Brand-Saberi, B., Huang, R.J., Zhi Q.X., Kontges, G., Ordahl, C. P. and Christ, B., 1995, *Developmental Dynamics* 202: 165-171.
40. Wilt, F. H., 1965, *Science* 147: 1588-1590.
41. Feinberg, R.N., Repo, M.A. and Saunders, J.W., 1983, *J. Exp. Zool.* 226: 391-398.
42. Wilson, D.J., Mcneill, J. and Hinchliffe, J.R., 1989, *Anat. Embryol.* 180: 383-391.
43. Asahara, T., Murohara, T., Sullivan, A., Silver, M., Van Der Zee, R., Li, T., Witzenbichler, B., Schattemen, G. and Isner, I.M., 1997, *Science* 275: 964-967.
44. Takahashi, T., Kalka, C., Masuda, H., Chen, D., Silver, M., Kearny, M., Magner, M., Isner, J.M. and Asahara, T., 1999, *Nature Medecine* 5: 434-438.
45. Caprioli, A., Jaffredo, T., Gautier, R., Dubourg, C. and Dieterlen-Lièvre, F., 1998, *Proc. Natl. Acad. Sci. Usa.*, 95: 1641-1646.

Regulation of Angiogenesis and Transduction Mechanisms Involved

THE ROLE OF PERICYTES IN CONTROLLING ANGIOGENESIS *IN VIVO*

S Egginton, A-L Zhou, [1]MD Brown, O Hudlická
Angiogenesis Research Group, Department of Physiology, and [1]School of Sport & Exercise Sciences, University of Birmingham, Birmingham B15 2TT, UK.

ABSTRACT In order to evaluate the interaction between endothelial cells and the perivascular pericytes during physiological angiogenesis, stereological analysis of fine structure was performed on samples of rat skeletal muscle where capillary growth was induced to a similar extent by three different interventions (indirect electrical stimulation, vasodilatation by α_1-blockade, stretch due to synergist extirpation). There was a significant reduction in the relative area of contact between pericytes and the capillary abluminal surface with stimulation, and withdrawal of pericyte processes coincided with an increase in anatomical capillary supply. These data indicate that pericytes may play an anti-angiogenic role *in vivo* in normal adult tissue similar to that proposed for *in vitro* models of angiogenesis, with their retraction during increased muscle activity possibly releasing endothelial cells from their contact inhibition. However, following long-term peripheral vasodilatation expansion of the capillary bed was accompanied by a co-ordinated increase in pericytes, such that coverage of capillaries was similar to that in control muscles. In addition, growth of capillaries following prolonged stretch resulted in a slightly greater increase in the pericyte population, suggesting they may be permissive for endothelial cell migration. Thus, the role of pericytes in controlling physiological angiogenesis is dependent on the nature of the initial stimulus, suggesting that *in vitro* data have to be interpreted with caution when discussing the mechanism of capillary growth *in vivo*.

Angiogenesis: From the Molecular to Integrative Pharmacology
Edited by Maragoudakis, Kluwer Academic / Plenum Publishers, New York, 2000

1. INTRODUCTION

We have previously demonstrated that mechanical factors acting primarily from the capillary luminal (shear stress and/or capillary wall tension) or abluminal side (stretch of muscle fibres and adjacent capillaries and/or long-term muscle contractions) represent an important stimulus for capillary growth (Hudlická *et al*, 1998; Zhou *et al*, 1998a). So far this has been demonstrated in tissues where endothelial cell proliferation is difficult to achieve, such as skeletal and cardiac muscles. Although the importance of some perivascular cell types in controlling endothelial proliferation has been studied in cell culture *in vitro*, there are few data on their role in the transduction of signals involved in capillary growth *in vivo*. Of particular interest is the potential role that pericytes (PCs) play in processes that are fundamental to *in vivo* angiogenesis, especially in regulating endothelial cell proliferation and differentiation.

It is important to realise that those cell lines studied in culture are, strictly speaking, pericyte-like cells whose phenotype may have been modified during extraction or passage. Indeed, as there are no unequivocal cytochemical markers for PCs, the only universally applicable criterion for their identification is based on structure. Pericytes are defined as those perivascular cells found under the basement membrane of capillaries, having a close structural relationship with the endothelial cells (Fig 1). They are synonymous with Rouget or mural cells mentioned in the older literature and are found in the vicinity of most continuous or fenestrated capillaries, although they appear to represent a heterogeneous family of cells possibly with different functions in various tissue (Sims, 1986; Tilton, 1991; Shepro & Morel, 1993; Nehls & Drenckhahn, 1993; Egginton *et al*, 1996). Pericytes are thought to originate from interstitial fibroblasts and give rise to vascular smooth muscle cells, with which they share many common structural features (Rhodin, 1967). A number of roles have been postulated for these apparently pluripotential mesenchymal cells (Fig 2) that range from providing structural support for capillaries (preventing excess dilatation), secretion of extracellular matrix components, providing a scaffold along which endothelial cells migrate during sprouting, control of vascular permeability, and regulation of capillary perfusion either directly by means of their contractility or indirectly *via* production of vasoactive agonists in a paracrine fashion (Nehls *et al*, 1992; Shepro & Morel, 1993; Tilton *et al*, 1979b).

Increases in the pericyte population appear to inhibit capillary

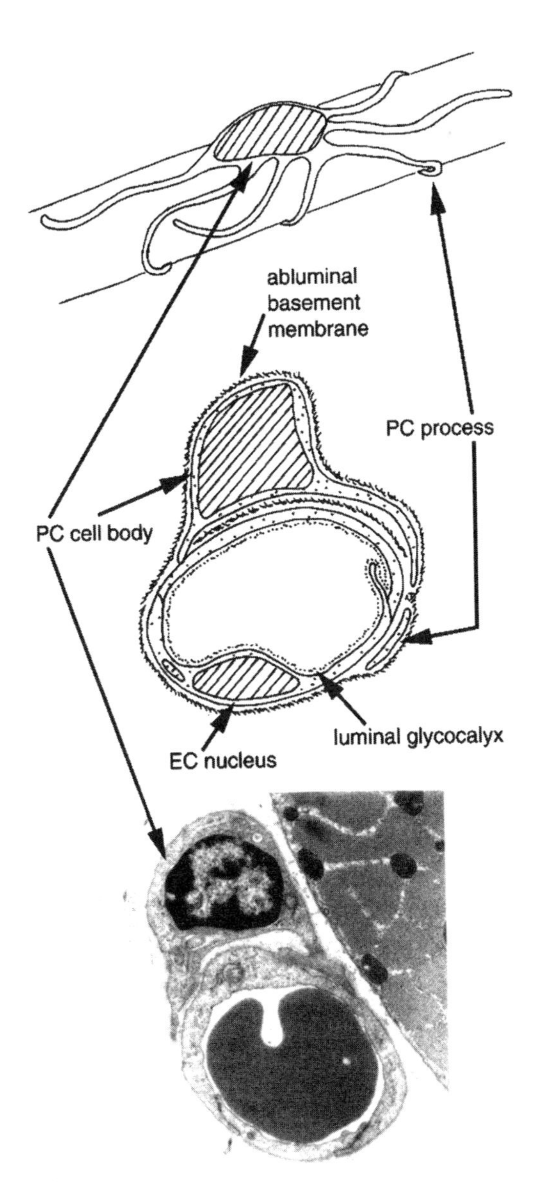

Figure 1. Pericyte morphology in mammalian skeletal muscle. Diagrammatic representation of the stellate cell in close association with a capillary in longitudinal aspect (top), and in a section through a pericyte cell body (middle). Electron micrograph showing the inclusion of both the PC cell body and processes within the basement membrane of the capillary (bottom).

proliferation in skin wounds (Crocker *et al*, 1970), and during developmental or pathological neovascularisation new vessels grow from capillaries without pericytes or where pericyte degeneration has occurred, whereas there is a decline in endothelial cell proliferation and capillaries stop growing when pericytes (re)appear (Kuwabara & Cogan, 1963; De Oliveira, 1966; Ausprunk & Folkman, 1977, Sims, 1986). In addition, pericytes inhibit endothelial cell growth when grown in co-culture, an effect similar to that of the vascular smooth muscle cells but opposite to their precursor (fibroblast) cell line, and a process that is apparently mediated by cell-cell contact (Orlidge & D'Amore, 1987).

However, a universal anti-angiogenic role for pericytes is unlikely, and they may even be permissive of capillary neoformation. For example, angiogenesis in the brain during embryonic development begins most commonly from capillaries rich in pericytes (Bär & Wolff, 1972), and within tumours or during wound healing pericytes are more numerous in vessels undergoing angiogenesis than in resting vessels (Schlingemann *et al*, 1990). The possibility clearly exists of alternative roles for pericytes under different circumstances and in different tissues. Endothelial cells secrete elements of

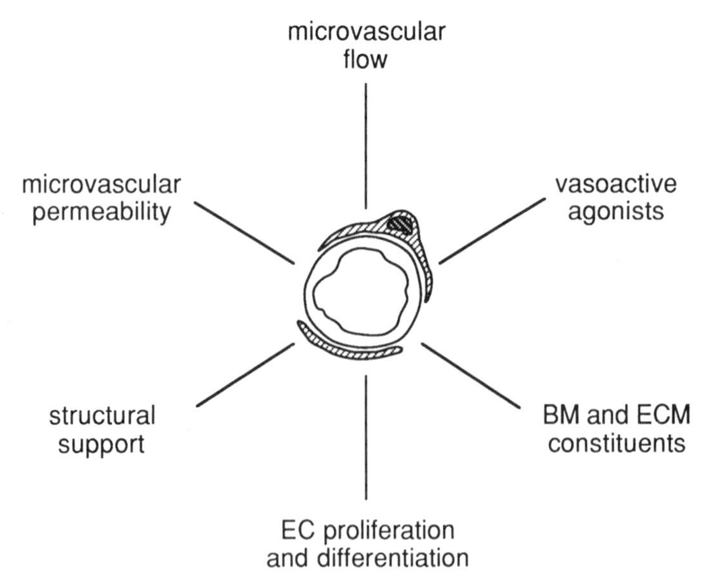

Figure 2. Postulated roles for pericytes in modifying capillary function. Note that most suggestions come from *in vitro* studies of pericyte-like cells, and that little evidence for their *in vivo* function is available.

the extracellular matrix (ECM) which may stimulate pericyte proliferation (Swinscoe & Carlson, 1992; Newcombe & Herman, 1993), while pericytes themselves synthesise many components of the ECM that are essential for EC migration. Mitotic division of pericytes during would healing has been proposed as a mechanism by which new capillary sprouts are guided in their migration through the ECM (Cliff, 1963). Indeed, the many similarities between ECs and PCs during angiogenesis may reflect their joint participation in the formation of capillary sprouts (Schor *et al*, 1992), possibly co-ordinated by reciprocal endothelial cell-pericyte interaction (Shepro & Morel, 1993; Egginton *et al*, 1996).

Most of the evidence indicating the possible role of pericytes in regulating angiogenesis has come from *in vitro* studies, and structural investigations of developmental or pathological changes. There are few data on the role of pericytes in controlling angiogenesis in mature, physiologically normal tissue when capillary growth has been induced experimentally (Egginton *et al*, 1996). Pericytes surround most capillaries in both skeletal and cardiac muscle (Tilton *et al*, 1979a), although their role may differ among tissues, e.g. contraction was demonstrated in the former but not the latter (Tilton *et al*, 1979b), and coverage is inversely related to capillary permeability (Shepro & Morel, 1993). We undertook morphometric analysis of PC structure to investigate the response of these cells during periods of active capillary growth, and their interaction with the underlying ECs, in order to address the controversy between anti- and pro-angiogenic roles proposed for these cells. In particular, we wished to establish whether they may play different roles when angiogenesis was elicited by different stimuli. The regulatory role of pericytes was therefore examined by quantifying the altered endothelial cell - pericyte interaction following angiogenic induction in skeletal muscle, under conditions where we have previously demonstrated capillary growth, viz: (1) sustained electrical stimulation; (2) long-term administration of the vasodilating drug prazosin; (3) prolonged stretch due to extirpation of a synergist. We conclude that the role played by pericytes is dependent on the nature of the initial angiogenic stimulus.

2. MATERIAL AND METHODS

All animals were treated in accordance with the UK Animals (Scientific Procedures) Act, 1986 and provided with food and water *ad libitum* during experimentation. Three experimental approaches were used that involved different forms of *in vivo* angiogenesis (Zhou *et al*, 1998a).

Stimulation

Experiments were performed on male Sprague-Dawley rats of around 150 g body mass from an in-house colony. All surgical procedures were performed under aseptic conditions and halothane anaesthesia. Five animals served as controls without any intervention (C), in 5 other animals electrodes were implanted in the vicinity of the common peroneal nerve to indirectly stimulate ankle flexors. Animals were monitored after surgery to ensure normal behaviour, and stimulation was begun about 24 h later. Muscles were stimulated for 8h/day at a frequency of 10 Hz, 0.3 ms pulse width, and with supramaximal voltage (Dawson & Hudlicka, 1989). Using this pattern of stimulation there was an increase in the anatomical capillary supply after 7 days of stimulation, but no change in muscles stimulated for 2 days and only a small increase after 4 days of stimulation (Brown *et al*, 1998). Thus, we sought to quantify structural changes once the growth process was well established (Egginton *et al*, 1996).

Dilatation

Experiments were carried out on male adult Sprague-Dawley rats, final body mass approximately 350 g. Twelve rats were divided into two groups, the first was given prazosin (gift of Pfizer plc) in their drinking water (50 mg/l) for 1 week (approximately 2 mg/day), while the second group had water only and was used as a control (Zhou *et al*, 1998c).

Stretch

Four male adult Wistar rats (final body weight approximately 310 g) were subjected to unilateral extirpation of the m. tibialis anterior (TA), performed under aseptic conditions and a combination of medetomidine and ketamine anaesthesia. This produced a sustained stretch of the remaining m. extensor digitorum longus (EDL) and m. extensor hallucis proprius (EHP), thus initiating compensatory overload hypertrophy. Muscles from both legs were removed 2 weeks later with the contralateral muscle acting as control (Zhou *et al*, 1998b).

Tissue preparation

Slices about 5 mm thick from EDL/EHPs of each animal were snap frozen in isopentane precooled in liquid nitrogen to be used for histochemical assessment of capillary supply, following staining for alkaline phosphatase (ALP) activity, which delineates capillary endothelium but not that of larger vessels. The number of capillaries and fibres was counted using an unbiased sampling frame in two fields from each muscle, and the results are expressed as capillary-to-fibre ratio (C:F) or capillary density (CD; mm^{-2}). We have previously demonstrated that the method used for ALP staining provides maximum counts for mature capillaries (Brown & Egginton, 1988), although it may underestimate the presence of very new capillaries as these stain only weakly. The influence of anastomoses and lateral sprouts on quantification of the capillary bed is accommodated by the counting protocol (Egginton, 1990).

The EHP was superfused with phosphate buffered gluteraldehyde *in situ* for ~ 5 min and subsequently fixed by immersion in the same fixative for around 30 min at 4°C; samples of EDL were fixed by immersion alone (Zhou *et al*, 1998b,c). The muscles were then sliced carefully into four blocks and placed in fresh fixative for a further 24 h at 4°C. Tissue blocks were then rinsed in buffer and post-fixed in osmium, dehydrated in alcohol, cleared in propylene oxide and vacuum-embedded in epoxy resin. One block per muscle was chosen at random from each animal, and 1.0 μm semithin sections were stained with toluidine blue to orientate the muscle fibres and capillaries for transverse sections. Ultrathin sections (60-70 nm) were double-stained with methanoic uranyl acetate and aqueous lead citrate, and viewed with a Jeol 100 CX11 electron microscope at an accelerating voltage of 60 kV. Some sections were mordanted with tannic acid prior to heavy metal staining in order to enhance the contrast of cell membranes and the basal laminae.

Morphological Observations

Ultrathin transverse sections (TS) were sampled in a systematic random manner by using a predefined square lattice with approximately 40 capillaries per animal to minimise inter-group variability and photographed onto 35-mm film for quantitative analysis (Egginton *et al*, 1996). Longitudinal sections (LS) were also examined in a subsample of both EHP and EDL muscles. Basement membrane and cytoplasmic interdigitations were visible with minimal goniometer tilt correction. Oblique capillary profiles were often obtained from collapsed (nonpatent) vessels; otherwise, only those with an aspect ratio (maximum:minimum diameter) close to unity (representing $<20^{\circ}$ axial deviation from true cross section) were used in the analyses.

Stereological Analysis

The fixative osmolarity was chosen to approximate physiological values, thereby emphasising any possible changes in capillary endothelium whilst adequately preserving ultrastructural integrity. Electron micrographs were projected at a final magnification of up to x16,000 onto a stereological counting grid *via* a microfilm reader. A lattice spacing (d) of 1.3 cm (equivalent to 0.8-1.5 μm) was used for quantification of individual capillary morphology by using standard point-count and line-intercept techniques for estimates of volume (Vv) and surface densities (Sv), respectively (Egginton *et al*, 1993). In practice, these represent the area and perimeter length of any structure as a proportion of capillary cross sectional area (see Egginton *et al*, 1996). Surface-to-volume ratio (S/V) and profile cross-sectional areas could be calculated in a similar manner, with mean cell thickness calculated as the individual cell area divided by the average of luminal and abluminal perimeters.

Our definition of pericytes (PC) is that they are those highly branched perivascular cells closely associated with capillaries and bounded by a common basal lamina (Fig 1). The relative contact area between cell types was estimated as the boundary length of pericyte membrane subtending the abluminal capillary surface, given as a fraction of the total capillary boundary length (abluminal circumference). Data for pericyte processes and cell bodies (defined by the presence of a nucleus) were combined for each capillary in this analysis. While this approach could underestimate PC volume relative to other studies that also include sub-adjacent PCs, the subtended perimeter will be comparable. The frequency of pericyte processes found in close association with inter-endothelial cell (EC) clefts was calculated as (number of clefts with PC processes/total number of clefts). Other indices of endothelial cell-pericyte interaction were scored on a nominal scale (Egginton *et al*, 1993), e.g. presence of cytoplasmic budding (projections), and interdigitations of projections within the boundary of the adjacent cell type.

Statistical Analysis

Single-factor analysis of variance (ANOVA) was used for comparison of values, with Scheffe's multiple-range F test to estimate significance between groups, set at $\alpha = 0.05$.

3. RESULTS

Control

Capillary morphology was similar to that previously described (Brown & Egginton, 1988; Egginton *et al*, 1993; 1996) with thin endothelial cells containing numerous cytoplasmic vesicles, bounded by distinct intercellular junctions (clefts) with no obvious signs of disturbance to the luminal glycocalyx, and surrounded by a clearly visible and intact basement membrane (BM) (Fig 1). Signs of abluminal endothelial activity such as endothelial cytoplasmic projections to the perivascular space were only occasionally seen in control muscle. Pericytes occurred within the BM envelope around virtually all capillaries, i.e. their apical BM (facing the interstitium) fused with that of the EC to form a contiguous layer. Pericyte basal and EC abluminal BMs formed a discontinuous structure within the intercellular gap, which was absent where the two cell types came into close contact. This was clearest at the margins of pericyte projections where junctions similar to those observed between two ECs could occasionally be found between PCs and ECs. On average, pericyte coverage of the capillary perimeter was around 30% in both control and sham operated skeletal muscle samples, when random cross-sections of different capillaries were analysed. However, <1% of capillary profiles were devoid of any associated pericyte cell body or processes, i.e. essentially the whole length of any individual capillary was to some extent associated with pericytes.

Stimulation

Chronic stimulation of hindlimb extensor muscles produced a 60% increase in capillary to fibre ratio (C:F) after 7d (Fig 3). Angiogenesis was accompanied by development of irregular luminal and abluminal EC surfaces, evidence of extensive EC activation, and the occurrence of small sprouts. A basement membrane was always present around capillaries, but became amorphous where PC projections were in close proximity to the capillary wall (Hansen-Smith *et al*, 1996).

After 7d of stimulation PC coverage declined by around 30% (Fig 4). Gross changes in pericyte morphology were not evident, although the volume density of PC nuclei was reduced following stimulation. As a greater proportion of capillaries from stimulated muscles had an adjacent pericyte cell body than did control muscle this suggests an increase in cytoplasmic volume due to withdrawal of cellular processes and rounding-up of existing PCs. Supporting evidence for this is that pericyte topology was altered, resulting in a decrease in surface: volume ratio (S/V; Fig 4). Consistent with this interpretation is an increase in the proportion of capillaries lacking any PC contact. Although no obvious change was evident in the nature of the close association between ECs and PCs there was a greater proportion of PC projections associated with EC clefts in stimulated muscles while interdigitation between cell types increased with stimulation, particularly of PC projections into ECs (Fig 5).

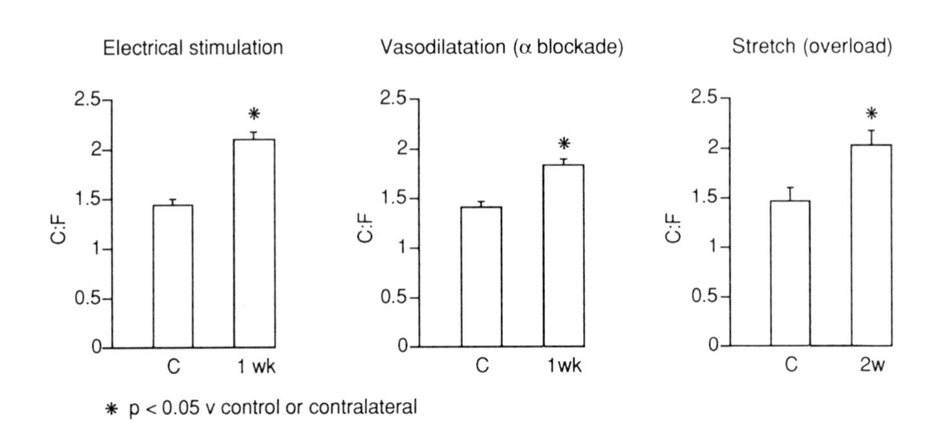

Figure 3. Animal models of physiological angiogenesis showing similar degrees of capillary growth (Brown *et al*, 1998; Zhou *et al*, 1998b & c), but involving different forms of angiogenesis (Zhou *et al*, 1998a)

Dilatation

Chronic administration of the vasodilator prazosin produced a 27% increase in C:F after 1 week (Fig 3). Angiogenesis occurred by luminal division (splitting of existing vessels), and possibly elongation of endothelial cells, leading to a progressive increase in the number of ECs per capillary after prazosin treatment, as a result of endothelial cell proliferation or longitudinal reorganisation of the capillary wall (Zhou *et al*, 1998c).

Following dilatation, however, there was no evidence of either a change in PC morphology (Fig 4) nor in the proportion of the capillary abluminal surface coverage by PCs (Fig 4). While the PC abundance was

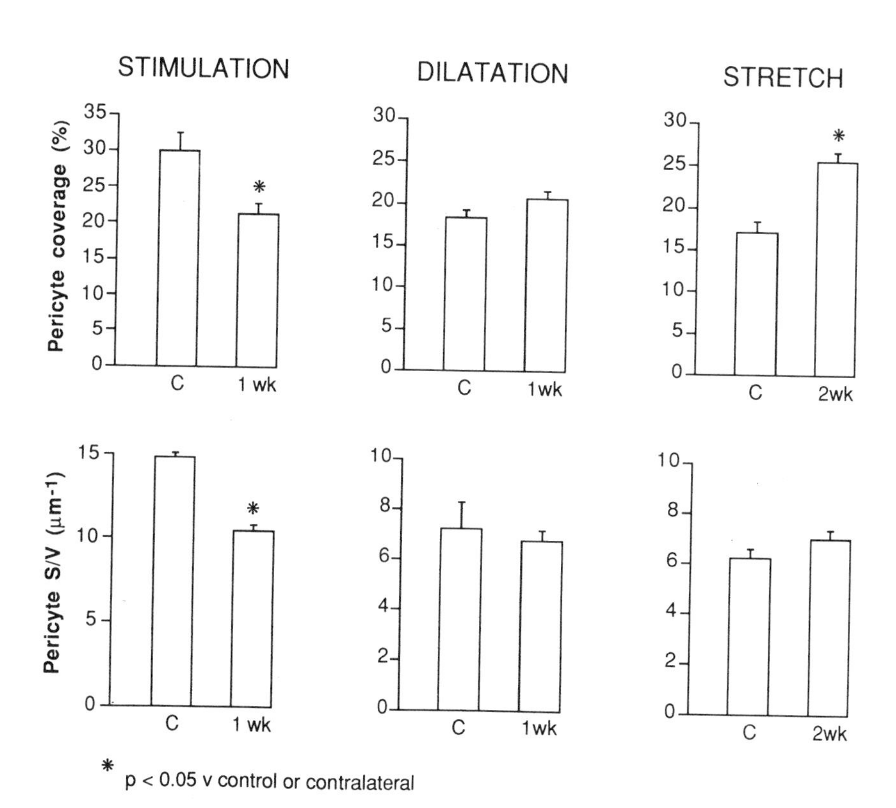

Figure 4. Top: pericyte coverage (percentage of the capillary circumference subtended by PC cell bodies and processes) in muscles during angiogenesis. Three different models give three different responses – withdrawal, maintenance and expansion for stimulation, dilatation and stretch, respectively. Bottom: pericyte surface to volume ratio (S/V) confirmed retraction of PC processes in stimulated muscle, with little change in PC morphology in response to dilatation or stretch.

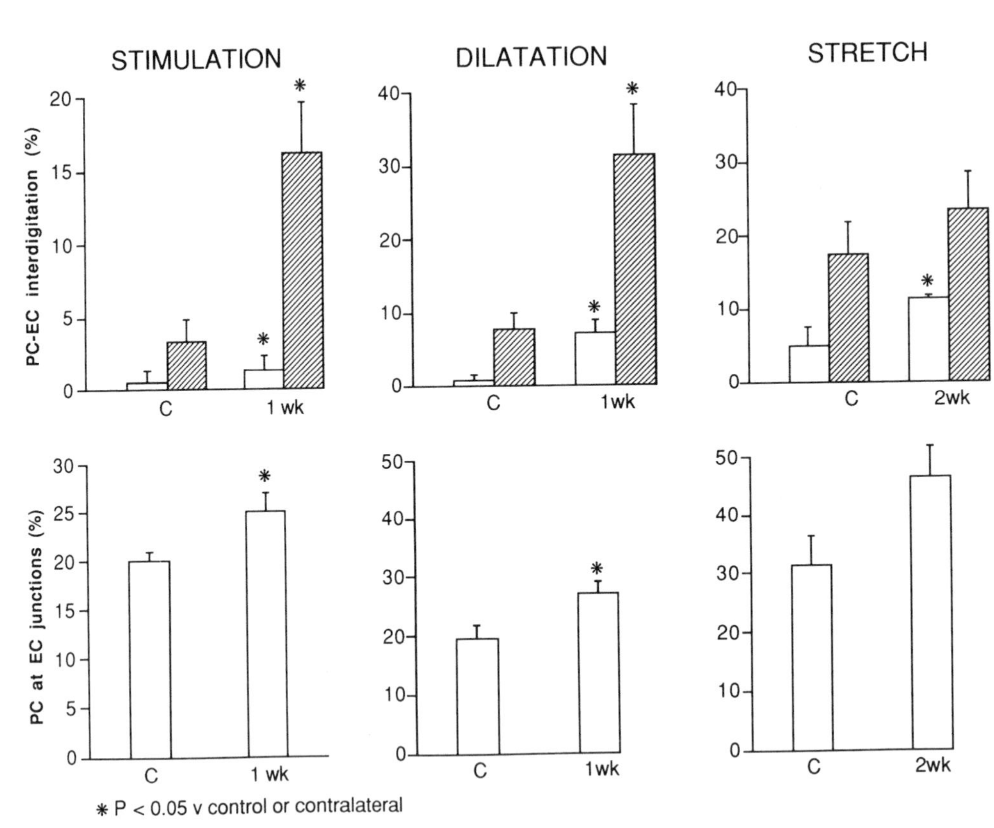

Figure 5. Top: proportion of capillaries with overt EC (open bars) or PC (hatched bars) protrusions towards the adjacent cell type, facilitating interdigitation or other forms of intimate contact. A significant increase in one or both forms of cellular interaction is seen during angiogenesis. Bottom: prevalence of PCs located at EC junctions increased in stimulated and vasodilated muscles, where blood flow was increased, but not in stretched muscle where blood flow was unaltered.

Table 1. Pericyte abundance around capillaries during *in vivo* angiogenesis (%)

	Capillaries lacking any PC contact	Capillaries with adjacent PC cell body
Control	< 1	17.0
1 wk stimulation	4.0	28.0
Control	5.0	5.8
1 wk dilatation	4.6	8.8
Control	10.0	3.8
2 wk stretch	6.3	2.5

unchanged following prazosin treatment (Table 1) some interaction between ECs and the adjacent pericytes (PCs) was evident, including an increase in both EC projections into PCs, and *vice versa* (Fig 5). There was also an increase in the proportion of intercellular junctions where PC processes were evident (Fig 5).

Stretch

Stretch-induced angiogenesis led to a significant increase in capillary supply only after 2 weeks, although signs of EC activation were evident at 1 week (Zhou et al, 1998b) including thickened capillary endothelial cells (ECs), increased organelle content (endoplasmic reticulum, ribosomes and mitochondria), and the presence of large vacuoles and irregular luminal capillary surface observed in other models of angiogenesis (Hansen-Smith *et al*, 1996; Zhou *et al*, 1998a). The proportion of capillaries with focal discontinuity of the basement membrane (BM) was higher, which is indicative of the first step in EC sprouting. Small sprouts usually had 1-2 ECs, a complete BM and associated pericytes (Zhou *et al*, 1998b).

As with dilatation, stretch induced no significant changes in the extent of PC coverage despite nearly a 40% increase in C:F (Fig 3). There was little change in PC abundance (Table 1) or morphology following synergist extirpation, with a similar S/V as control muscle (Fig 4). However, PC topology was markedly different with a substantial increase in cellular processes that was reciprocal between ECs and PCs (Fig 5), as well as a realignment of PC processes towards EC clefts (Fig 5).

4. DISCUSSION

Electrical stimulation of skeletal muscle is a very effective way of inducing capillary proliferation (Hudlicka, 1991), while angiogenesis has previously been demonstrated following administration of prazosin (Dawson & Hudlicka, 1990) and imposition of stretch (Zhou *et al*, 1998b). We examined muscles after suitable intervals that produced a comparable degree of capillary growth following each intervention (Fig 3).

In vivo studies on the possible interaction between endothelial cells and pericytes have tended to concentrate on pathological states such as wound healing and tumour growth, with considerable attention being addressed toward retinopathies. Some workers even consider that all angiogenesis is a result of an inflammatory response to altered metabolic demand or physical environment. However, little attention has been paid to the role of perivascular pericytes in mature tissue in which *in vivo* angiogenesis has been induced as a physiological response. In comparing *in vivo* models of angiogenesis, both inter-organ and inter-specific variations are to be expected. For the present purposes, however, these appear not to influence the degree of pericyte coverage in normal tissue to any great extent, with our data being similar to that reported for rat skeletal muscle (Tilton *et al*, 1979a). Small differences may be associated with different size or strain of rats used in these studies, although the comparison of experimental data with their respective controls allow robust conclusions to be drawn. Capillary morphology alters little after stimulation, other than a thickening of endothelial cells and signs of activation such as increased mitochondrial and endoplasmic reticulum content. Although newly formed capillaries tend to be smaller than pre-existing vessels, with slit-like lumen and an increased proportion (volume density) of endothelial cells (Hansen-Smith *et al*, 1996; Zhou *et al*, 1998a), their influence on mean capillary size is small. Similar changes in EC morphology were noticed following stretch and chronic dilatation, but with new vessels formed by external sprouting or internal division, respectively (Zhou *et al*, 1998b,c).

What can morphometry tell us about the function of pericytes? We chose to use morphometric techniques in order to avoid the ambiguities inherent in cell culture approaches and the uncertainties of cytochemical identification. Unambiguous identification of PCs on the basis of definitive structural criteria is particularly important when tissue integrity is disturbed, e.g., when fibroblasts may otherwise be easily mistaken for PCs.

Pericyte coverage

Pericyte coverage of the capillary abluminal surface is consistent with one of the earliest postulated roles for pericytes, that of a physical restraint on capillary proliferation. Variation in extent is unlikely to result from inter-tissue differences in the pattern of microvessel growth, where angiogenesis occurs either by sprout formation or by internal division of existing capillaries. The degree of tissue-specificity reported, with lung having less and brain more coverage (Weibel, 1974; Allsopp & Gamble, 1979), may indicate other potential roles for pericytes, than that of controlling microvascular permeability. The reduction in coverage coincident with angiogenesis as a result of stimulation can be most readily explained by withdrawal of pericyte processes, which is consistent with a reduction in S/V ratio in skeletal muscle as PCs became more rounded in appearance. It would also explain the data on PC abundance, where a greater proportion of the capillary length lacks any adjacent PC processes, and the apparent increase in the probability of finding a PC cell body next to a capillary. This interpretation relies on pericytes displaying some degree of contractility (Murphy & Wagner, 1994) for which there is no direct evidence *in vivo* (Nehls & Drenckhahn, 1993), although there are morphometric data that are consistent with such a phenomenon (Tilton *et al*, 1979b). As a consequence, the constraint on EC proliferation apparently imposed by the proximity of PCs (Orlidge & D'Amore, 1987) may be removed.

However, angiogenesis induced by chronic dilatation or stretch involves a maintenance, or even increased coverage by pericytes. In the former case PC abundance data suggests that a co-ordinated growth of PCs and capillaries occurs. This would be consistent with their proposed role in modification of the extracellular matrix to assist EC migration (Nehls *et al*, 1992) or acting as a scaffold along which ECs may be guided (Cliff, 1963). In stretched muscle, increased coverage may result from the migration of fibroblasts and their transformation into PCs (Rhodin & Fujita, 1989) prior to their transformation into smooth muscle cells during capillary arterialisation (Hansen-Smith *et al*, 1998). In both situations, the role of PCs would appear to be permissive for angiogenesis.

Pericyte topology

Pericytes show extensive branching with PC processes forming an incomplete layer around the capillary abluminal surface. The proportion of capillaries with pericyte profiles having close cytoplasmic contact with endothelial cells, which are reported to include tight junctions (Schulze &

Firth, 1993), was similar to that observed in lung and heart by Eppling (1966). The increased likelihood of finding a pericyte process adjacent to or covering an endothelial cleft may reflect slippage of the pericyte around the capillary. Whatever the cause, the position of a PC process relative to an EC cleft clearly has the potential to alter capillary permeability by either increasing EC cleft width or exposing the junction surfaces following PC contraction. This would be most likely *via* PC contraction (Tilton, 1991) if interdigitations of EC and PC projections provided firm anchorage points.

Endothelial cell communication with pericytes raises the possibility of flow regulation, supported by morphological observations showing pericytes spanning two capillary sprouts (Nehls *et al* , 1992) or mature vessels (Weibel, 1974; Gaudio *et al*, 1990). This may be a result of a reduced capillary diameter if PCs are wrapped around the vessel, or by adjusting the haemodynamics of flow at vessel bifurcations (increasing resistance and/or turbulance). Direct communication between endothelial cells and pericytes is suggested by the appearance of tight junctions (above), but there are also close cytoplasmic interdigitations ('peg and socket' relationships; Egginton et al, 1996) which may act as physical unions by which pericyte contractility exerts its influence on capillary diameter (Matsusaka, 1975; Gaudio *et al*, 1990). Interestingly, such structures were absent from the myocardial capillary bed (Schulze & Firth, 1993; Egginton *et al*, 1996), while we observed a significant increase in the occurrence of such structures when skeletal muscle was stimulated. The coordinated increase in both EC and PC populations following dilatation and, to a lesser extent, stretch also involved an increase in EC-PC interaction which would be consistent with inter-cellular communication providing a means of co-ordinating cellular proliferation.

5. CONCLUDING REMARKS

The explanation that perhaps best fits the present morphometric data is that both sprouting endothelial cells and pericytes may participate in sprout formation during angiogenesis (Schor *et al*, 1992). As samples in the present study were taken at random, the morphometric analyses provide average values from these pericytes both located close to, and remote from, the site of capillary neoformation. Hence, pericytes may perform a dual role of preventing EC proliferation at the proximal end, and providing a scaffold for EC migration at the distal end of capillary sprouts (Rhodin & Fujita, 1989). Whether this is a consequence of pericytes inducing endothelial tube formation (Shepro & Morel 1993) and/or acting as guiding structures (Cliff, 1963) awaits definitive *in vivo* studies.

These results support the suggestion that an inverse correlation exists between endothelial cell proliferation and the extent of pericyte coverage, and demonstrates co-ordinated regulation of capillaries and pericytes during physiological angiogenesis. Thus, in skeletal muscle there is reciprocal change in relative pericyte coverage and capillary supply accompanying or preceding *in vivo* angiogenesis following increased muscle activity. In contrast, where capillary growth was induced by primarily mechanical stimuli, the data suggests that there is co-ordinated growth of capillaries and pericytes during physiological angiogenesis.

Table 2. Pericyte involvement in physiological angiogenesis

	Stimulation	Dilatation	Extirpation
Capillary growth	✓	✓	✓
PC withdrawal	✓	X	X
PC interdigitation	✓	✓	✓
PC clelf coverage	✓	X	✓
EC sprouts	✓	X	✓
EC number	✓	✓	✓

ACKNOWLEDGMENTS

This work was supported by the MRC (UK) and NATO.

REFERENCES

Allsopp, G and Gamble, H.J. (1979) An electron microscopic study of the pericytes of the developing capillaries in human fetal brain and muscle. J. Anat. 128: 155-168

Ausprunk, D.H. & Folkman, J. (1977) Migration and proliferation of endothelial cells in preformed and newly formed vessels during tumour angiogenesis. Microvasc. Res. 14: 53-65

Bär, T., & Wolff, J.R. (1972) The formation of capillary basement membranes during internal vascularization of the rat's cerebral cortex. Z. Zellforsch. 133: 231-248

Brown, M.D., Egginton, S. (1988) Capillary density and fine structure in rabbit papillary muscles after a high dose of norepinephrine. Microvascular. Res. 36: 1-12

Brown, M.D., Walter, H., Hansen-Smith,F.M., Hudlicka, O., Egginton, S. (1998) Lack of involvement of basic fibroblast growth factor (FGF-2) in capillary growth in skeletal muscles exposed to long-term contractile activity. Angiogenesis 2: 81-91

Cliff, W.J. (1963) Observations on healing tissue: A combined light and electron microscopic investigation. Philos. Trans. R. Soc. London 246: 305-325

Crocker, D.J., Murad, T.M. & Geer, J.C. (1970) Role of the pericyte in wound healing. An ultrastructural study. Exp. Mol. Pathol. 13: 51-56

Dawson, J.M., Hudlická, O. (1990) The effect of long-term administration of prazosin on the microcirculation in skeletal muscle. Cardiovasc. Res. 23: 913-920

De Oliveira, F. (1966) Pericyctes in diabetic retinopathy. Br. J. Opthal. 50: 134-143

Egginton, S. (1990) Morphometric analysis of tissue capillary supply. In: Boutilier, R.G. (ed) Vertebrate Gas Exchange from Environment to Cell. Advances in Comparative and Environmental Physiology 6: 73-141

Egginton, S., Hudlická, O., Glover, M. (1993) Fine structure of capillaries in ischaemic and non-ischaemic muscles in rat striated muscle. Effect of torbafylline. Int. J. Microcirc. Clin. Exp. 12: 33-44

Egginton, S., Hudlická, O., Brown, M.D., Graciotti, L., Granata, A-L. (1996) *In vivo* pericyte-endothelial cell interaction during angiogenesis in adult skeletal muscle. Microvasc. Res. 51: 213-228

Eppling, G.P., (1966) Electron microscopic observations of small blood vessels in the lungs and hearts of normal cattle and swine. Anat. Rec. 155 513-530

Gaudio, E., Pannarale, L., Caggiati, A., & Marinozzi, G. (1990) A three-dimensional study of the morphology and topography of pericytes in the microvascular bed of skeletal muscle. Scanning Microsc. 4: 491-500

Hansen-Smith, F., Hudlická, O., Egginton, S. (1996) *In vivo* angiogenesis in adult rat skeletal muscle: early changes in capillary network architecture and ultrastructure. Cell Tiss. Res. 286: 123-136

Hansen-Smith, F.M., Egginton, S., Hudlická, O. (1998) Growth of arterioles in chronically stimulated adult rat skeletal muscle. Microcirculation 5: 49-59

Hudlická, O. (1991) What makes blood vessels grow? J. Physiol. 444: 1-24

Hudlická, O., Brown, M.D., Egginton, S. (1992) Angiogenesis in skeletal and cardiac muscle. Physiol. Rev. 72: 369-117

Hudická, O., Brown, M.D., Egginton, S. (1996) Angiogenesis in skeletal muscle. In: Maragoudakis, M.E. (ed.) 'Molecular, cellular and clinical aspects of angiogenesis'. Plenum Press, New York. pp141-150.

Hudlicka, O., Egginton, S., Brown, M.D. (1998) Angiogenesis in heart and skeletal muscle – models for capillary growth. In: Maragoudakis, M.E. (ed) Angiogenesis: Models, modulators and clinical applications. Plenum Press, New York. pp19-33

Kuwabara, T., Cogan, D.G. (1963) Retinal vascular patterns VI. Mural cells of the retinal capillaries. Arch Ophthalmol. 69: 492-502

Matsusaka, T. (1975) Tridimensional views of the relationship of pericytes to endothelial cells of capillaries in the human choroid and retina. *J. Electron. Microsc.* 24: 13-18

Murphy, D.D. & Wagner, R.C. (1994) Differential contractile response of cultured microvascular pericytes to vasoactive agents. Microcirculation. 1: 121-128

Nehls, V., Denzer, K., Drenckhan, D. (1992) Pericyte involvement in capillary sprouting during angiogenesis *in situ*. Cell Tissue Res. 270: 469-474.

Nehls, V. and Drenckhaln, D. (1993) The versitility of microvascular pericytes: From mesenchyme to smooth muscle? Histochemistry 99, 1-12.

Newcomb, P.M. & Herman, I.M. (1993) Pericyte growth and contractile phenotype : Modulation by endothelial-synthesized matrix and comparison with aortic smooth muscle. J. Cell Physiol. 155: 385-393

Orlidge, A., D'Amore, P.A. (1987) Inhibition of capillary endothelial cell growth by pericytes and smooth muscle cells. J. Cell Biol. 105: 1455-1462.

Rhodin, J.A.G. (1967) The ultrastructure of mammalian arterioles and precapillary sphincters. J. Ultrastruct. Res. 18, 181-223

Rhodin, J.A.G. and Fujita, H. (1989) Capillary growth in the mesentary of normal young rats. Intravital video and electron microscope analyses. J. Submicrosc. Cytol. Pathol. 21, 1-34

Schlingemann, R.O., Rietveld, F.J.R., deWaal, R.M.W., Ferrone, S. & Ruiter, D.J. (1990) Expression of the high molecular weight melanoma-associated antigen by pericytes during angiogenesis in tumours and in healing wounds. Am. J. Pathol. 136: 1393-1405

Schor, A.M., Canfield, A.E., Sutton, A.B., Allen, T.D., Sloan, P. & Schor, S.L. (1992) The behaviour of pericytes *in vitro*: Relevance to angiogenesis and differentiation. In "Angiogenesis: Key Principles" (P.B. Weisz & R Langer, Eds) Birkhauser Verlag, Basel.

Schulze, C., and Firth, J.A. (1993) Junctions between pericytes and the endothelium in rat myocardial capillaries: A morphometric and immunogold study. Cell Tissue Res. 271, 145-154

Shepro, D., and Morel N.M.L. (1993) Pericyte physiology FASEB. J. 7, 1031-1038

Sims, D.E. (1986) The pericyte – a review. Tissue Cell 18: 153-174.

Swinscoe, J.C. & Carlson, E.C. (1992) Capillary endothelial cells secrete a heparin-binding mitogen for pericytes J. Cell Sc. 103: 453-461

Tilton, R.G., Kilo, C., and Williamson, J.R. (1979a) Pericycte-endothelial relationship in cardiac and skeletal muscle capillaries. Microvasc. Res. 18, 325-335

Tilton, R.G., Kilo, C., Williamson, J.R., and Murch, D.W. (1979b) Differences in pericyte contractile function in rat cardiac and skeletal muscle microvasculatures. Microvasc. Res. 18, 336-352

Tilton, R.G. (1991) Capillary pericyctes: Perspectives and future trends. J. Electron Microsc. Tech. 19, 327-344

Weibel, E.R. (1974) On pericytes, particularly their existance on lung capillaries. Microvasc. Res. 8: 218-235

Zhou, A-L., Egginton, S., Hudlická, O. (1998a) Ultrastructural study of three types of physiological angiogenesis in adult rat skeletal. In: Maragoudakis, M.E. (ed.) 'Angiogenesis. Models, modulators and clinical applications'. Plenum Press, New York. pp556-557.

Zhou, A-L., Egginton, S., Brown, M.D., Hudlická, O. (1998b) Capillary growth in overloaded, hypertrophic adult rat skeletal muscle: an ultrastructural study. Anat. Rec. 252: 49-63.

Zhou, A-L., Egginton, S., Hudlická, O., Brown, M.D. (1998c) Internal division of capillaries in rat skeletal muscle in response to chronic vasodilator treatment with α_1 antagonist prazosin. Cell Tiss. Res. 293: 293-303.

INSIGHTS INTO THE VASODILATION OF RAT RETINAL VESSELS EVOKED BY VASCULAR ENDOTHELIAL GROWTH FACTOR$_{121}$ (VEGF$_{121}$)

HASSÉSSIAN, H.M.
Centre de Recherche Guy-Bernier, Hôpital Maisonneuve-Rosemont, Department of Ophthalmology, Université de Montréal

1. INTRODUCTION

We have previously reported that cultured retinal capillary endothelial cells express VEGFR-2, whereas VEGFR-1 protein could not be detected (Hasséssian, 1997). Both types of VEGF receptor are tyrosine kinases (Ferrara and Davis-Smyth, 1997). Furthermore, we have demonstrated that VEGF$_{121}$ is more potent than VEGF$_{165}$ to evoke vasodilation in the retina, and that the VEGF$_{121}$ evoked vasodilation is due to the release of nitric oxide (NO) as well as prostacyclin from the endothelium (Hasséssian et al, 1998). Concurrent with the vasodilator effect, we observed increases of intracellular Ca^{2+}, in endothelial cells, which were almost entirely due to influx across the plasma membrane (Hasséssian, 1997). Both VEGF$_{165}$ and VEGF$_{121}$ were found to evoke Ca^{2+} influx (Hasséssian, 1997). We initiated the current set of experiments to determine if the Ca^{2+} influx is necessary for the vasodilator response to VEGF$_{121}$ and to investigate further the cellular mechanisms responsible for the vasodilation.

Angiogenesis: From the Molecular to Integrative Pharmacology
Edited by Maragoudakis, Kluwer Academic / Plenum Publishers, New York, 2000

2. METHODS

2.1 Vasomotor Video Imaging

Eye cup preparations were used to study the retinal vasculature essentially as described by Lahaie et al (1998). Male rats (200 g) were sacrificed by guillotine and enucleated. The eyes were immediately placed in ice cold Krebs buffer (pH 7.4) consisting of the following ingredients (mM) : 120 NaCl, 4.5 KCl, 3.3 $CaCl_2$, 1.0 $MgSO_4$, 27 $NaHCO_3$, 1.0 KH_2PO_4 , 10 glucose and 1.0 U/ml heparin. A circular incision was made 3-4 mm posterior to the ora serrata, to remove the anterior segment and the vitreous without handling the retina. The remaining eye cup, containing the retina, was fixed with pins to a wax base in a 200 μl well containing Krebs buffer (pH 7.4) equilibrated with 95% O_2 and 5% CO_2 and maintained at 37^0C. The preparation was allowed to stabilize for 30-45 min before the start of experiments. Non perfused vessels of 100-200 μm diameter were then examined from fresh tissue. The outer vessel diameter was recorded with a video colour camera (3CCD-DC330, DAGE-MTI inc, Michigan City, USA) mounted on a microscope (MZ12, Leica), and responses were quantified by a digital image analyzer (Sigma Scan Software, Jandel Scientific, Corte Madera, CA). Vessels were constricted using 9,11-dideoxy-11α, 9α epoxy-methanoprostaglandin $F_{2\alpha}$ (U-46619, Sigma-Aldrich Canada, Ltd) to provide initial tone necessary for the study of vasodilation. A single application of 1 mM U-46619 was effective for more than 90 min. Vascular diameter was recorded 10 min after topical application of each concentration of agent, at which time a stable response had been reached. The responses are expressed as a percentage of the surface area, constituted by a chosen length of vessel, when compared to the resting vessel prior to the application of U-46619.

2.2 Statistical Analysis

Data were analysed, as required, with a paired or non-paired Student's t-test, where a probability of $P<0.05$ was considered significant, and n represents the number of retinas which were used. Data are expressed as means ± standard error of the mean.

3. RESULTS AND DISCUSSION

3.1 Tyrosine Kinase Mediated, Ca^{2+} Dependent Vasodilation

Concentrations of 10 pM-100 nM $VEGF_{121}$ produced a dose dependent vasodilation with an ED_{50} of 50 pM, and the maximal effect was obtained with 100 pM $VEGF_{121}$ (Figure 1A). This ED_{50} is similar to the K_d that has been reported for the VEGFR-2 (Ferrara and Davis-Smyth, 1997), the only VEGF receptor type present on retinal capillary endothelial cells (Hassésian, 1997), hence the response is consistent with a VEGFR-2 mediated effect. A 10 mM concentration of $GdCl_3$, which is known to block Ca^{2+} influx through non-selective cation channels in non excitable cells such as those in endothelium (Egee et al., 1998; Fernando and Barritt, 1994), was able to block the vasodilation evoked by $VEGF_{121}$. This demonstrates that the Ca^{2+} influx produced by $VEGF_{121}$ in endothelial cells is necessary for the vascular response. Herbamycin (1 μM), an inhibitor of tyrosine kinase, was without effect on resting vessel diameter, whereas it blocked the vasodilation response to $VEGF_{121}$. Furthermore, retinal vessels treated with herbamycin could vasodilate in response to sodium-nitroprusside (Figure 1B), therefore the inhibition by herbamycin is not due to a non-specific effect. Clearly the vasodilation is tyrosine kinase mediated, and dependent on endothelial cell Ca^{2+} influx.

3.2 Are G-Proteins Required?

Activation of G-proteins with 100 μM of the hydrolysis resistant GTPγS, vasodilated retinal vessels (Figure 2A), demonstrating that G-proteins are part of the cellular mechanism for vasodilation in the retina. In order to specify which G-protein type is involved in $VEGF_{121}$ evoked vasodilation, we used ADP-ribosylating toxins. Application of, 0.1 μM cholera toxin, which catalyses the ADP-ribosylation of the α-subunit of G-proteins and is known to be sufficient to activate only G-proteins of the G_s, G_{olf} and G_t variety (Di Rita et al., 1991; Moss and Vaughan, 1992), was without significant effect on vessel diameter (Figure 2B). However, 0.1 μM pertussis toxin, which is known to catalyse the ADP-ribosylation of G-proteins and thus inactivates G_i , G_o and G_t –proteins (Hazes et al., 1996;

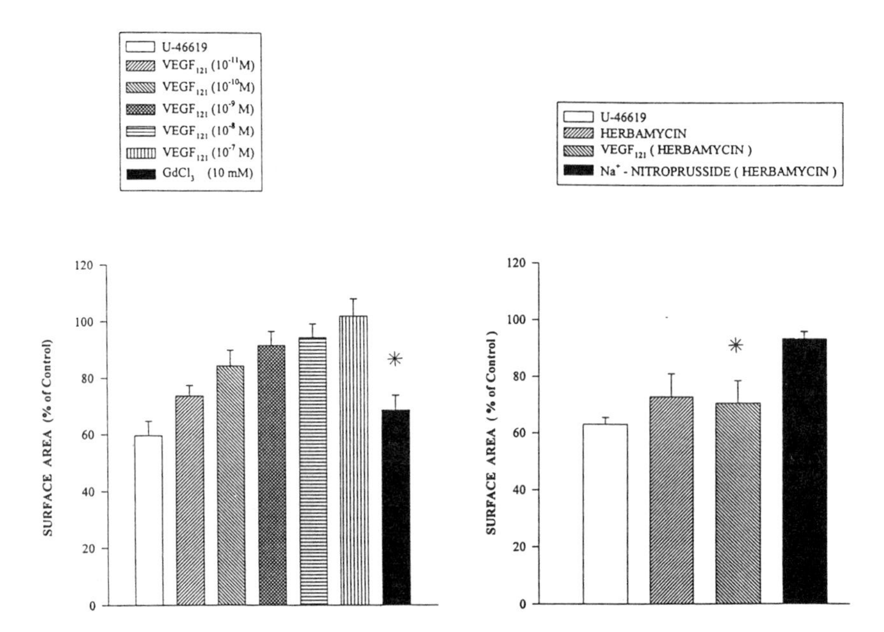

Figure 1. Antagonism of the Vasodilation by Inhibition of Ca^{2+} Influx or Tyrosine Kinase. A) Vasodilation evoked by increasing concentrations of $VEGF_{121}$, and the antagonism of the vasodilator response to 10 nM $VEGF_{121}$ produced by 10 mM $GdCl_3$. * $P<0.05$ (n=8) when the response to 10 nM $VEGF_{121}$ is compared in the presence and absence of 10 mM $GdCl_3$ using a paired Student's t-test. B) The effect of herbamycin on the vasodilation produced by 10 nM $VEGF_{121}$ and sodium nitroprusside. * $P<0.05$ (n=6) treatment with herbamycin significantly antagonised the response to 10 nM $VEGF_{121}$ when compared using a non-paired Student's t-test with the group shown in Figure 1A.

Moss and Vaughan, 1992), was able to block the effect of 1 nM $VEGF_{121}$ on vascular tone (Figure 2C), indicating that G-proteins of the G_i or G_o variety are involved. Where between endothelial cell receptor activation and smooth muscle relaxation, are G_i or G_o proteins involved remains to be established.

3.3 Adenylate Cyclase Derived Effects on Vascular Tone

Forskolin produced a concentration dependent vasodilation with an ED_{50} of 0.01μM (Figure 3A), demonstrating that cAMP mediates vasodilation of retinal vessels. This is consistent with an effect on adenylate

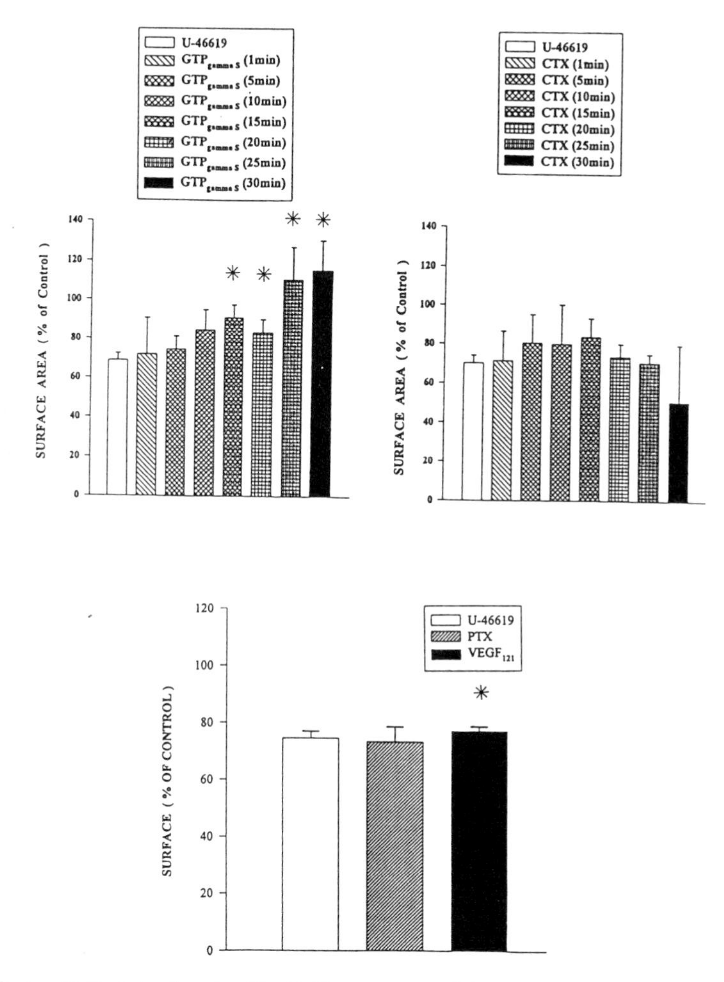

Figure 2. Are G-Proteins Involved? A) Time course of the vasodilation evoked by 100 μM GTPγS. ∗ $P<0.05$ (n=7) when compared, using a paired Student's t-test, to column derived from U-46619 alone. B) The vessel diameter over time after administration of 0.1 μM cholera toxin (CTX). There is no significant effect ($P>0.05$, n=13) over the time period which was studied. C) Vessel diameter after treatment with 0.1 μM pertussis toxin and antagonism of the vasodilation evoked by 10 nM $VEGF_{121}$.∗ $P<0.05$ (n=11) when compared, using a paired Student's t-test to the column derived from pertussis toxin alone.

cyclase, and not on MAP kinase since the concentration we used is 2500 times lower than what would be required to inhibit MAP kinase (Galli et al., 1995; Laurenza et al., 1989; Li et al., 1995; Lomo et al., 1995). Inhibition of adenylate cyclase with 500 μM MDL-12,330A (Correia-de-Sa and Robeiro, 1994; Lippe and Ardizzone, 1991) did not produce an effect on resting vessel diameter, whereas it did block the vasodilation response to $VEGF_{121}$ (Figure 3B). Hence, cAMP production is necessary for the vasodilation in response to $VEGF_{121}$. In contrast, stimulation of protein kinase A with 1 μM Sp-cAMP was without effect on the diameter of endothelium denuded vessels, demonstrating that stimulation of smooth muscle protein kinase A will not produce vasodilation. When 1 μM Rp-cAMP was used, to inhibit protein kinase A, the vasodilation response to $VEGF_{121}$ could not be blocked (Figure 3B). Clearly, although cAMP is required for $VEGF_{121}$ evoked vasodilation, protein kinase A does not mediate this response. G-proteins differentially regulate adenylyl cyclase isozymes, hence conclusions on how and which, if any, G-proteins are involved in the regulation of adenylyl cyclase activity required for $VEGF_{121}$ mediated vasodilation in the rat retina can not be made at this moment.

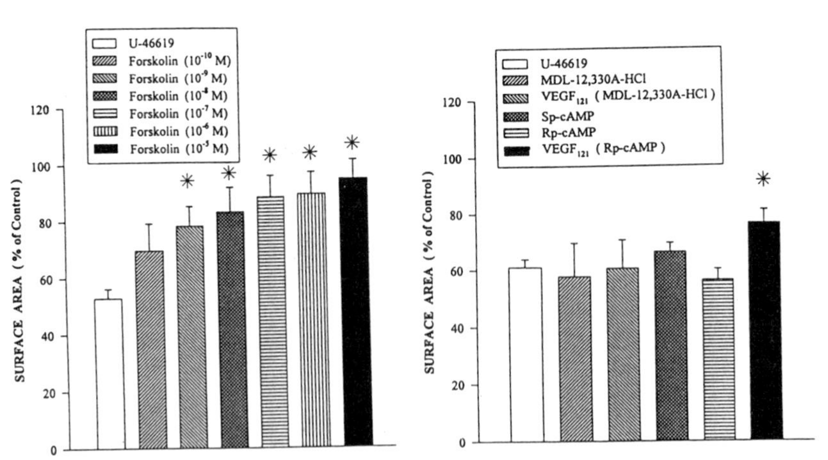

Figure 3. Adenylate Cyclase Mediated Vasodilation. A) Vasodilation produced by increasing concentrations of forskolin. * $P<0.05$ (n=6) when compared to the column derived from U-46619 using a paired Student's t-test. B) Effect of various agents on vessel diameter and on the response to 10 nM $VEGF_{121}$. * $P<0.05$ (n=9) when compared, using a paired Student's t-test, to the vessel diameter produced by the column with Rp-cAMP alone.

4. CONCLUSION

The results demonstrate that Ca^{2+} influx is necessary for the vasodilation evoked by $VEGF_{121}$. The G-proteins which mediate $VEGF_{121}$ evoked vasodilation are pertussis toxin sensitive, and therefore likely to be of either G_i or G_o type. Furthermore, cAMP production is necessary for $VEGF_{121}$ evoked vasodilation, and protein kinase A does not appear to be involved, although tyrosine kinase activity is required for this response. The source of these mediators and the link between the various elements in this response will now have to be established.

ACKNOWLEDGMENTS

This work was supported by the Medical Research Council of Canada (MRCC), the Heart and Stroke Foundation of Canada (HSFC) and the Fondation J-L Levesque. H.M.H. is a Scholar of the Fonds de la Recherche en Santé du Québec (FRSQ).

REFERENCES

DiRita, V.J., Parsot, C., Jander, G. and J.J. Mekalanos, Regulatory Cascade Controls Virulence in Vibrio Cholerae, PNAS 88:5403-5407, 1991.

Egee, S, Mignen, O, Harvey, B.J. and S. Thomas, Chloride and Non-selective Cation Channels in Unstimulated Trout Red Blood Cells, J. Physiol (Lond), 511:213-224, 1998.

Fernando, K.C. and G.J. Barritt, Characterisation of the Inhibition of the Hepatocyte Receptor-Activated Ca^{2+} Inflow System by Gadolinium and SK&F969365, Biochim Biophys Acta, 1994.

Ferrara, N. and T. Davis-Smyth, The Biology of Vascular Endothelial Growth Factor, Endocrine Reviews, 18(1):4-25, 1997.

Galli, C., Meucci, O., Scorziello, A., Werge, T.M., Calissano, P. and G. Schettini, Apoptosis in Cerebellar Granule Cells is Blocked by High KCl, Forskolin, and IGF-1 through Distinct Mechanisms of Action: The Involvement of Intracellular calcium and RNA Synthesis, J. Neurosci. 15:1172-1179, 1995.

Hasséssian, H., Vaca, L. and D.L. Kunze, Blockade of the Inward Rectifier Potassium Current by the Ca^{2+}- ATPase Inhibitor 2',5'-di(tert-butyl)-1,4-benzohydroquinone (BHQ), Br. J. Pharmacol., 112:1118-1122, 1994.

Hasséssian, H., The Mechanism of Ca^{2+} Influx Evoked by VEGF Acting on Bovine Retinal Capillary Endothelial Cells, Proceedings of the NATO Advanced Studies Institute, Angiogenesis: Models, Modulators and Clinical Applications, 1997.

Hasséssian, H., El-Dwairi, Q., Abran, D.A. Belgaçd, N., Kazanjian, A. and S. Chemtob, Differential Effects of $VEGF_{165}$ and $VEGF_{121}$ on the Nitric Oxide System of Retinal Vascular Endothelial Cells, Investigative Ophthalmology and Visual Science (ARVO) Abs # 4230, 1998.

Hazes, B., Boodhoo, A., Cockle, S.A. and R.J., Read, Crystal Structure of the Pertussis Toxin-ATP Complex: A Molecular Sensor, J. Mol. Biol. 258:661-671, 1996.

Lahaie, I, Hardy, P., Hou, X., Hasséssian, H., Asselin, P., Lachapelle, P., Almazan, G., Varma, D.R., Morrow, J.D., Roberts II, L.J. and S. Chemtob, A Novel Mechanism for Vasoconstrictor Action of 8-Isoprostaglandin $F_{2\alpha}$ on Retinal Vessels, Am J Physiol 274:R1406-R1416, 1998.

Laurenza, A., Sutkowski, E.M. and K.B., Seamon, Forskolin: A Specific Stimulator of Adenylyl Cyclase or a Diterpene with Multiple Sites of Action, Trends Pharmacol Sci. 10:442-447, 1989.

Li, X., Zarinetchi, F., Schrier, R.W. and R.A., Nemenoff, Inhibition of MAP Kinase by Prostaglandin E2 and Forskolin in Rat Renal Mesangial Cells, Am. J. Physiol. 269:C986-991, 1995.

Lomo, J., Blomhoff, H.K., Beiske, K., Stokke, T. and E.B., Smeland, TGF-beta 1 and cyclicAMP Promote Apoptosis in Resting Human B Lymphocytes, J. Immunol. 154:1634-1643, 1995.

Lückhoff, A. and D.E., Clapham, Inositol 1,3,4,5-tetrakisphosphate Activates an Endothelial Ca^{2+}-Permeable Channel, Nature 355:356-358, 1992.

Moss J. and M., Vaughan, Activation of Cholera Toxin by ADP-ribosylation Factors, 20-kDa Guanine Nucleotide-Binding Proteins, Curr. Top. Cell Regul. 32:49-72, 1992.

Nilius, B, Ion Channels in Nonexcitable Cells. In : Cell Physiology, Edited by Nicholas Sperelakis, Academic Press, 1995.

SIGNAL TRANSDUCTION AND TRANSCRIPTIONAL REGULATION OF ANGIOGENESIS

YASUFUMI SATO, MAYUMI ABE, KATSUHIRO TANAKA, CHIKA IWASAKA, NOBUYUKI ODA, SHINICHI KANNO, MANAMI OIKAWA, TOHRU NAKANO, TAKAYUKI IGARASHI

Department of Vascular Biology, Institute of Development, Aging and Cancer, Tohoku University

Abstract When quiescent endothelial cells (ECs) are exposed to angiogenic factor such as VEGF,' ECs express proteases to degrade extracellular matrices, migrate, proliferate and form new vessels. However, the molecular mechanism of these events is not fully characterized yet. We are studying the signal transduction and transcriptional regulation of angiogenesis. We investigated the properties of two VEGF receptors, Flt-1 and KDR, by using two newly developed blocking monoclonal antibodies (mAbs), i.e., anti-human Flt-1 mAb and anti-human KDR mAb. VEGF elicited induction of transcription factor Ets-1 in human umbilical vein endothelial cells (HUVECs). This induction was mediated by the KDR/Flt-1 heterodimer and the KDR homodimer. The role of transcription factor Ets-1 in angiogenesis was further clarified. We established both high and low Ets-1 expressing EC lines, and compared angiogenic properties of these cell lines with a parental murine EC line, MSS31. The growth rate was almost identical among three cell lines. It appeared that gene expressions of matrix metalloproteinases (MMP-1, MMP-3, and MMP-9) as well as integrin beta3 were correlated with the level of Ets-1 expression. As a result, the invasiveness was enhanced in high Ets-1 expressing cells and reduced in low Ets-1 expressing cells compared with parental cells, and high Ets-1 expressing cells made more tube-like structures in type 1 collagen gel. These results indicate that Ets-1 is a principle transcription factor converting ECs to the angiogeneic phenotype.

Angiogenesis: From the Molecular to Integrative Pharmacology
Edited by Maragoudakis, Kluwer Academic / Plenum Publishers, New York, 2000

Vascular endothelium is a continuous monolayer of endothelial cells (ECs) lining on the inner surface of blood vessels. ECs are normally quiescent and play an important role in maintaining the integrity of blood vessels. However, they have the ability to form neo-vessels. Angiogenesis is a process by which new blood vessels are formed from pre-existing ones, and is a fundamental process in reproduction and development. The vascular system is the first functional organ that develops in the vertebrate embryo. Vasculature of the entire body is formed via two distinctive processes; vasculogenesis and angiogenesis (for review, see Risau, 1997). The initial process of vascular development is vasculogenesis. In this process, EC precursors so-called hemangioblasts or angioblasts of mesoderm origin differentiate into ECs and form primitive vascular plexus. In a subsequent process of angiogenesis, neovessels form from extant primitive vessels and distribute to the entire body. In the adult, physiological angiogenesis is observed only in restricted places such as in endometrium and ovarian follicle, and this process is normally transient. However, persistent angiogenesis appears to play a crucial role in pathological states including solid tumors, diabetic retinopathy, rheumatoid arthritis, and atherosclerosis.

Angiogenesis is a complex phenomenon which includes at least four distinct properties of ECs; degradation of vascular basement membrane and interstitial matrices by proteases, migration, proliferation, and tube formation. Among them, coordinated reinforcement of matrix degradation and cell migration is characteristic of the angiogenic phenotype of ECs (see Risau, 1997 for a review). A number of proteases are required for the degradation of extracellular matrices. Among them, matrix metalloproteinases (MMPs) are indispensable for the degradation of collagen, a major component of extracellular matrices. ECs express various MMPs including MMP-1 (interstitial collagenase; Iwasaka et al., 1996), MMP-2 (gelatinase A; Herron et al., 1986; Vacca et al., 1997), MMP-3 (stromelysin 1; Herron et al., 1986; Duhamel-Clérin et al., 1997), MMP-9 (gelatinase B; Herron et al., 1986; Vacca et al., 1997) and MT1-MMP (Zucker et al., 1995). Integrins, α and β heterodimeric adhesion molecules, are requisite for cell migration. ECs express various integrins as well. Among them, integrin $\alpha_v\beta_3$ plays an important role in angiogenesis. Relevant studies demonstrated that the blocking of the binding of endothelial integrin $\alpha_v\beta_3$ by a monoclonal antibody or a synthetic antagonistic peptide inhibited angiogenesis and forced neovascular ECs into apoptosis (Brooks et al., 1994a; Brooks et al., 1994b).

Ets-1 is a prototype of the Ets family transcription factors. Ets family transcription factors are known to regulate the expression of numerous genes including proteases (Gutman et al., 1990; Wasylyk et al., 1991; Nerlov et al., 1992; Gum et al., 1996) and integrins (Block et al., 1996; Doubeikovski et al., 1997). Ets domain is a DNA-binding domain composed with 85 amino acids. Ets domain binds to Ets binding motif, GGAA/T, in the cis-acting element of target genes as a monomer. Characteristic feature of Ets family transcription factors is its cooperation with other transcription factors for their transactivation activity. For example, Ets binding motif and AP-1 binding site are often combined to generate a responsive complex. Pointed domain, which is homologous to drosophila Ets family transcription factor Pointed, contains a target amino acid sequence for MAP kinase. When this sequence is phosphorylated by MAP kinase, transactivation activity of Ets-1 is markedly enhanced. The *ets-1* gene is transiently expressed in ECs during vascular development in embryo (Vandenbunder et al., 1989; Kola et al., 1993; Pardanaud et al., 1993; Maroulakou et al.; 1994) as well as angiogenesis in adult (Wernert et al., 1992; Wernert et al., 1994). In spite of these observations, the roles of Ets-1 in vascular development and angiogenesis were ill defined. Therefore, we examined the role of Ets-1 in angiogenesis. First, we demonstrated that Ets-1 is induced in ECs in response to representitive angiogenic growth factors such as VEGF and bFGF (Iwasaka et al., 1996; Tanaka et al., 1999).

VEGF, a dimeric endothelial cell (EC)-specific growth factor, is thought to play a principal role by stimulating migration and proliferation of ECs and the expression of angiogenesis-related genes in ECs (for review, see Ferrara and Davis-Smyth, 1997). VEGF binds to two receptor-type tyrosine kinases, Flt-1 (VEGF receptor-1)(de Vries et al., 1992) and KDR/Flk-1 (VEGF receptor-2)(Quinn et al., 1992), and to membrane protein neuropilin-1, which does not contain a tyrosine kinase domain (Soker et al., 1998). Flk-1 is the murine homologue of human KDR, which shares 85% homology with KDR (Matthews et al., 1991). Both Flt-1 and KDR/Flk-1 have seven immunoglobulin-like domains in the extracellular domain, a single transmembrane region, and a consensus tyrosine kinase domain in the intracellular domain, which is interrupted by a kinase-insert domain. It has been shown that Flt-1 and KDR/Flk-1 have distinct functions in vascular development in embryos. Flk-1 knockout mice, which die by embryonic day 8.5 (E8.5), lack both ECs and hematopoietic cells (Shalaby et al., 1995). In contrast, Flt-1 knockout mice, which also die around E8.5, have abundant ECs, which do not assemble in functional vessels (Fong et al., 1995). In spite of these observations in knockout mice, the properties of KDR/Flk-1 and Flt-1 in the signal transduction of VEGF in differentiated ECs is largely unknown. We investigated the roles of two VEGF receptors in the expression of *ets-1* mRNA in human umbilical vein endothelial cells (HUVECs) by using newly developed blocking monoclonal antibodies (mAbs) against Flt-1 and KDR. VEGF-mediated induction of *ets-1* was completely inhibited by anti-KDR mAb, whereas anti-Flt-1 mAb exhibited no inhibition. These results indicate that KDR-medimated signal is responsible for the induction of *ets-1* mRNA in HUVECs (submitted for publication).

ECs cultured on the surface of type 1 collagen gel invaded and formed tube-like structures in the gel, and the specific elimination of Ets-1 synthesis with phosphorothioate *ets-1* antisense oligodeoxynucleotides (ODNs) inhibited this process (Iwasaka et al., 1996). Therefore, the expression of Ets-1 in ECs was required for angiogenesis. In order to understand the precise role of Ets-1 in angiogenesis, we transfected murine ets-1 sense cDNA or antisense cDNA and established high or low Ets-1 expression EC lines, and compared angiogenic properties of these cell lines with those of the parental murine EC line, MSS-31. The growth rate was almost identical for each cell line, indicating that Ets-1 did not play a major role in proliferation. In contrast, the invasiveness was markedly enhanced in high Ets-1 expression cells and reduced in low Ets-1 expression cells compared with that of parental cells. Cell invasion requires proteases for matrix degradation and integrins for cell migration. The gene expressions of matrix metalloproteinases (MMP-1, MMP-3, and MMP-9) as well as gelatinolytic activity of MMP-9 were significantly increased in high Ets-1 expression cells. Low Ets-1 expression cells could not spread on a vitronectin substratum, and the phosphorylation of focal adhesion kinase was markedly impaired because of the reduced expression of integrin β3. These results indicate that Ets-1 is a principal regulator converting ECs to the angiogenic phenotype (Oda et al., 1999).

In case of endothelial denudation, remaining ECs at the wound edge migrate, proliferate, and repair the denuded area. This process, termed "re-endothelialization", provides an important mechanism for maintaining the normal vascular wall. While the migration and proliferation of ECs are required for re-endothelialization, cell migration is the initiating and rate-limiting event as it releases ECs from contact-inhibition and leads to subsequent proliferation. We observed that Ets-1 is induced in ECs at the edge of the wound. An immunohistological analysis revealed that the induction of Ets-1 protein was restricted to migrating ECs at the edge of wound and that the protein returned to the basal level when re-endothelialization was accomplished (Tanaka et al., 1998). Although re-endothelialization and angiogenesis are distinct phenomena with respect to initiators (injury vs. angiogenic factors) and functional purposes (repair vs. neo-vessel formation), the molecular mechanism that regulates these processes may be common.

A number of genes are expresssed in ECs during angiogenesis. Thus, the regulation of the expression of these genes in ECs has become an important research subject. Ets-1 is found to be one of the principle transcription factors which regulates angiogenesis by inducing proteases and integrin in ECs. However, our understanding about the role of Ets-1 in ECs is fragmentary, since Ets binding motifs are found in cis-element of numerous genes. Further study is required to clarify the entire role of Ets-1 in the biology of ECs.

REFERENCES

Block, K.L., Shou, Y., and Poncz, M. 1996 An Ets/Sp1 interaction in the 5'-flanking region of the megakaryocyte-specific alpha IIb gene appears to stabilize Sp1 binding and is essential for expression of this TATA-less gene. *Blood* 88:2071-2080.

Brooks, P.C., Clark, R.A.F., and Cheresh, D.A. 1994a Requirement of vascular integrin $\alpha_v\beta_3$ for angiogenesis. *Science* 264:569-571.

Brooks, P.C., Montgomery, A.M.P., Rosenfeld, M., Reisfeld, R.A., Hu, T., Klier, G., and Cheresh, D.A. 1994b Integrin $\alpha_v\beta_3$ antagonists promote tumor regression by inducing apoptosis of angiogenic blood vessels. *Cell* 79:1157-1164.

de Vries, C., Escobedo, J.A., Ueno, H., Houck, K., Ferrara, N. & Williams, L.T. 1992 The fms-like tyrosine kinase, a receptor for vascular endothelial growth factor. *Science* 255:989-991.

Doubeikovski, A., Uzan, G., Doubeikovski, Z., Prandini, M.H., Porteu, F., Gisselbrecht, S., and Dusanter-Fourt, I., 1997 Thrombopoietin-induced expression of the glycoprotein IIb gene involves the transcription factor PU.1/Spi-1 in UT7-Mpl cells. *J. Biol. Chem.*, 272:24300-24307.

Duhamel-Clérin, E., Orvain, C., Lanza, F., Cazenave, J, and Klein-Soyer, C. 1997 Thrombin receptor-mediated increase to two matrix metalloproteinase, MMP-1 and MMP-3, in human endothelial cells. *Arterioscler. Thromb. Vasc. Biol.* 17:1931-1938.

Fong, G.H., Rossant,J., Gertsenstein,M. & Breitman,M.L. 1995 Role of the Flt-1 receptor tyrosine kinase in regulating the assembly of vascular endothelium. *Nature* 376:66-70.

Gum, R., Lengyel, E., Juarez, J., Chen, J.H., Sato, H., Seiki, M., and Boyd, D. 1996 Stimulation of 92-kDa gelatinase B promoter activity by ras is mitogen-activated protein kinase kinase 1-independent and requires multiple transcription factor binding sites including closely spaced PEA3/ets and AP-1 sequences. *J. Biol. Chem.* 271:10672-10680.

Gutman, A., and Wasylyk, B. 1990) The collagenase gene promoter contains a TPA and oncogene-responsive unit encompassing the PEA3 and AP-1 binding sites. *EMBO J.* 9:2241-2246.

Herron, G.S., Werb, Z., Dwyer, K., and Banda, M.J. 1986 Secretion of metalloproteinases by stimulated capillary endothelial cells. *J. Biol. Chem.* 261:2810-2813.

Iwasaka, C., Tanaka, K., Abe, M. & Sato, Y. 1996 Ets-1 regulates angiogenesis by inducing the expression of urokinase-type plasminogen activator and matrix metalloproteinase-1 and the migration of vascular endothelial cells. *J. Cell. Physiol.* 169:522-531.

Kola, I., Brookes, S., Green, A. R., Garber, R., Tymms, M., Papas, T. S. and Seth, A. 1993 The Ets1 transcription factor is widely expressed during murine embryo development and is associated with mesodermal cells involved in morphogenetic processes such as organ formation. *Proc. Natl. Acad. Sci. USA*. 90:7588-7592.

Maroulakou, I.G., Papas, T.S. and Green, J.E. 1994 Differential expression of ets-1 and ets-2 proto-oncogenes during murine embryogenesis. *Oncogene*. 9:1551-1565.

Matthews, W., Jordan, CT., Gavin, M., Jenkins, N.A., Copeland, N.G. & Lemischka, I.R. 1991 A receptor tyrosine kinase cDNA isolated from a population of enriched primitive hematopoietic cells and exhibiting close genetic linkage to c-kit. *Proc. Natl. Acad. Sci. USA* 88:9026-9030.

Nerlov, C., Rorth, P. Blasi, F., and Johnsen, M. 1991 Essential AP-1 and PEA3 binding elements in the human urokinase enhancer display cell type-specific activity. *Oncogene* 6: 1583-1592.

Oda, N., Abe, M., and Sato, Y. 1999 ETS-1 converts endothelial cells to the angiogenic phenotype by inducing the expression of matrix metalloproteinases and integrin β3. *J. Cell. Physiol.* 178:121-132,.

Pardanaud, L. and Dieterlen-Lievre, F. 1993 Expression of c-ets-1 in early chick embryo mesoderm: relationship to the hemangioblastic lineage. *Cell Adhes. Commun.* 1:151-160.

Quinn, T.P., Peters, K.G., de Varies, C., Ferrara, N. & Williams, L.T. 1993 Fetal liver kinase 1 is a receptor for vascular endothelial growth factor and is selectively expressed in vascular endothelium. *Proc. Natl. Acad. Sci. USA* 90:7533-7537.

Risau, W. 1997 Mechanisms of angiogenesis. *Nature* 386:671-674.

Shalaby, F., Rossant,J., Yamaguchi, T.P., Gertsenstein, M., Wu, X.F., Breiman, M.L. & Schuh, A.C. 1995 Failure of blood-island formation and vasculogenesis in Flk-1-deficient mice. *Nature* 376:62-66.

Soker, S., Takashima, S., Miao, H.Q., Neufeld, G. & Klagsbrun, M. 1998 Neuropilin-1 is expressed by endothelial and tumor cells as an isoform-specific receptor for vascular endothelial growth factor. *Cell* 92:735-745.

Tanaka, K., Oda, N., Iwasaka, C., Abe, M. & Sato, Y. 1998 Induction of Ets-1 in endothelial cells during reendothelialization after denuding injury. *J. Cell. Physiol.* 176:235-244.

Tanaka, K., Abe, M., and Sato, Y. 1999 Roles of ERK1/2 and p38 MAP kinase in the signal transduction of bFGF in endothelial cells during angiogenesis. *Jpn. J. Cancer Res.* 90:647-654.

Vacca, A., Moretti, S., Ribatti, D., Pellegrino, A., Pimpinelli, N., Bianchi, B., Bonifazi, E., Ria, R., Serio, G., and Dammacco, F. 1997 Progression of mycosis fungoides is associated with changes in angiogenesis and expression of the matrix metalloproteinase 2 and 9. *Eur. J. Cancer,* 33:1685-1692.

Vandenbunder, B., Pardanaud,L., Jaffredo, T., Mirabel, M. A. and Stehelin, D. 1989 Complementary patterns of expression of c-ets-1, c-myb and c-myc in the blood-forming system of the chick embryo. *Development* 106:265-274.

Wasylyk, C., Gutman, A., Nicholson, R., and Wasylyk, B. 1991 The c-Ets oncoprotein activates the stromelysin promoter through the same elements as several non-nuclear oncoproteins. *EMBO J.* 10:1127-1134.

Wernert, N., Raes, M., Lassalle, P., Dehouck, M., Gosselin, B., Vandenbunder, B. and Stehelin, D. 1992 c-ets-1 proto-oncogene is a transcription factor expressed in endothelial cells during tumor vascularization and other forms of angiogenesis in humans. *Am. J. Pathol.* 140:119-127.

Wernert, N., Gilles, F., Fafeur, V., Bouali, F., Raes, M.-B., Pyke, C., Dupressoir, T., Seitz, G., Vandenbunder, B. and Stehelin, D. 1994 Stromal expression of c-Ets1 transcription factor correlates with tumor invasion. *Cancer Res.* 54:5683-5688.

Zucker, S., Conner, C., Dimassmo, B.I., Ende, H., Drews, M., Seiki, M., Bahou, W.F. 1995 Thrombin induces the activation of progelatinase A in vascular endothelial cells. Physiologic regulation of Angiogenesis. *J. Biol. Chem.* 270:23730-23738.

DIFFERENTIAL CONTRIBUTION OF BRADYKININ RECEPTORS IN ANGIOGENESIS

Lucia Morbidelli, Astrid Parenti, Sandra Donnini, #Harris J. Granger, Fabrizio Ledda and *^Marina Ziche

*Department of Pharmacology and *C.I.M.M.B.A., University of Florence, Viale G. Pieraccini 6, 50139, Florence, and ^Institute of Pharmacological Science, University of Siena, Siena Italy; #Microcirculation Research Institute and Department of Physiology, Texas A&M University System Health Science Center, College Station, TX 77843-1114, USA*

Bradykinin (BK) contributes to the inflammatory response inducing vasodilation of postcapillary venules and has been demonstrated to induce neovascular growth in subcutaneous rat sponges. In this study the ability of BK to stimulate cell growth and migration in cultured endothelium from coronary postcapillary venules (CVEC) has been investigated. BK promotes growth of endothelial cells and its mitogenic activity involves c-Fos expression. Only the B1 receptor appears to be responsible for the proliferation induced by BK and suggests that this type of receptor might be implicated in favouring angiogenesis of coronary venules. Phospholipase C (PLC) activation and the endogenous upregulation of nitric oxide synthase (NOS) were monitored in response to B1 and B2 receptor activation. BK-induced IP turnover was triggered by B2 activation, while NOS activation was linked to the B1 receptor. NOS inhibition in CVEC prevented BK/B1 induced cell growth. These results substantiate the protective and trophic role of BK on microvascular endothelium by demonstrating that B1 receptor stimulation on venular endothelium is mandatory for the angiogenic effect of BK through its coupling to NOS activation.

1. INTRODUCTION

Kinins are potent vasodilating peptides that are released from precursor kininogens by kallikreins. Kinins bind to receptors and mediate a broad spectrum of biological effects including vasodilation, smooth muscle contraction and relaxation, pain and inflammation. Bradykinin (BK) is a potent inflammatory mediator capable of inducing vasodilation and dramatically enhancing the transport of water and proteins across postcapillary venules by its action on endothelial cells.

Two subtypes of bradykinin receptors, B1 and B2 have been defined based on their pharmacological properties (Hall, 1992). In physiological conditions the B2 receptor is constitutively expressed at vascular level. The B1 receptor expression, undetectable in physiological conditions, is found strongly up-regulated following tissue injury and inflammation (Ahluwalia and Perretti, 1999). The most BK effects are reported to be linked to B2 receptor activation. However an increasing number of evidence suggests that the B1 receptor might be relevant for the cardioprotective effect of BK (Bouchard et al., 1998).

Fan and coworkers (Hu and Fan, 1993; Hu et al., 1995) reported on angiogenesis induced in vivo by BK in synergism with interleukin-1 (IL-1). The receptor subtype involved in BK-induced neovascular response and the mechanism by which BK induced angiogenesis however were not elucidated.

The aim of the present study was to assess whether BK could directly affect endothelial cell growth and migration, which are necessary steps for angiogenesis. Endothelial cells form postcapillary venules were selected for this study since these vessels are the primary site of BK effects as well as of neovascular growth. Furthermore, we attempted to characterize the BK receptor subtype present in cultured coronary venular endothelial cells and its role in mediating BK effects.

2. MATERIALS AND METHODS

Cell line and culture conditions

The coronary venular endothelial cells (CVEC) were isolated and characterized as previously described (Schelling et al., 1998). Cells were maintained in culture in DMEM supplemented with 10% bovine calf serum (CS) and antibiotics (100 U/ml penicillin and 100 μg/ml streptomycin) on gelatin coated dishes. Cells were cloned and each clone was subcultured up to a maximum of 25 passages. Passages between 15 and 20 were used in these experiments.

Proliferation studies

Cell proliferation was quantified by total cell number as reported (Ziche et al., 1994). Briefly, $1x10^3$ cells resuspended in 10% CS were seeded in each well of 96 multiwell plates. After adherence (4 h) the medium was replaced with 1% CS DMEM containing test substances. After 48 h incubation, the supernatants were removed from the multiwell plates and cells were fixed with methanol and stained with Diff-Quik. Cell numbers were obtained by counting in 7 random fields at a magnification of 100x with the aid of an ocular grid (21 mm^2). Data are expressed as percent increase over basal response as well as cell number/well.

c-Fos immunohistochemistry

2500 cells were seeded onto 96 multiwell plates in DMEM containing 10 % CS and let adhere overnight. Cells were starved from serum for 3 days. Following stimulation cells were washed with PBS and fixed in 2% formaldehyde for 20 min. The cells were washed with PBS containing 10 mM glycine, permeabilized for 5 min with 1% Nonidet P-40 and then incubated at 4°C overnight with polyclonal sheep c-Fos peptide antiserum (Genosys, 1:500 dilution in PBS containing 1% of normal horse serum). The bound antibody was detected with biotinylated anti-sheep IgG and the avidin-biotin peroxidase (ABC, Vector Laoratories) according to the manifacturer's instructions and visualized by incubation with 3,3'-diaminobenzidine tetrahydrochloride (0.3 mg/ml) and hydrogen peroxide (0.015%) for 4-8 min. The number of labelled nuclei was counted in a blinded manner at 100 X magnification in 3 random fields/well with the aid of an ocular grid.

Migration assay

The Boyden Chamber procedure was used to evaluate cell migration (Ziche et al., 1994). The method is based on the passage of endothelial cells across porous filters against a concentration gradient of the migration effector. The two wells of the Neuro Probe 48-well micro-chemotaxis chamber were separated by a PVP-free polycarbonate filter, 8 μm pore size, coated with type I collagen (100 μg/ml) and fibronectin (10 μg/ml). Test solutions were dissolved in 0.1% CS medium and placed in the lower wells. 50 μl of cell suspension ($1.2x10^4$ cells) were added to each upper well. The chamber was incubated at 37°C for 4 hr and the filter was then removed and fixed in methanol. Cells migrated on the lower surface of the filter were stained with Diff-Quik (Dade International) and counted using a light microscope at a magnification of 400 in 10 random fields per each well. Migration was expressed as the number of total cells counted per experiment.

Determination of NOS activity

NOS activity was measured on adherent cells as previously reported (Parenti et al., 1998). Briefly, subconfluent CVEC were serum-starved overnight, and incubated for 20 min with 1 μCi of [^{3}H]-L-arginine plus 10 μM L-arginine. Test substances were added for 5 min at 37°C. The reaction was stopped by cold HEPES buffer containing 4 mM EDTA and supernatant removed. 0.5 ml ethanol added to each monolayer was allowed to evaporate and 2 ml of 10 mM HEPES-Na were added for 20 min. The supernatant was collected and applied to 0.8 ml Dowex AG50WX-8 (Na-form) and vigorously shacked for 45 min. Then 0.5 ml was collected and added to 3 ml of scintillation liquid. NOS activity was expressed as cpm/mg protein and as well as percent increase over basal activity.

Inositol phosphate activation

Inositol monophosphate (IP1) levels were measured as previously reported (Parenti et al., 1998). Briefly, [^{3}H]-myo-inositol (2 μCi/ml)-labelled CVEC were incubated for 10 min with 20 mM LiCl to block myo-inositol-1-phosphatase and then with test compounds for the designed times. Reaction was stopped by ice-cold methanol for 30 min. Cells were scraped and cell associated inositols extracted by chloroform-methanol (1:1). Water-soluble fractions were applied to anion-exchange columns (Resin AG-X8, 200-400 mesh, formate form) and water-soluble eluted IP_1 levels were measured as recovered radioactivity.

3. RESULTS

BK induces proliferation and c-Fos activation of microvascular endothelial cells

CVEC proliferation was evaluated as the total number of cells recovered after 48 hr exposure to BK. BK (1-100 nM) sustained cell growth in a

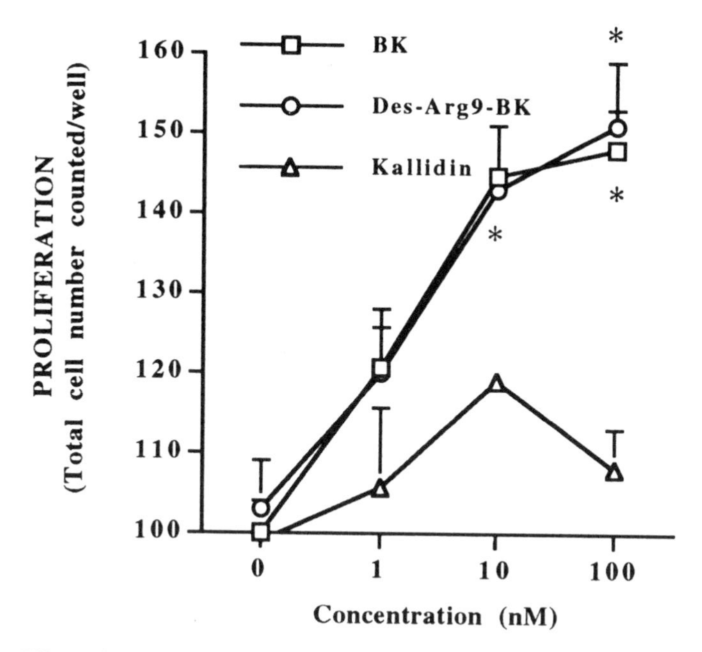

Figure 1: Effect of BK, selective B1 and B2 receptor agonists on cell growth. The total number of cells recovered after 48 hr exposure to BK, the selective B1 (Des-Arg9-BK) and B2 receptor (kallidin) agonists was counted. Data represent the results of at least 4 experiments run in duplicate. $*P<0.01$ vs basal.

concentration-dependent manner with maximal effect at 100 nM (148±3 counted cells/well vs a basal value of 100±4.3, $P<0.05$) (Fig. 1).

BK was then evaluated for its ability to activate early gene expression. BK (100 nM) induced c-Fos activation producing a significant accumulation of protein immunoreactivity in the nucleus within 15 min (3.4 fold of basal control condition). Maximal activation was observed at 1 hr, producing a 18 fold increase of positive nuclei number, and after 3 hr stimulation the immunoreactivity was still significantly different from basal levels (4 fold). FGF-2 (10 nM) activated c-Fos to a much smaller extent than BK, producing its maximal effect at 1 hr (2 fold increase) (Fig. 2).

Receptor subtype characterization

The receptor subtype involved in cell growth induced by BK was assessed by the use of the selective B1 and B2 receptor agonists, Des-Arg9-BK and kallidin, respectively. Cells were treated with increasing

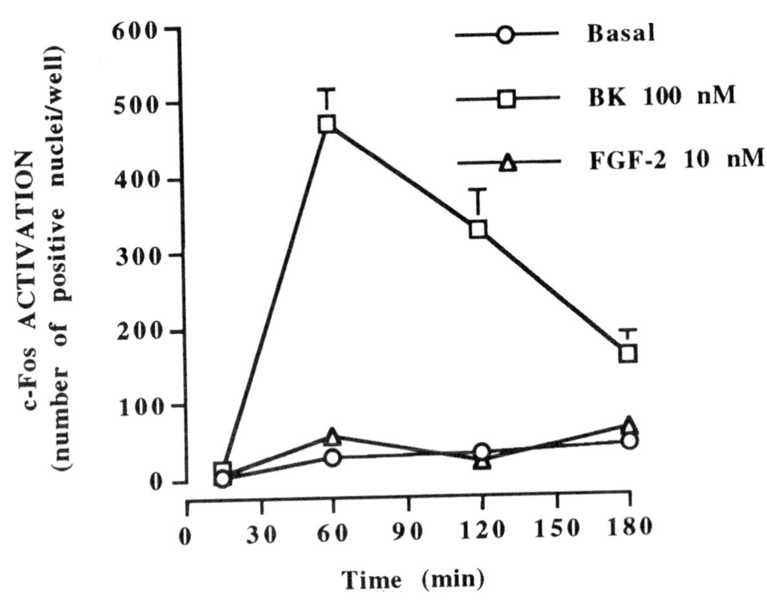

Figure 2: Effect of BK in stimulating c-fos gene expression. c-Fos protein accumulation in the nucleus was detected by immunohistochemistry in postcapillary endothelial cells exposed to 100 nM BK and 10 nM FGF-2 for different times. Data are expressed as total positive nuclei counted/well (n=4 in triplicate).

concentrations of the two peptides (1-100 nM) and cell growth was measured as total number of cells counted after 48 hr incubation. The selective B1 receptor agonist Des-Arg9-BK reproduced the proliferative effect induced by BK, while the selective B2 receptor agonist kallidin did not (Fig. 1).

To further characterize the receptor subtype activated by BK on CVEC, cells were treated with selective B1 and B2 receptor antagonists and then exposed to BK. The effect of BK on CVEC was reduced in a concentration-dependent manner by the pretreatment with the B1 receptor antagonist Des-Arg9-Leu8-BK. The maximal effective concentration of the antagonist able to completely block BK-induced growth was 1 μM (Fig. 3). In parallel experiments, the selective B2 receptor antagonist HOE 140 (Hock et al., 1991) used at the concentration of 0.1 and 1 μM, did not significantly affect BK-induced proliferation. The specificity of B1 antagonist was assessed on the growth of CVEC induced by a different endothelial cell mitogen, FGF-2. Des-Arg9-Leu8-BK at the dose of 1 μM did not modify FGF-2-induced proliferation.

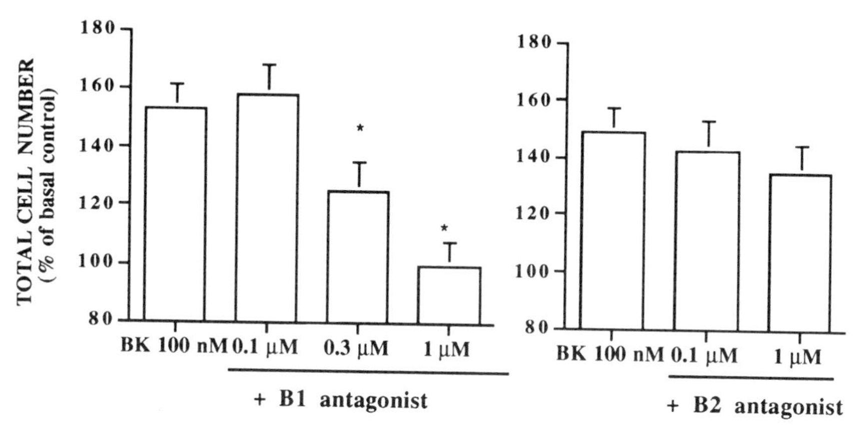

Figure 3: Effect of selective B1 and B2 receptor antagonists on BK-induced proliferation. The total number of cells was counted after 48 hr exposure to BK in cells treated with different concentrations of the selective B1 (Des-Arg9-Leu8-BK) and B2 (HOE 140) receptor antagonist. Data are expressed as % of the effect of antagonists alone, and represent the results of at least 4 experiments run in duplicate. Control basal value was 110±6 counted cells/well. *P<0.01 vs BK alone.

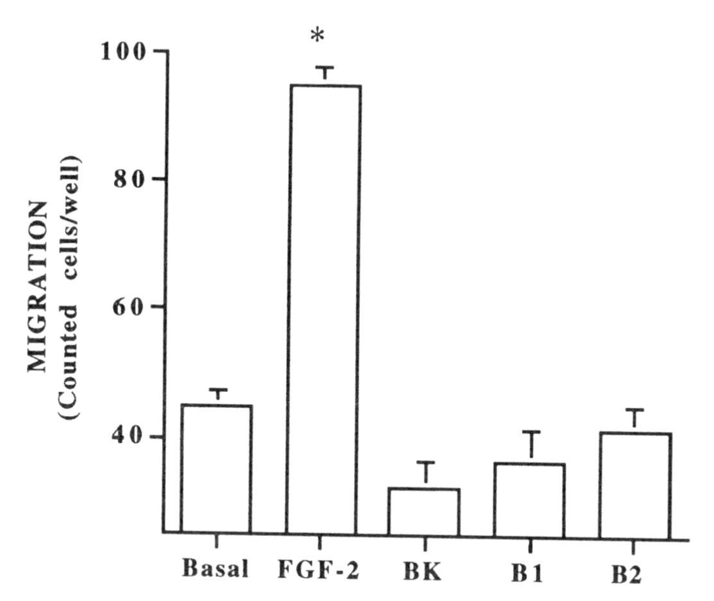

Figure 4: Effect of 100 nM BK, selective B1 and B2 receptor agonists and FGF-2 on cell migration. The 48-well microchemotaxis chamber was used to assess migration. The chamber was incubated at 37°C for 4 hr. Data are means ± SEM from at least 4 experiments run in triplicate. *P<0.01 vs basal.

BK does not promote migration

During angiogenesis endothelial cells migrate and acquire an invasive capacity by increasing proteolytic activity. The effect of BK was then assayed on migration. BK had no effect of CVEC chemotaxis, while endothelial cells were highly responsive to FGF-2 in both assays (Fig. 4). Moreover, neither the B1 nor the B2 receptor agonist stimulated CVEC mobilization.

BK/B1 receptor activates ecNOS

We have recently demonstrated that ecNOS activation is essential to transduce the angiogenic activity of vasoactive peptides, such as substance P (SP) and vascular endothelial growth factor (VEGF) (Ziche et al., 1994, 1997). BK is a known activator of the NOS pathway in endothelial cells which mediates the vasodilating properties of the kinin (Schlemper and Calixto, 1994).

We then investigated whether BK could stimulate ecNOS in CVEC, and the type of receptor involved. The ability to stimulate ecNOS activity was

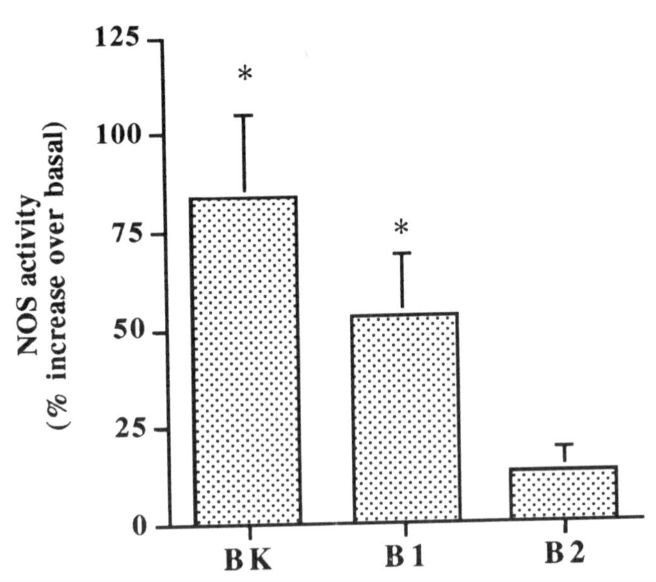

Figure 5: BK activates NOS activity. ecNOS activity was measured following 5 min exposure to 100 nM BK and B1 and B2 agonists. In unstimulated CVEC ecNOS activity was 7792 ± 1400 cpm/mg protein. Mean ± SE of 3 experiments. Data are expressed as percent increase over basal. *$P<0.05$ vs basal activity.

assessed by measuring the conversion of [^{3}H]L-arginine in [^{3}H]L-citrulline. Within 5 min from exposure, BK and the selective B1 agonist (100 nM) significantly increased ecNOS activity, while the B2 receptor agonist did not (Fig. 5).

We then assessed whether the proliferative effect of BK in cultured endothelium could be regulated by the NOS pathway. Thus CVEC were stimulated with BK and B1 agonist (100 nM) in the presence of the NOS inhibitor L-NMMA. The treatment with L-NMMA (200 μM) completely prevented CVEC proliferation in response to either BK or the B1 agonist (Table 1).

Table 1: Effect of NOS inhibition on BK/B1 agonist induced proliferation.

Stimuli	Control	+ L-NMMA
BK	145±5 **	98±6 @
B1 agonist	160±7 **	102±3 @

CVEC proliferation in response to 100 nM BK and Des-Arg9-BK in the presence of 200 μM L-NMMA. Data are reported as % of basal proliferation. Mean ± SE of 3 experiments in duplicate. **P<0.01 vs basal proliferation; @P<0.001 vs BK or Des-Arg9-BK-induced proliferation.

The BK/B2 receptor activates PLC

We have previously demonstrated that BK induces a rapid turnover of IPs and increases calcium levels in CVEC (Ziche et al., 1993). We assessed the receptor subtype involved by pretreating CVEC with B1 and B2 receptor antagonists. After 15 min stimulation, BK (100 nM) significantly increased IP1 accumulation (2.5 ± 0.3 fold, P<0.01, n=8). HOE140, the selective B2 receptor antagonist, was able to concentration-dependently inhibit IP1 accumulation in response to 100 nM BK (50 and 98% inhibition with 0.1 and 1 μM HOE140, n=3), while Des-Arg9-Leu8-BK did not modify BK-induced PLC activation.

4. DISCUSSION

This study was aimed to investigate by which mechanism BK contributed to angiogenesis. We demonstrate for the first time that BK exerts a specific and direct growth promoting effect on coronary postcapillary endothelial cells while it does not induce their migration. Early gene expression of c-fos is substantially increased by BK exposure (Morbidelli et al., 1998).

The receptor involved in BK-induced endothelial cell proliferation was characterized. Our findings demonstrate that in postcapillary venular endothelial cells both B1 and B2 receptors are present and that activation of only the B1 receptor is required to promote endothelial cell growth. The B1 receptor agonist Des-Arg9-BK is able to mimic BK-induced proliferation, while the B2 receptor agonist kallidin is devoid of any activity. Consistent with these data the selective B1 receptor antagonist Des-Arg9-Leu8-BK blocks endothelial cell proliferation promoted by BK while it does not affect FGF-2 action on endothelial cell growth. Conversely, the highly specific B2 receptor antagonist, HOE140, does not modify BK-induced responses (Morbidelli et al., 1998).

The angiogenic effect of BK is due to the direct activation of B1 receptor on capillary endothelial cells which transduces the autocrine upregulation of the NOS pathway. BK/B2 receptor activation does not involve endothelial cell proliferation and is preferentially linked to PLC activation. Despite the presence of a functionally B2 receptor, other functions than proliferation have to be ascribed to the B2 receptor of CVEC, possibly linked to hemodynamic changes and permeability (Cambridge and Brain, 1995, Wahl et al., 1996, Braun et al., 1997).

Angiogenesis plays a crucial role in different pathological and physiological processes including wound healing and the recovery from myocardial infarction. Proliferation of coronary endothelial cells in the border zone of the ischemic myocardium contributes to limit the damage to the myocardium and the infarct size by favouring angiogenesis (Granger et al., 1994). It has been reported that brief periods of myocardial ischemia are cardioprotective (Yellon et al., 1993). Increased levels of bradykinin are produced as a result of the ischemic damage together with other vasoactive agents (as adenosine, angiotensin, endothelin) which can affect the neovascular process (Cohen and Downey, 1996). Our results suggest that on the coronary endothelium of postcapillary venules kinins exert a trophic effect which may result in a protective action for ischemic heart disease (Starkopf et al., 1997). The direct effect of BK on coronary endothelium shown here provides further evidence in support of the cardioprotective effect of ACE inhibitors in vivo through their action on endogenous kinins (Noda et al., 1993; Schriefer et al., 1997).

In conclusion our findings indicate a role for BK on coronary postcapillary endothelial cell proliferation relevant for the promotion of angiogenesis. Pharmacological agents mimicking BK effect linked to B1 receptor activation such as B1 agonist or PKC inducer, and favouring BK action such as ACE inhibitor, have the potential to confer cardioprotection by inducing angiogenesis in ischemic heart diseases or during revascularization surgery.

ACKNOWLEDGMENTS

This work was supported by funds from the National Council of Research (CNR), the Italian Ministry of University, Scientific Research and Technology (MURST) and Programma Nazionale di Ricerca sui Farmaci (seconda fase) of MURST (to M.Z) and the Texas Advanced Technology and Research Program (to H.J.G.).

REFERENCES

Ahluwalia A, and Perretti M. (1999). B1 receptors as a new inflammatory target. Could this B the 1?. *TiPS* 20,100-104

Bouchard, J. F., Chouinard, J., and Lamontagne, D. (1998) Role of kinins in endothelial protective effect of ischaemic preconditioning. *Br. J. Pharmacol.* 123, 413-420

Braun, C., Ade, M., Unger, T., van der Woude, F. J., and Rohmeiss, P. (1997) Effects of bradykinin and icatibant on renal hemodynamics in conscious spontaneously hypertensive and normotensive rats. *J. Cardiovasc. Pharmacol.* 30, 446-454

Cambridge, H., and Brain, S. D. (1995) Mechanism of bradykinin-induced plasma extravasation in the rat knee joint. *Br. J. Pharmacol.* 115, 641-647

Cohen, M.V., and Downey, J.M. (1996). Myocardial preconditioning promises to be a novel approach to the treatment of ischemic heart disease. *Annu. Rev. Med.*, 47, 21-29.

Granger, H. J., Ziche, M., Hawker, J.R. Jr, Meininger, C. J., Czisny, L. E., and Zawieja, D. C. (1994). Molecular and cellular basis of myocardial angiogenesis. *Cell. Mol. Biol. Res.,* 40, 81-85.

Hall, J. M. (1992) Bradykinin receptors: pharmacological and biological roles. *Pharmac Ther.* 56, 131-190

Hock, F.J. Wirth, K., Albus, U., Linz, W., Gerhards, H.J., Wiemer, G., Henke, St. Breipohl, G., Konig, W., Knolle, J., and Scholkens, B.A. (1991). Hoe 140 a new potent and long acting bradykinin-antagonist: in vitro studies. *Br. J. Pharmacol.*, 102, 796-773.

Hu, D.-E., and Fan, T.-P. D. (1993) [Leu8]des-Arg9-bradykinin inhibits the angiogenic effect of bradykinin and interleukin-1 in rats. *Br. J. Pharmacol.* 109, 14-17

Hu, D.-E., Hiley, C.R., Smither, R.L., Greesham, G.A., and T.-P.D. Fan. (1995). Methods in laboratory investigation. Colleration of ^{133}Xe clearance, blood flow and histology in rat sponge model of angiogenesis. *Lab. Invest.*, 72, 601-610.

Morbidelli, L., Parenti, A., Giovannelli, L., Granger, H. J., Ledda, F., and Ziche, M. (1998) B1 receptor involvement in the effect of bradykinin on venular endothelial cell proliferation and potentiation of FGF-2 effects. *Br. J. Pharmacol.* 124, 1286-1292

Noda, K., Sasaguri, M., Ideishi, M., Ikeda, M., and Arakawa, K. (1993). Role of locally formed angiotensin II and bradykinin in the reduction of myocardial infarct size in dogs. *Cardiovasc. Res.*, 27, 334-340.

Parenti, A., Morbidelli, L., Cui, X. L., Douglas, J. G., Hood, J., Granger, H. J., Ledda, F., and Ziche, M. (1998) Nitric oxide is an upstream signal for vascular endothelial growth factor-induced extracellular signal-regulated kinases$_{1/2}$ activation in postcapillary endothelium. *J. Biol. Chem.* 273, 4220-4226

Schelling, M. E., Meininger, C. J., Hawker, J. R., and Granger, H. J. (1988) Venular endothelial cells from bovine heart. *Am. J. Physiol.* 254, H1211-H1217

Schlemper, V., and Calixto, J. B. (1994) Nitric oxide pathway-mediated relaxant effect of bradykinin in the guinea-pig isolated trachea. *Br. J. Pharmacol.* 111, 83-88

Schriefer, J. A., Broudy, E. P., and Hassen, A. H. (1997). Endopeptidase inhibitors decrease myocardial ischemia/reperfusion injury in an in vivo rabbit model. *J. Pharmacol. Exp. Ther.*, 278 (3), 1034-1039.

Starkopf, J., Bugge, E., and Ytrehus, K. (1997). Preischemic bradykinin and ischaemic preconditioning in functional recovery of the globally ischaemic rat heart. *Cardiovasc. Res.*, 33 (1), 63-70.

Wahl, M., Whalley, E. T., Unterberg, A., Shilling, L., Parsons, A. A., Baethmann, A., and Young, A. R. (1996) Vasomotor and permeability effects of bradykinin in cerebral microcirculation. *Immunpharmacol.* 33, 257-263

Yellon, D.M., Alkhulaifi, A.M., and Pugsley, W.B. (1993). Preconditioning the human myocardium. *Lancet*, 342, 276-277.

Ziche, M., Morbidelli, L., Choudhuri, R., Zhang, H.-T., Donnini, S., Granger, H. J., and Bicknell, R. (1997) Nitric oxide-synthase lies downstream of vascular endothelial growth factor but not basic fibroblast growth factor induced angiogenesis. *J. Clin. Invest.* 99, 2625-2634

Ziche, M., Morbidelli, L., Masini, E., Amerini, S., Granger, H. J., Maggi, C. A., Geppetti, P., and Ledda L. (1994) Nitric oxide mediates angiogenesis in vivo and endothelial cell growth and migration in vitro promoted by substance P. *J. Clin. Invest.* 94, 2036-2044

Ziche, M., Zawieja, D., Hester, R. K., and Granger, H. J. (1993) Calcium entry, mobilization and extrusion in postcapillary venular endothelium exposed to bradykinin. *Am. J. Physiol.* 34, H569-H580

SENSITIVITY OF DIFFERENT VASCULAR BEDS IN THE EYE TO NEOVASCULARIZATION AND BLOOD-RETINAL BARRIER BREAKDOWN IN VEGF TRANSGENIC MICE

S.A. VINORES, N.L. DEREVJANIK, M.A. VINORES, N. OKAMOTO, AND P.A. CAMPOCHIARO

The Wilmer Eye Institute, Johns Hopkins University School of Medicine, 825 Maumenee Bldg., 600 N. Wolfe St., Baltimore, MD 21287-9289, USA

Abstract: Neovascularization (NV) causes visual deficits in ocular disorders such as diabetic retinopathy, age-related macular degeneration, and retinopathy of prematurity. An understanding of the angiogenic factors promoting this abnormal vascular growth is necessary to devise a therapeutic approach to inhibit NV. One factor known to promote NV is vascular endothelial growth factor (VEGF), which can also induce a breakdown of the blood-retinal barrier (BRB) leading to macular edema, another major cause of visual loss in a variety of ocular disorders. To investigate the role of VEGF on ocular NV, transgenic mice have been produced that overexpress VEGF in the photoreceptors under control of the rhodopsin promoter. Eyes from these mice and from littermates not expressing the transgene were examined using immunohistochemistry, griffonia simplicifolia isolectin-B4 (GSA) staining to clearly visualize vessels, and electron microscopy. Levels of transgene expression were determined by the polymerase chain reaction. In normal mice, retinal vessels are organized into a superficial and a deep capillary bed with some vessels forming a shunt between both beds. In a transgenic line of mice that overexpresses VEGF (V-6), NV originates from the deep capillary bed at about postnatal day 10 (P10) and extends through the photoreceptor layer to form vascular complexes in the subretinal space with BRB breakdown occurring only in the area of NV. The superficial capillary bed and the choroidal vasculature are unaffected. In another line of transgenic mice with a higher expression rate of VEGF (V-24), photoreceptor degeneration begins at P7-8, soon after the onset of transgene expression, without widespread NV, as was observed in V-6 mice. In conclusion, overexpression of VEGF in transgenic mice is sufficient to cause retinal NV, but only the deep capillary bed is responsive. Increasing the expression of VEGF does not necessarily increase the amount of NV. A better understanding of the specific factors and conditions that result in a particular pattern of ocular NV may provide clues regarding the pathogenesis of ocular neovascular disease.

1. VEGF

NV is a serious ocular complication associated with a number of ocular disorders, but the angiogenic factors operating in these various disorders and their interactions with each other are not clearly understood. VEGF is an endothelial cell mitogen that has been shown to stimulate angiogenesis associated with neoplasia (Connolly et al 1989a, b, Leung et al 1989, Kondo et al 1993, Senger 1993). It can also promote ocular NV (Adamis et al 1994, Aiello et al 1994, Miller et al 1994, Murata et al 1995, Pierce et al 1995, Stone et al 1995, Ozaki et al 1997) and may be operative in ocular disease processes. VEGF is induced under hypoxic conditions (Shweiki et al 1992, Plate et al 1992, Goldberg and Schneider 1994, Hashimoto et al 1994, Minchenko et al 1994a, b, Levy et al 1995, Pierce et al 1995) and its upregulation has been demonstrated in ischemic retinopathies, such as diabetic retinopathy (Murata et al 1996, Mathews et al 1997, Vinores et al 1997), retinopathy of prematurity, and retinal vascular occlusive disease (Vinores et al 1997). However, VEGF has also been shown to be upregulated in a number of disorders in which hypoxia is not a feature, such as experimental autoimmune uveoretinitis (EAU), which shows a striking upregulation of VEGF in the inner retina (Vinores et al 1997). This finding suggests that factors other than hypoxia may be capable of inducing VEGF in the retina in pathological conditions, as has been demonstrated in other systems. A number of other factors, such as interleukin-1β (Ben-Av et al 1995, Li et al 1995, Jackson et al. 1997, Ristimaki et al 1998), prostaglandins E_1 and E_2 (Harada et al 1994), TNFα (Ryuto et al 1996), EGF, PDGF-BB, bFGF (Tsai et al 1995), and inflammatory cytokines from activated T-cells (Samaniego et al 1998) have been shown to induce the production and secretion of VEGF in other systems and the same thing is likely to occur in the eye.

VEGF expression is not limited to the vascular endothelium, but can be demonstrated in neurons in areas of cerebral infarction and in tumor cells from a wide variety of neoplasms, where its expression appears to be correlated with angiogenesis and tumor growth (Takano et al 1996, Provias et al 1997, Salven et al 1997, Westphal et al 1997). VEGF controls vasculogenesis by acting directly on the vascular endothelial cells through the flk-1 receptor (Shalaby et al 1995). Within tumors, VEGF appears to be a survival factor for new vessels, but it is not required for mature vessels (Benjamin and Keshet 1997). Within the retina, VEGF is temporally and spatially correlated with angiogenesis and vasculogenesis (Miller et al 1994, Stone et al 1995, Murata et al 1996) and with BRB breakdown (Vinores et al 1995, Mathews et al 1997) and it can be localized to ganglion cells, glia, retinal pigmented epithelial (RPE) cells, and photoreceptors. During developmental vasculogenesis in the eye, VEGF is transiently expressed by astrocytes as they migrate across the nerve fiber layer, preceding the formation of the superficial retinal vessels. It appears to be likewise transiently expressed by Muller cells in the inner nuclear layer preceding the formation of the adjacent deep layer of the retinal vasculature (Stone et al 1995). In experimental ischemic retinopathy, which was designed to serve as a model for retinopathy of prematurity, neonatal rats are exposed to elevated levels of oxygen for 14 days and then returned to the relative hypoxia of room air (Penn et al 1993, 1994). In these rats, increased VEGF was localized in the vessels of the inner retina (Vinores et al 1998), from which NV originates, but it was more prominently induced in the retinal ganglion cells (Vinores et al. 1997).

Increased expression of VEGF in the retina can result, not only in NV and vascular leakage leading to macular edema, but in the recruitment and activation of inflammatory cells and their adhesion to the vascular endothelium (Clauss et al 1996, Melder et al 1996, Lu et al 1999). VEGF administered intravitreally by injection (Tolentino et al 1996) or by slow release implants (Ozaki et al 1997) induced NV and retinal vascular leakage. When slow release vinyl acetate pellets containing VEGF were implanted into rabbit eyes (Ozaki et al 1997), marked retinal NV resulted 2 weeks after implantation and many of the retinal vascular endothelial (RVE) cells were positive for proliferating cell nuclear antigen (PCNA) indicating that they were actively proliferating. When similar pellets were implanted into monkey eyes, however, grossly dilated vessels were observed in the inner retina with widespread leakage revealed by immunohistochemical staining for albumin, but the RVE cells were negative for PCNA, indicating that despite the greatly increased retinal vascular component, NV had not occurred. Ultrastructural localization of albumin revealed that VEGF mediates BRB breakdown by opening some of the interendothelial cell junctions and by increasing transendothelial vesicular transport (Luna et al 1997).

The most striking upregulation of VEGF was visualized in the neurons and glia of the inner retina of rats and mice developing EAU resulting from immunization with the photoreceptor-specific S-antigen (Vinores et al 1997, 1998). Normal Lewis rats did not have demonstrable VEGF within the retina, but 8 days after immunization with S-antigen, which is prior to any pathological changes, a marked upregulation of VEGF had occurred in the inner retina. Immunohistochemical staining for VEGF intensified by 11 days, when the early stages of inflammation were observed in half of the rats. Despite intense staining for VEGF in the retinas of rats developing EAU, NV does not occur. This could not be accounted for by sub-threshold levels of VEGF, since this disorder demonstrated the most intense staining for VEGF that was seen, or by an insufficient number of VEGF receptors, since no decrease in flt-1 or flk-1 receptors was observed. It is, therefore, likely that NV was prevented in rats developing EAU by an angiogenesis inhibitor, such as TGFβ. In support of this hypothesis is the observation that, coincident with the upregulation of VEGF in rats developing EAU, there was increased immunoreactivity for TGFβ1 and β2 within the same layers of the retina, but rats with ischemic retinopathy and a more modest upregulation of VEGF, developed NV and showed little or no increase in TGFβ. In addition, in mice developing EAU, which does not have an abrupt onset as it does in rats and may involve NV, a comparable upregulation of VEGF was seen in the inner retina, but there was no increase in TGFβ.

2. GENERATION OF TRANSGENIC MICE

VEGF and other growth factors can be induced by other factors, as mentioned above, and may therefore represent part of a cascade initiated by an inflammatory mediator or other factor and not the primary event. To study the effect of increased VEGF on the retina as a primary event, transgenic mice were generated that overexpress VEGF in the photoreceptors, therefore allowing retina-specific expression. Transgenes were constructed using human cDNA for VEGF coupled to the rhodopsin promoter (Okamoto et al 1997). Transgene expression was first detected on postnatal day 6 (P6), as determined by reverse transcriptase polymerase chain reaction (Tobe et al 1998).

3. VEGF TRANSGENIC MICE

Three lines of VEGF transgenic mice were generated. In the first line, V-6, VEGF transgene expression was first detected at P6 and increased to reach a constant level between P14 and P21 (Okamoto et al, 1997, Tobe et al 1998). The second line, V-24, reached a higher level of expression at P14, but did not maintain this level as the expression decreased over the next 2 weeks. The third line, V-27, did not show expression of the transgene and had a normal phenotype; therefore, the line was not studied further. A comparison of the V-6 and V-24 lines showed that V-6 mice contained 3-5 transgene copies and V-24 mice contained 13-15 copies. The earliest phenotypic change in V-6 mice was the onset of retinal NV, which was evident at P10-14. Griffonia simplicifolia isolectin-B4 (GSA), a marker that can be used to visualize endothelial cells (Schulte and Spicer 1983, Sahagun et al 1989), was used to examine vascular profiles on 50 μm sections. In normal mice, GSA labeling clearly revealed the superficial and deep

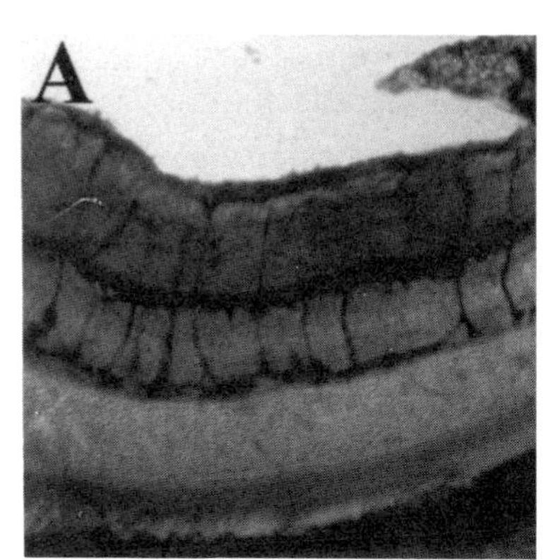

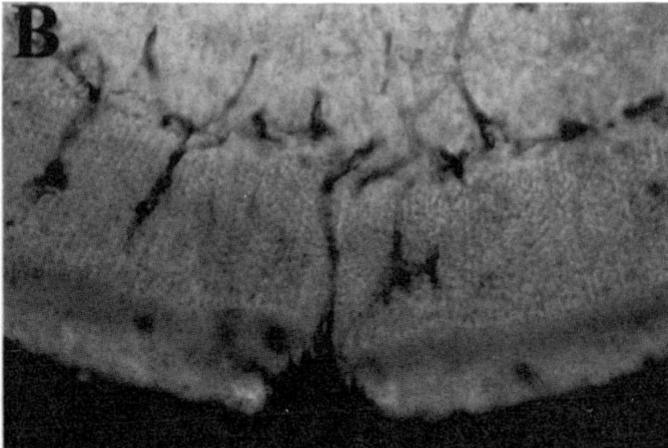

Figure 1. Retinal Neovascularization in the V-6 Line of VEGF Transgenic Mice. GSA labelling of vascular endothelial cells shows normal retinal vasculature (A) in a 2 week old mouse with the superficial (top) and deep (middle) capillary beds and some vessels forming shunts between the two vascular beds. No labelling is seen between the deep capillary bed and Bruch's membrane (bottom). (B) In a 2 week old V-6 mouse, retinal NV originating from the vessels in the deep capillary bed and extending across the photoreceptor and RPE layers is demonstrated.

capillary beds with some vessels forming shunts between the two vascular beds. Vessels were never observed between the deep capillary bed and Bruch's membrane. In V-6 mice, prominent NV originated from vessels in the deep capillary bed and extended across the photoreceptors to the RPE (Figure 1). The superficial capillary bed and the choroidal vasculature appeared unaffected. Immunohistochemical staining for albumin showed that widespread vascular leakage occurred in the outer retina (the area of NV), but the inner retinal vessels retained their BRB properties. GSA labeling was used at the ultrastructural level to visualize areas of NV, where it revealed vascular sprouting from pre-existing vessels in the deep capillary bed and newly-formed vessels extending across the photoreceptor layer. Some of the vessels were patent and contained red blood cells, but some clusters of proliferating endothelial cells had no lumen and would have been difficult to recognize without the GSA labelling.

In V-24 mice, the first phenotypic change was the onset of retinal degeneration, which commenced at P7 and progressed until there was a complete absence of photoreceptors. This was an unexpected finding and it is not clear whether the retinal degeneration is attributable to the overexpression of VEGF or to an insertional mutation caused by integration of the transgene into an essential retinal gene. Widespread NV was not seen, as it was in V-6 mice. These results show that VEGF expression induces retinal NV in the proximity of its source, in this case the photoreceptors, and increased expression of VEGF does not necessarily lead to increased NV, as seen when comparing V-6 and V-24 mice.

4. CONCLUSIONS

The retina is supplied by three different vascular beds: the superficial capillary bed, the deep capillary bed, and the choroidal vasculature. An angiogenic factor, under the appropriate conditions, can selectively target a specific vascular bed. When the source is the photoreceptors and the onset of expression is P6, VEGF overexpression induces NV only in the deep capillary bed unless the expression rate is higher and retinal degeneration is occurring, in which case proliferation from the deep capillary bed is reduced and the choroidal vasculature is also targeted. NV does not originate from the superficial retinal vessels under these conditions, but vascular leakage can occur from these vessels in association with high levels of VEGF expression. A better understanding of the angiogenic factors and the necessary conditions to provide a particular pattern of NV may provide valuable information about the mechanisms of NV in pathologic ocular disease.

5. ACKNOWLEDGMENTS

This study was supported by grants EY10017 and EY05951 from the National Eye Institute, National Institutes of Health, by Lew R. Wasserman Merit Awards (SAV, PAC), by a Juvenile Diabetes Foundation fellowship grant (NO), and by an unrestricted grant from Research to Prevent Blindness.

6. REFERENCES

Adamis, A.P., Miller, J.W., Bernal, M.-T., D'Amico, D.J., Folkman, J., Yeo, T.-K., and Yeo, K.-T., 1994, Increased vascular endothelial growth factor levels in the vitreous of eyes with proliferative diabetic retinopathy. Amer. J. Ophthalmol. 118: 445-450.

Aiello, L.P., Avery, R.L., Arrigg, P.G., Keyt, B.A., Jampel, H.D., Shah, S.T., Pasquale, L.R., Thieme, H., Iwamoto, M.A., Park, J.E., Nguyen, H.V., Aiello, L.M., Ferrara, N., and King, G.L., 1994, Vascular endothelial growth factor in ocular fluid of patients with diabetic retinopathy and other retinal disorders. New Engl. J. Med. 331: 1480-1487.

Ben-Av, P., Crofford, L.J., Wilder, R.L. Hla, T., 1995, Induction of vascular endothelial growth factor expression in synovial fibroblasts by prostaglandin E and interleukin-1: a potential mechanism for inflammatory angiogenesis. FEBS Lett. 372: 82-87.

Benjamin, L.E. and Keshet, E., 1997, Conditional switching of vascular endothelial growth factor (VEGF) expression in tumors - induction of endothelial cell shedding and regression of hemangioblastoma-like vessels by VEGF withdrawal. Proc. Natl. Acad. Sci. USA 94: 8761-8766.

Clauss, M., Weich, H., Breier, G. Knies, U., Rockl, W., Waltenberger, J., and Risau, W., 1996, The vascular endothelial growth factor receptor Flt-1 mediates biological activities. Implications for a functional role of placenta growth factor in monocyte activation and chemotaxis. J. Biol. Chem. 271: 17629-17634.

Connolly, D.T., Heuvelman, D.M., Nelson, R., Olander, J.V., Eppley, B.L., Delfino, J.J., Siegel, N.R., Leimgruber, R.M., and Feder, J., 1989a, Tumor vascular permeability factor stimulates endothelial cell growth and angiogenesis. J. Clin. Invest. 84: 1470-1478.

Connolly, D.T., Olander, J.V., Heuvelman, D., Nelson, R., Monsell, R., Siegel, N., Haymore, B.L., Leimgruber, R., and Feder, J., 1989b, Human vascular permeability factor. J. Biol. Chem. 264: 20017-20024.

Goldberg, M.A. and Schneider, T.J., 1994, Similarities between the oxygen-sensing mechanisms regulating the expression of vascular endothelial growth factor and erythropoietin. J. Biol. Chem. 269:4355-4359.

Harada, S., Nagy, J.A., Sullivan, K.A., Thomas, K.A., Endo, N., Rodan, G.A., and Rodan, S.B., 1994, Induction of vascular endothelial growth factor expression by prostaglandin E2 and E1 in osteoblasts. J. Clin. Invest. 93: 2490-2496.

Hashimoto, E., Kage, K., Ogita, T., Nakaoka, T., Matsuoka, R., and Kira, Y., 1994, Adenosine as an endogenous mediator of hypoxia for induction of vascular endothelial growth factor mRNA in U-937 cells. Biochem. Biophys. Res. Comm. 204: 318-324.

Jackson, J.R., Minton, J.A., Ho, M.L., Wei, N., and Winkler, J.D., 1997, Expression of vascular endothelial growth factor in synovial fibroblasts is induced by hypoxia and interleukin 1beta. J. Rheumatol. 24: 1253-1259.

Kondo, S., Asano, M., and Suzuki, H., 1993, Significance of vascular endothelial growth factor/vascular permeability factor for solid tumor growth, and its inhibition by the antibody. Biochem. Biophys. Res. Commun. 194: 1234-1241.

Leung, D.W., Cachianes, G., Kuang, W.-J., Goeddel, D.V., and Ferrara, N., 1989, Vascular endothelial growth factor is a secreted angiogenic mitogen. Science 246: 1306-1309.

Levy, A.P., Levy, N.S., Wegner, S., and Goldberg, M.A., 1995, Transcriptional regulation of the rat vascular endothelial growth factor gene by hypoxia. J. Biol. Chem. 270: 13333-13340.

Li, J., Perrella, M.A., Tsai, J.-C.,Yet, S.F., Hsieh, C.M., Yoshizuma, M., Patterson, C., Endege, W.O., Schlegel, R., and Lee, M.E., Induction of vascular endothelial growth factor gene expression by interleukin-1β in rat aortic smooth muscle cells. J. Biol. Chem. 279: 308-312.

Lu, M., Perez, V.L., Ma, N., Miyamoto, K., Peng, H.-B., Liao, and Adamis, A.P., 1999, VEGF increases retinal vascular ICAM-1 expression in vivo. Invest. Ophthalmol. Vis. Sci. 40: 1808-1812.

Luna, J.D., Chan, C.-C., Derevjanik, N.L., Mahlow, J., Chiu, C., Peng, B., Tobe, T., Campochiaro, P.A., and Vinores, S.A., 1997, Blood-retinal barrier (BRB) breakdown in experimental autoimmune uveoretinitis: Comparison with vascular endothelial growth factor, tumor necrosis factor α, and interleukin-1β-mediated breakdown. J. Neurosci. Res. 49: 268-280.

Mathews, M.K., Merges, C., McLeod, D.S., and Lutty, G.A., 1997, Vascular endothelial growth factor and vascular permeability changes in human diabetic retinopathy. Invest. Ophthalmol. Vis. Sci. 38: 2729-2741.

Melder, R.J., Koenig, G.C., Witwer, B.P., Safabakhsh, N., Munn, L.L., and Jain, R.K., (1996), During angiogenesis, vascular endothelial growth factor and basic fibroblast growth factor regulate natural killer cell adhesion to tumor endothelium. Nature Med. 2: 992-997.

Miller, J.W., Adamis, A.P., Shima, D.T., D'Amore, P.A., Moulton, R.S., O'Reilly, M.S., Folkman, J., Dvorak, H.F., Brown, L.F., Berse, B., Yeo, T.-K., and Yeo, K.-T., 1994, Vascular endothelial growth factor/vascular permeability factor is temporally and spatially correlated with ocular angiogenesis in a primate model. Amer. J. Pathol. 145: 574-584.

Minchenko, A., Bauer, T., Salceda, S., and Caro, J., 1994a, Hypoxic stimulation of vascular endothelial growth factor expression in vitro and in vivo. Lab. Invest. 71: 374-379.

Minchenko, A., Salceda, S., Bauer, T., and Caro, J., 1994b, Hypoxia regulatory elements of the human vascular endothelial growth factor gene. Cell. Mol. Biol. Res. 40: 35-39.

Murata, T., Nakagawa, K., Ishibashi, T., Ohnishi, Y., Inomata, H., and Sueishi, K., 1995, Temporal and spatial correlation between VEGF expression and retinal angiogenesis in neonatal rats. Invest. Ophthalmol. Vis. Sci. 36: S895.

Murata, T., Nakagawa, K., Khalil, A., Ishibashi, T., Inomata, H., and Sueishi, K., 1996, The relation between expression of vascular endothelial growth factor and breakdown of the blood-retinal barrier in diabetic rat retinas. Lab. Invest. 74: 819-825.

Ozaki, H., Hayashi, H., Vinores, S.A., Moromizato, Y., Campochiaro, P.A., and Oshima, K., 1997, Intravitreal sustained release of VEGF causes neovascularization in rabbits and breakdown of the blood-retinal barrier in rabbits and primates. Exp. Eye Res. 64: 505-517.

Penn, J.S., Tolman, B.L., and Henry, M.M., 1994, Oxygen-induced retinopathy in the rat; relationship of retinal nonperfusion to subsequent neovascularization. Invest. Ophthalmol. Vis. Sci. 35: 3429-3435.

Penn, J.S., Tolman, B.L., and Lowery, L.A., 1993, Variable oxygen exposure causes preretinal neovascularization in the newborn rat. Invest. Ophthalmol. Vis. Sci. 34: 576-585.

Pierce, E.A., Avery, R.L., Foley, E.D., Aiello, L.P., and Smith, L.E.H., 1995, Vascular endothelial growth factor/vascular permeability factor expression in a mouse model of retinal neovascularization. Proc. Natl. Acad. Sci. USA 92: 905-909.

Plate, K.H., Breier, G., Welch, H.A., and Risau, W., 1992, Vascular endothelial growth factor is a potential tumor angiogenesis factor in human gliomas in vivo. Nature 359: 845-848.

Provias, J., Claffey, K., delAguila, L., Lau, N., Feldkamp, M., and Guha, A., 1997, Meningiomas: role of vascular endothelial growth factor/vascular permeability factor in angiogenesis and peritumoral edema. Neurosurg. 40: 1016-1026.

Ristimaki, A., Narko, K., Enholm, B., Joukov, V., and Alitalo, K., 1998, Proinflammatory cytokines regulate expression of the lymphatic endothelial mitogen vascular endothelial growth factor-C. J. Biol. Chem. 273: 8413-8418.

Ryuto, M., Ono, M., Izumi, H., Yoshida, S., Weich, H.A., Kohno, K., and Kuwano, M., 1996, Induction of vascular endothelial growth factor by tumor necrosis factor alpha in human glioma cells. Possible roles of SP-1. J. Biol. 271: 28220-28228.

Sahagun, G., Moore, S.A., Fabry, Z., Shelper, R.L., and Hart, M.N., 1989, Purification of murine endothelial cell cultures by flow cytometry using fluorescein-labeled Griffonia simplicifolia agglutinin. Amer. J. Pathol. 134: 1227-1232.

Salven, P., Heikkilä, P., and Joensuu, H., 1997, Enhanced expression of vascular endothelial growth factor in metastatic melanoma. Brit. J. Canc. 76: 930-934.

Samaniego, F., Markham, P.D., Gendelman, R., Watanabe, Y., Kao, V., Kowalski, K., Sonnabend, J.A., Pintus, A., Gallo, R.C., and Ensoli, B., 1998, Vascular endothelial growth factor and basic fibroblast growth factor present in Kaposi's sarcoma (KS) are induced by inflammatory cytokines and synergize to promote vascular permeability and KS lesion development. Amer. J. Pathol. 152: 1433-1443.

Schulte, B.A. and Spicer, S.S., 1983, Histochemical evaluation of mouse and rat kidneys with lectin-horseradish peroxidase conjugates. Amer. J. Anat. 168: 345-362.

Senger, D.R., Van DeWater, L., Brown, L.F., Nagy, J.A., Yeo, K.-T., Yeo, T.-K., Berse, B., Jackman, R.W., Dvorak, A.M., and Dvorak, H.F., 1993, Vascular permeability factor (VPF, VEGF) in tumor biology. Cancer Metast. Rev. 12: 303-324.

Shalaby, F., Rossant, J., Yamaguchi, T.P., Gertsenstein, M., Wu, X.-F., Breitman, M.L., and

Schuh, A.C., 1995, Failure of blood-island formation and vasculogenesis in Flk-1-deficient mice. Nature 376: 62-66.

Shweiki, D., Itin, A. Soffer, D., and Keshet, E., 1992, Vascular endothelial growth factor induced by hypoxia may mediate hypoxia-initiated angiogenesis. Nature 359: 843-845.

Stone, J., Itin, A., Alon, T., Pe'er, J., Gnessin, H., Chan-Ling, T., and Keshet, E., 1995, Development of retinal vasculature is mediated by hypoxia-induced vascular endothelial growth factor (VEGF) expression by neuroglia. J. Neurosci. 15: 4738-4747.

Takano, S., Yoshii, Y., Kondo, S., Suzuki, H., Maruno, T., Shirai, S., and Nose, T., 1996, Concentration of vascular endothelial growth factor in the serum and tumor tissue of brain tumor patients. Canc. Res. 56: 2185-2190.

Tobe, T., Okamoto, N., Vinores, M.A., Derevjanik, N.L., Vinores, S.A., Zack, D.J., and Campochiaro, P.A., 1998, Evolution of neovascularization in mice with overexpression of vascular endothelial growth factor in photoreceptors. Invest. Ophthalmol. Vis. Sci. 39: 180-188.

Tolentino, M.J., Miller, J.W., Gragoudas, E.S., Jakobiec, F.A., Flynn, E., Chatzistefanou, K., Ferrara, N., and Adamis, A.P., 1996, Intravitreous injections of vascular endothelial growth factor produce retinal ischemia and microangiopathy in an adult primate. Ophthalmology 103: 1820-1828.

Tsai, J.-C., Goldman, C.K., and Gillespie, G.Y., 1995, Vascular endothelial growth factor in human glioma cell lines: induced secretion by EGF, PDGF-BB, and bFGF. J. Neurosurg. 82: 864-873.

Vinores, S.A., Chan, C.-C., Vinores, M.A., Matteson, D.M., Chen, Y.-S., Klein, D.A., Shi, A., Ozaki, H., and Campochiaro, P.A., 1998, Increased vascular endothelial growth factor (VEGF) and transforming growth factor β (TGFβ) in experimental autoimmune uveoretinitis: upregulation of VEGF without neovascularization. J. Neuroimmunol. 89: 43-50.

Vinores, S.A., Küchle, M., Mahlow, J., Chiu, C., Green, W.R., and Campochiaro, P.A., 1995, Blood-retinal barrier breakdown in eyes with ocular melanoma: a potential role for vascular endothelial growth factor. Amer. J. Pathol. 147: 1289-1297.

Vinores, S.A., Youssri, A.I., Luna, J.D., Chen, Y.-S., Bhargave, S., Vinores, M.A., Schoenfeld, C.-L., Peng, B., Chan, C.-C., LaRochelle, W., Green, W.R., and Campochiaro, P.A., 1997, Upregulation of vascular endothelial growth factor in ischemic and non-ischemic human and experimental retinal disease. Histol. Histopathol. 12: 99-109.

Westphal, J.R., van't Hullenaar, R.G., van der Laak, J.A., Cornelissen, I.M., Schalkwijk, L.J., van Muijen, G.N., Wesseling, P., de Wilde, P.C., Ruiter, D.J., and de Waal, R.M., 1997, Vascular density in melanoma xenografts correlates with vascular permeability factor expression but not with metastatic potential. Brit. J. Canc. 76: 561-570.

REVASCULARIZATION OF ISCHEMIC TISSUES WITH SIKVAV AND NEUROPEPTIDE Y (NPY)

Derrick S. Grant and Zofia Zukowska

The Cardeza Foundation for Hematologic Research, Jefferson Medical College, Thomas Jefferson University, 1015 Walnut St., Philadelphia PA.; Departments of Physiology and Biophysics, Georgetown University Medical Center, Washington, D.C..

Abstract: Angiogenesis, the process of new vessel growth, is necessary for many normal physiological and pathological processes such as tumor growth, wound healing and ischemia. We have recently examined *in vitro* and *in vivo* the ability of two potent angiogenic compounds, SIKVAV (a peptide derived from the alpha chain of laminin-1) and Neuropeptide Y (NPY) to revascularize ischemic tissue. These compounds were tested in an ex vivo capillary sprouting angiogenesis assay that uses rat aortic rings. Both NPY and SIKVAV in the presence of VEGF, stimulated the formation of long sprouts at concentrations of 1 ng NPY (0.2 pmole/L) and 100 μg SIKVAV. In comparison very little sprouting occurred in the control rings and 50 ng of VEGF alone was required to induce equivalent number of sprouts as NPY. SIKVAV and NPY were further tested in vivo in a rat hindlimb ischemic model. Both compounds (500 μg SIKVAV and 10 ng of NPY) were embedded in the rat hind limb following unilateral ligation of the femoral artery 1 cm proximal to the adductor hiatus. After two weeks control peptides show little or no revascularization of the hindlimb distal to the ligation; however, both SIKVAV and NPY demonstrated a two-fold increase in new vessels in the region proximal to the ligation. Histological sections of latex perfused hindlimb demonstrated that ligated limbs had very few latex-filled dermal capillaries. Limbs treated with SIKVAV and NPY, however, demonstrated normal distribution in the dermal capillary beds. These data indicate that both SIKVAV and NPY are potent angiogenic factors that show promising potential clinical application to the revascularization of ischemic tissue.

Angiogenesis: From the Molecular to Integrative Pharmacology
Edited by Maragoudakis, Kluwer Academic / Plenum Publishers,, New York, 2000

1. INTRODUCTION

The process of angiogenesis is essential to wound healing following injury or physiological insufficiencies such as an ischemic myocardium. We have recently examined *in vivo* the ability of SIKVAV (a peptide derived from the alpha chain of laminin-1) to revascularize ischemic tissue. In this manuscript, we review the work done with SIKVAV and its role in angiogenesis as it pertains to the revascularization of ischemic tissue. In addition we will also present data that introduces the new angiogenic agent Neuropeptide Y (NPY) (Zukowska-Grojec et al., 1995) to revascularize ischemic tissue. In this report, we review the structure and activities of SIKVAV and NPY as they pertain to their role in ischemia-driven angiogenesis.

2. TISSUE ISCHEMIA

The vascular supply to the heart, limbs and peripheral tissues is essential for normal physiological functions. Under certain pathologic conditions, however, vascular supply may be reduced to such an extent that it leads to necrosis of the tissue (Jennings et al., 1975). Angiogenesis, the process of new vessel growth, is necessary for many normal physiological (development, reproductive cycle) and pathological processes such as tumor growth, wound healing and ischemia (Banai et al., 1994; Folkman, 1995; Folkman and Shing, 1992). Investigators have explored the use of angiogenic agonists to revascularize ischemic tissue(Banai et al., 1994; Harada et al., 1994). We have recently examined *in vitro* and *in vivo* the ability of two potent angiogenic compounds, SIKVAV (a peptide derived from the alpha chain of laminin-1) (Grant et al., 1992; Kibbey et al., 1992) and Neuropeptide Y (Zukowska-Grojec et al., 1995).

3. NORMAL VASCULAR DISTRIBUTION AND ANGIOENIC THERAPY IN THE LIMBS

The peripheral microcirculation is comprised largely of an extensive network of arterioles, capillaries and postcapillary venules, which allow a rich distribution of nutrients to the muscles, cutaneous and subcutaneous tissues. Blood supply to the lower limb is primarily provided by the femoral artery and its branches (profundus, tibial and peroneal). Closely associated with these vessels are nerves, some of the sensory origin and most of them being sympathetic, providing the vessels with resting vasoconstrictor tone. Sympathetic nerve activity is, in fact, the main regulator of limb blood flow at rest but it also contributes to redistribution of flow to the skeletal muscle during exercise. Blockage of arterial supply by a thrombus or a constriction results in tissue hypoxia which release several angiogenic factors, including bFGF, VEGF and IL-8. Additionally nerve plexus may secrete trophic factors that may stimulate vascular growth. In most cases, however, the endogenous supply of angiogenic factors is insufficient, and does not produce enough of an inductive effect to fully revascularize the ischemic tissues. Isner has shown that the application of exogenous growth factors to patent vessels surrounding ischemic regions can result in increased collateralization and a subsequent increase in blood flow (Asahara et al., 1995; Isner, 1997; Isner et al., 1996; Isner et al., 1995; Rohovsky et al., 1996; Takeshita et al., 1996). Under this treatment, local arterioles and capillaries branch and become angiogenic especially with the use of angiogenic factors such as bFGF and VEGF(Isner et al., 1996). More recently Isner's group have shown that gene therapy using bFGF or VEGF to stimulate new blood vessel growth is an effective way of bypassing occluded arteries and reestablishing blood flow to ischemic tissues (Isner, 1999). Isner raises the issue however, as to the reliability of this approach. He indicates that further clinical investigation of both recombinant proteins and alternative dosing regimens for gene therapy will be necessary to define the optimal therapeutic strategy (Isner, 1999).

While the *in vivo* potency and angiogenic efficacy of exogenous bFGF and VEGF are undisputed, it is still unclear what role they play endogenously, and to what extent their actions involve interactions with other growth factors, proteases and extracellular matrix proteins (components of a complex network of systems participating in angiogenesis). It has been the intent of our laboratories to determine both *in vitro* and *in vivo* if new factors such as the laminin-1-derived peptide SIKVAV or the sympathetic nerve-derived neurotransmitter, neuropeptide Y, are angiogenic and could be used to revascularize ischemic tissue.

4. VASCULAR GROWTH FACTORS AND SIKVAV IN THE REVASCULARIZATION OF ISCHEMIC TISSUE

A possible treatment for the prevention of tissue infarction might involve the local administration of a potent angiogenic compound (without a short halflife) in the area of ischemia. This local administration would stimulate revascularization of the ischemic tissue and could rescue it from the irreversible damage caused by infarction. As indicated above several studies have investigated the efficacy of enhanced revascularization of ischemic tissue stimulated by local release of heparin dependant angiogenic factors such as bFGF and VEGF. The laminin derived peptide SIKVAV seems not to require heparin and has a direct effect on endothelial cell behavior. The SIKVAV peptide is smaller than the other growth factors and can be synthesized; therefore it has potential clinical applications and may be practical to use alone or in combination with VEGF in myocardial revascularization. Therefore, it was our intent to determine both *in vitro* and *in vivo* whether SIKVAV could be used in the revascularization of ischemic tissue.

5. ROLE OF SIKVAV IN ISCHEMIC HINDLIMB REVASCULARIZATION

We have used a rat model where longitudinal incision was performed, extending inferiorly from the inguinal ligament to a point 7 mm proximal to the patella of both limbs. Through this incision of one limb, using surgical loops, the femoral artery was dissected free, along its length in this region and ligated proximal to the adductor hiatus. A slow release LVAX pellet impregnated either with no peptide (placebo), a control peptide (100 μg/day), VEGF(VEGF-1, 165) (50 ng/day) or SIKVAV (100 μg/day) was placed in the popliteal fossa. These pellets were small approximately 5 mm (containing calcium alginate) remained throughout the experiment and dissolved by 21 days. With this method, 50-100 μg of peptide per day was released during the first week. The incision was then sutured and the animal was allowed to recover for two weeks. Evaluation of angiogenesis in the ischemic limb was first done by opening the abdomen of each rat, and inserting a catheter into the abdominal aorta. Then injected radiopaque dye

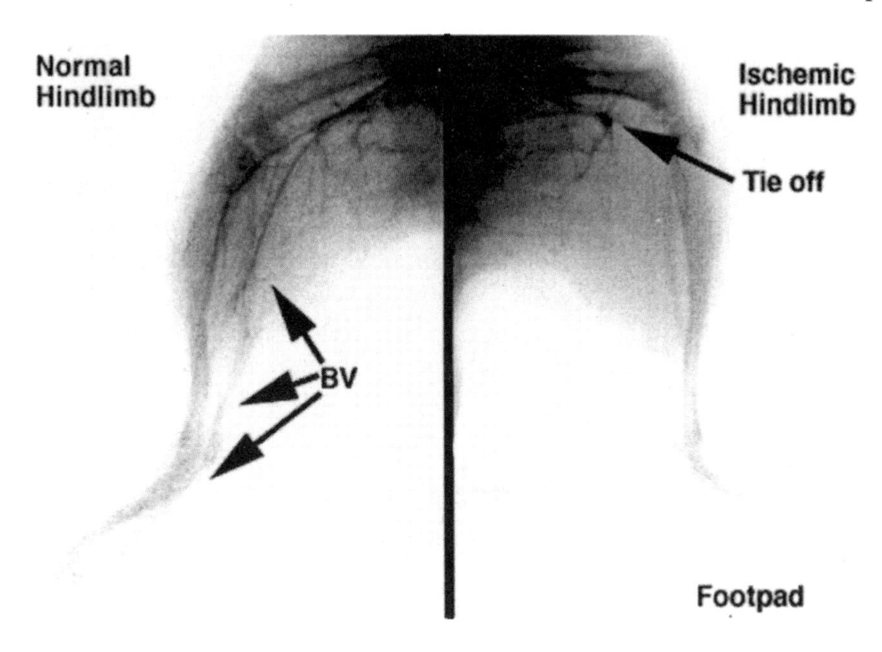

Figure 1a. Radiographs of the blood flow to hindlimb of the rat following tie off of the femoral artery. Little blood flow is seen to the hindlimb with no treatment whereas collaterals can be observed with the SIKVAV treated animal.

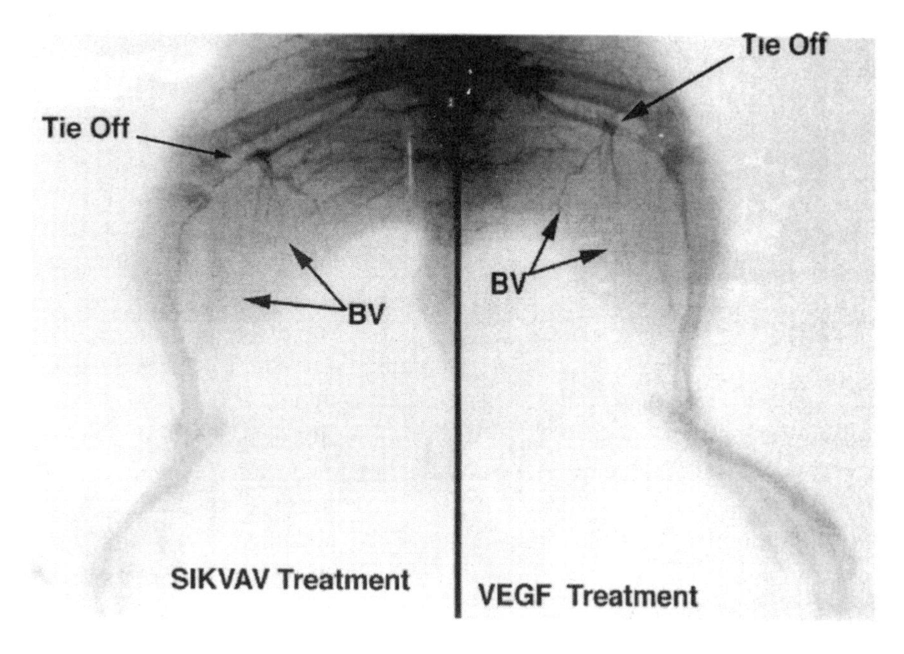

Figure1b. Angiogram of the blood supply to he hindlimb of the rats following treatment with SIKVAV or with VEGF.

into the lower limbs. Radiographs were then taken to evaluate the degree of blood flow to the limb (see figure 1). Limbs which did not have a ligated femoral artery (sham operated) showed normal blood flow to the adductor muscles and calf muscles. Additionally, the planter vessels of the footpads could clearly be seen (figure 1).

In contrast limbs where the femoral artery was tied off showed markedly reduced blood flow to the thigh and calf and little or no flow to the foot. Limbs that contained a VEGF (VEGF-1, 165) or SIKVAV slow release pellet demonstrated remarkable revascularization of the limb including blood flow to the adductor and calf muscles (figure 1). Note; limbs with control peptide appeared similar to ligated untreated limbs, however, the control peptide had a slight increase in blood flow to the adductor compartment but little or no flow to calf and foot (not shown).

We used a second method to evaluate the degree of blood flow to the tissues and muscles of the hindlimb. Following the surgery and 14-21 day recovery, the rat's aorta was cannulated as above, however this time latex was injected into the vascular bed and allowed to harden. Tissue samples from the adductor and calf muscles and the footpad were excised and processed for H & E light microscopic histological sections. Evaluation of normal sections of the rat footpad (figure 2) showed the presence of latex in the capillary bed adjacent to the dermal papilla in the skin of the footpad. Little or no latex was seen in the limbs with ligated femoral artery indicating that little blood flow was reaching the skin in this area. This condition mimics the situation observed with diabetic patients.

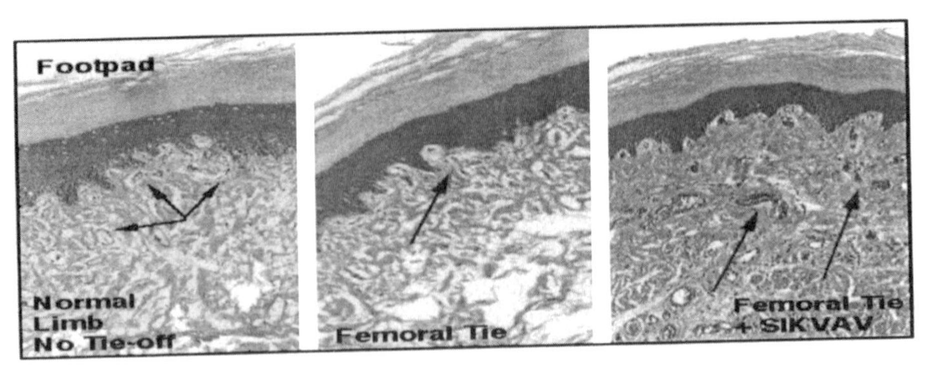

Figure 2. Latex perfusion of vessel in the rat footpad.

When SIKVAV or VEGF-treated limbs were examined all the capillaries in the dermal region of the skin were filled with latex indicating blood flow to the region. We examined the adductor muscle from the normal limb and compared it to the limb treated with SIKVAV and found that there was little difference in latex content in the capillaries surrounding the muscle fibers (not shown). These data indicate that SIKVAV has an effect on inducing revascularization of the hindlimb similar to VEGF. We do not know, however, whether the increase in blood flow to the ischemic limb is due solely to neovascularization or a combination of new angiogenesis and collateralization. Secondly it would be interesting to determine the effect of SIKVAV on angiogenesis.

6. THE ROLE OF NEUROPEPTIDE Y IN ANGIOGENESIS

Sympathetic nerves have been considered to provide a trophic action on blood vessel development and this action is believed to be mediated by catecholamines. However, catecholamines, at physiological concentrations, have weak or no direct mitogenic activities for either VSMCs or endothelial cells (Blaes and Boissel, 1983; Okazaki et al., 1994). Neuropeptide Y (NPY), a sympathetic non-adrenergic transmitter, and is the most abundant of all peptides in the heart and the brain (reviewed in (Zukowska-Grojec and Wahlestedt, 1993). It is released within cardiovascular tissues during exercise, stress and ischemia, and causes vasoconstriction by acting at Y1 receptor, one of the five of its Gi-coupled receptors (Zukowska-Grojec and Wahlestedt, 1993). More recently, our (Zukowska-Grojec et al., 1993) and other (Erlinge et al., 1994) laboratories have discovered that NPY is also a potent mitogen for vascular smooth muscle cells. Unlike the NPY's vasoconstrictor activity (which is restricted to the Y1 receptor activation and requires a full molecule of the peptide at concentrations of at least 10-fold higher than the circulating levels) the mitogenic effect of NPY occurs over a wide range of concentrations beginning with sub-pM (Figure 3), and involves both Y1 and non-Y1- receptors (Zukowska-Grojec et al., 1998). This has led us to propose that NPY's physiological function is foremost that of a vascular growth factor and only in pathophysiological conditions would the peptide additionally become a vasoconstrictor .

To determine whether NPY fulfills the criteria of a trophic vascular factor, we undertook further studies to examine its potential angiogenic

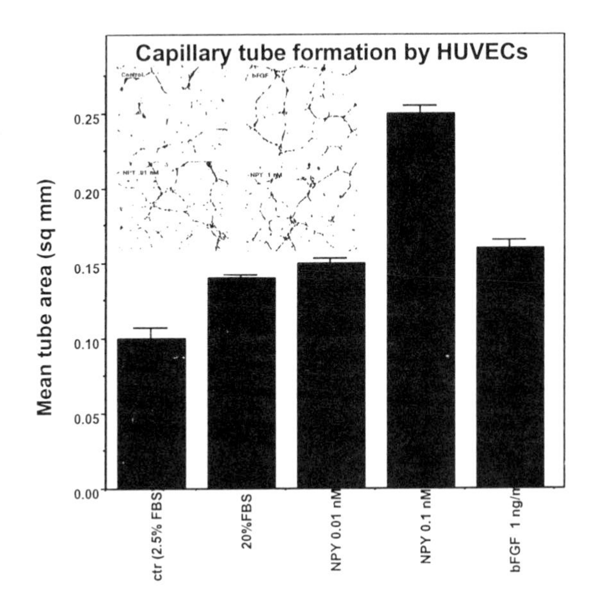

Figure 3. Effect of NPY on tube formation in vitro.

Table 1. Compounds and their activities

Compound	Name	Function
NPY	Neuropeptide Y	Vasoconstrictor Neuromodulator
Y1-Y6	Neuropeptide Y Receptor	Binds NPY and related proteins; Gi/o-coupled
NPY3-36	Enzyme processed NPY	Binds selectively to Y2 receptor
DPPIV	Dipeptidyl peptidase IV: (enzyme) cd26	Cleaves NPY to $NPY_{3\text{-}36}$
LpNPY	$Leu_{31}Pro_{34}NPY$	Y1 receptor agonist
BIBP3226	®-N2-(diphenacyl)-N-[(4-hydroxyphenyl)methyl]-D-arginineamide	Y1 antagonist

activity. In several *in vivo* and *in vitro* assays, NPY was found to be angiogenic. At low physiological concentrations, *in vitro,* NPY promoted vasodilation, adhesion, migration, proliferation and capillary tube formation by human umbilical vein endothelial cells (HUVECs, Figure 3), and enhanced sprouting of the rat aorta, with potencies comparable to those of VEGF and bFGF (Zukowska-Grojec et al., 1998). *In vitro* and *in vivo,* in murine and chick angiogenic assays, NPY was angiogenic and as efficacious as bFGF (Figure 3). Under these conditions *in vitro* NPY's action involved activation of both Y1 and Y2 receptors. Of the two, the Y1 receptor responded with more immediate up-regulation of expression within the first 6 hrs of capillary tube formation on Matrigel, but its expression subsided by 20 hrs and its activation was unable to fully mimic the NPY's effect (Zukowska-Grojec et al., 1998). Conversely, the Y2 receptor expression was initially less dramatic but increased to comparable levels and remained sustained for 20 hrs of cell differentiation on Matrigel. Additionally, the Y2 receptor agonist, NPY3-36, fully mimicked and Y2 receptor antagonist, T4 [NPY27-36]4), blocked the NPY's effect (Zukowska-Grojec et al., 1998). Thus, it is the Y2 which appears to be the main NPY angiogenic receptor (Figure 4-5).

When examined, the endothelium was found to express not only Y1 and Y2 receptors and be a site of peptide action, but also to be a source of the NPY's "converting enzyme", dipeptidyl peptidase IV (DPPIV) (Mentlein et al., 1993; Zukowska-Grojec et al., 1998). This endothelial membrane-bound serine-protease, is also present in activated T and B lymphocytes (as CD26) (Mentlein et al., 1993), and has been implicated in wound healing, AIDS (Callebaut et al., 1993) and metastasis formation either directly or in association with gelatinase (Pineiro-Sanchez et al., 1997), another protease involved in angiogenesis. Interestingly, DPPIV terminates NPY's Y1 activity and cleaves a dipeptide Tyr_1-Pro_2 from NPY1-36 to form a Y2 agonist, NPY_{3-36} (Mentlein et al., 1993). Processed NPY3-36 loses its ability to activate vasoconstrictor Y1 receptor but is strongly angiogenic (Zukowska-Grojec et al., 1998).

Our recent data indicate that blockade of DPPIV with inhibitors such as Diprotin A and aminoacylpyrrolidine-2-nitriles (Li et al., 1992) or monoclonal antibodies which neutralize its enzymatic activity (Pineiro-Sanchez et al., 1997) attenuates growth-promoting effects of NPY1-36 in vascular smooth muscle cells (Wesely et al., 1998) and abolish the NPY1-36- (but not NPY3-36-) induced endothelial cell migration in a wound

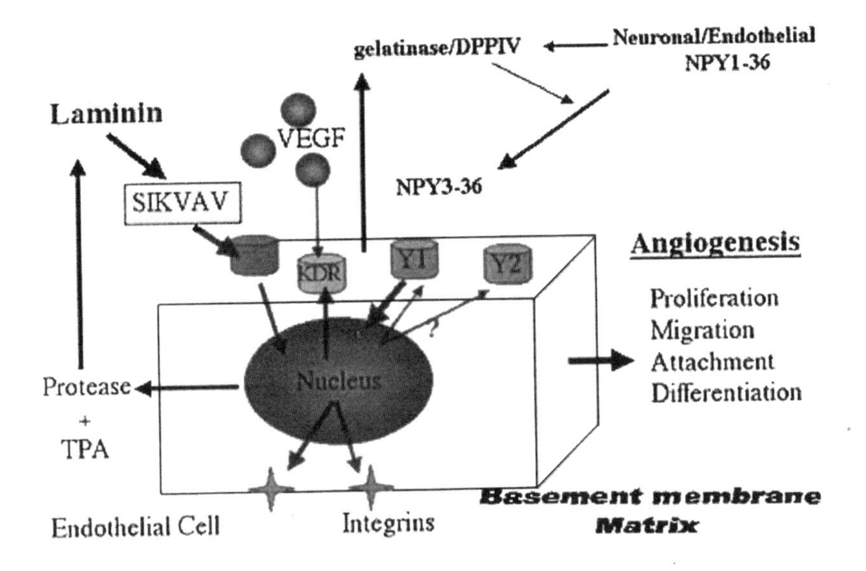

healing assay (unpublished observation from our laboratory). Thus, DPPIV appears to be an important regulator of NPY's angiogenic activity, and may determine whether or not NPY release in the ischemic tissues such as the heart leads to vasoconstriction via Y1 receptors or angiogenesis via Y2 receptors (Figure 3-5).

This enzyme may be also a link between the two systems under studies in our laboratories, laminin and NPY, via its association with other proteases (elastase and gelatinase) whose activation leads to generation of angiogenic fragments from laminin such as SIKVAV. Thus, a proposed cascade of events initiated by angiogenic stimuli may involve activation of elastase, DPPIV and gelatinase, degradation of extracellular matrix and liberation of angiogenic laminin-derived peptides such as SIKVAV, and formation of NPY-derived fragment, NPY3-36, which become inactive as a vasoconstrictor but is potently angiogenic (Figure 4-5).

Initially we have assumed all angiogenic activity of NPY to come from the sympathetic nerves. However, much to our surprise, we have found that human endothelial cells possess their own autocrine NPY, containing its mRNA and abundant intracellular stores of the peptide, co-localizing with DPPIV (Zukowska-Grojec et al., 1998). Thus, we have demonstrated the existence of two complete NPY systems: an autocrine endothelial and

Table 2. NPY Activities in Angiogenic Models

Angiogenic Model	Effect	Relative change
Matrigel in vitro	Increase in tube formation	++++
Matrigel in vivo Mouse skin	Increased capillary formation and invasion into Matrigel.	+++++
Chicken Yolk sac	Increased branching of capillaries at edge of yolk sac	++
Aortic Sprouts	Increased capillary sprouting in rat aortic ring	++++

paracrine neuronal system, both of which may be important in angiogenesis during tissue repair (reviewed in (Zukowska-Grojec, 1997).

7. NEUROPEPTIDE Y AND ITS ROLE IN LIMB ISCHEMIA

Sympathetic nerve activity is an important regulator of blood flow to the limbs by providing vasoconstrictor tone to blood vessels and this tone increases in states such as stress, exercise, and also ischemia. Sympathectomy has therefore been used to relieve the symptoms of peripheral vascular disease, the pain of claudication in the ischemic limb. The beneficial effects of sympathectomy is assumed to be due to vasodilation, however is only temporary and the procedure has largely been abandoned. Considering that sympathetic neurotransmitter NPY is angiogenic and its release may be increased during ischemic states (Jensen et al., 1994; Kaijser et al., 1994), we postulated that the peptide plays a pivotal role in revascularization of the ischemic limb. However, its beneficial effects may be reduced by sympathetic vasoconstrictor tone primarily provided by catecholamines but presumably also by NPY's Y1 receptor-mediated vasoconstriction.

NPY has been found to be released from the heart during ischemic conditions such as angina pectoris, myocardial infarction, and congestive heart failure (Kaijser et al., 1990), and from the skeletal muscle during exercise in humans and animals (Jensen et al., 1994; Kaijser et al., 1994).

The exercise-induced elevation in circulating NPY levels is further augmented by hypoxia (Jensen et al., 1994) but, interestingly, this response of NPY is lost in older patients (Jensen et al., 1994). Exercise is also implicated to stimulate angiogenesis in the heart and muscles, and the effect appears to be synergistic with ischemia (White et al., 1998). Furthermore, the ability to respond with angiogenesis appears to be reduced with age. To test whether NPY could provide an angiogenic stimulus to the ischemic limb, we used the rat model of chronic (14 days) occlusion of the femoral artery (as above) with or without unilateral sympathectomy. Also a local administration of NPY to the ischemic limb in a slow release pellet was placed in the popliteal fossa (Zukowska-Grojec, 1997). At 14 days, femoral artery occlusion-induced ischemia increased venous output of NPY by 25%, and markedly depleted NPY-immunoreactivity from the femoral artery in the ischemic compared to the non-ischemic limb, with no changes in NPY receptor expression in the skeletal muscle. Administration of NPY (1 μg/21 day in a slow release pellet, Innovative Research of America, Rockville, MD) into the ischemic limb increased venous NPY outflow 3-fold, prevented vascular tissue depletion of NPY, and induced expression of NPY Y1, Y2 and Y5 receptors (Zukowska-Grojec, 1997). The slow release NPY pellet resulted in slight elevation of systemic plasma NPY levels and also induced expression of Y1 and Y2 but not Y5 receptors in the contralateral non-ischemic leg. Thus, NPY itself promotes the induction of its own receptors, independently of ischemia. Similar NPY-mediated induction of Y1 and Y2 receptor expression were observed *in vitro* in rat vascular smooth muscle cells, and implicated in promoting cell growth (Zukowska-Grojec et al., 1998). These receptors, therefore, may play an angiogenic role, in particular, the Y5 receptor since its expression was specific for ischemia. Indeed, along with its receptor induction, NPY markedly improved vascularization of the ischemic limb, increasing capillary density in the soleus and gastroccnemius muscles two-fold as compared to the untreated ischemic limb, and restored capillary density to the level present in the non-ischemic muscles (Zukowska-Grojec et al., 1998).

Similarly NPY was also able to restore blood flow to the rat footpad following femoral ligation and ischemia. Interestingly, the effectiveness of NPY in a local delivery system both in terms of its ability to raise venous plasma NPY outflow and increasing capillary density in the ischemic limb was reduced in rats with streptozotocin-induced diabetes (unpublished observation from our laboratory). This indicates that NPY processing to

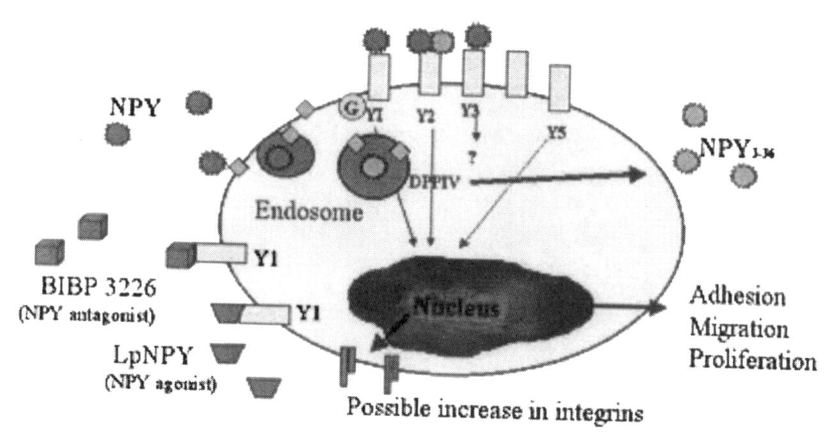

Figure 5. Schematic diagram of the effect and mechanism of action of NPY and SIKVAV on the endothelial cell during the induction of angiogenesis.

angiogenic fragments, detectable by our assay, is impaired in this condition.

Thus, our initial studies have supported the notion that the NPY system plays a role in angiogenesis in the ischemic limb. This is the first demonstration that NPY or NPY-derived peptides may be useful as angiogenic therapy for the peripheral arterial insufficiency and limb ischemia. Several factors position NPY favorably as compared to other angiogenic molecules such as VEGF and bFGF: it is a smaller molecule. Therefore it is easier to deliver and less expensive to produce, and is angiogenic at extremely low concentrations at which it is devoid of vasoconstrictor activities, it is endogenously released, thus amenable to pharmacological manipulations, and may be synergistic with other angiogenic factors and heparin. Also, by being a mitogen for both vascular smooth muscle and endothelial cells, similarly to bFGF but unlike VEGF, NPY may be able to stimulate growth of not just capillaries but of fully developed vessels containing both cell elements. We postulate that it is the Y5 and/or a combination of Y2 and Y5-type NPY receptors which are angiogenic. Thus, specific agonists activating these two receptors might be more beneficial that NPY itself, an agonist for the Y1-Y5 receptors, since they would angiogenic but devoid of unwanted Y1 vasoconstrictive activity.

We are currently examining the effect of combining NPY and SIKVAV in the hind limb ischemic model to evaluate if the effect of these compounds is additive or synergistic. This may be an effective approach to revascularizing of chronic ischemic tissues by providing a local infusion to enhance angiogenesis. The current therapeutic options to improve tissue blood flow include surgical bypass, balloon angioplasty or endoluminal recanalisation; treatments which are all hampered by significant failure rate. We hope to apply this approach to treatment of diabetic patients in an effort to alleviate ischemia and necrosis of limbs.

REFERENCES

Asahara, T., C. Bauters, L.P. Zheng, S. Takeshita, S. Bunting, N. Ferrara, J.F. Symes, and J.M. Isner. 1995. Synergistic effect of vascular endothelial growth factor and basic fibroblast growth factor on angiogenesis in vivo. *Circulation*. **92**:II365-371.

Banai, S., M.T. Jaklitsch, M. Shou, D.F. Lazarous, M. Scheinowitz, S. Biro, S.E. Epstein, and E.F. Unger. 1994. Angiogenic-induced enhancement of collateral blood flow to ischemic myocardium by vascular endothelial growth factor in dogs. *Circulation*. **89**:2183-2189.

Blaes, R.D., and J.P. Boissel. 1983. Growth-promoting effects of catecholamines on rat aortic smooth muscle cells in culture. *Journal of Cellular Physiology*. **116**:167-172.

Callebaut, C., B. Krust, E. Jacotot, and A.G. Hovanessian. 1993. T cell activation antigen, CD26, as a cofactor for entry of HIV in CD4+ cells [see comments]. *Science*. **262**:2045-2050.

Erlinge, D., J. Brunkwall, and L. Edvinsson. 1994. Neuropeptide Y stimulates proliferation of human vascular smooth muscle cells: cooperation with noradrenaline and ATP. *Regulatory Peptides*. **50**:259-265.

Folkman, J. 1995. Angiogenesis in cancer, vascular, rheumatoid and other disease. *Nat Med*. **1**:27-31.

Folkman, J., and Y. Shing. 1992. Angiogenesis. *J Biol Chem*. **267**:10931-10934.

Grant, D.S., J.L. Kinsella, R. Fridman, R. Auerbach, B.A. Piasecki, Y. Yamada, M. Zain, and H.K. Kleinman. 1992. Interaction of endothelial cells with a laminin A chain peptide (SIKVAV) in vitro and induction of angiogenic behavior in vivo. *J Cell Physiol*. **153**:614-625.

Harada, K., W. Grossman, M. Friedman, E.R. Edelman, P.V. Prasad, C.S. Keighley, W.J. Manning, F.W. Sellke, and M. Simons. 1994. Basic fibroblast growth factor improves myocardial function in cronically ischemic porcine hearts. *J. Clin. Invest.* **94**:623-630.

Isner, J.M. 1997. Angiogenesis for revascularization of ischaemic tissues [editorial]. *Eur Heart J*. **18**:1-2.

Isner, J.M. 1999. Manipulating angiogenesis against vascular disease. *Hospital Practice*. **June 15**:69-86.

Isner, J.M., A. Pieczek, R. Schainfeld, R. Blair, L. Haley, T. Asahara, K. Rosenfield, S. Razvi, K. Walsh, and J.F. Symes. 1996. Clinical evidence of angiogenesis after arterial gene transfer of phVEGF165 in patient with ischaemic limb [see comments]. *Lancet*. **348**:370-374.

Isner, J.M., K. Walsh, J. Symes, A. Pieczek, S. Takeshita, J. Lowry, S. Rossow, K. Rosenfield, L. Weir, E. Brogi, and et al. 1995. Arterial gene therapy for therapeutic angiogenesis in patients with peripheral artery disease [news]. *Circulation*. **91**:2687-2692.

Jennings, R.B., C.E. Ganote, and K.A. Reimer. 1975. Ischemic tissue injury. *Am J. Pathol.* **81**:179-198.

Jensen, E.W., K. Espersen, I.L. Kanstrup, and N.J. Christensen. 1994. Exercise-induced changes in plasma catecholamines and neuropeptide Y: relation to age and sampling times. *J Appl Physiol.* **76**:1269-1273.

Kaijser, L., J. Pernow, B. Berglund, J. Grubbstrom, and J.M. Lundberg. 1994. Neuropeptide Y release from human heart is enhanced during prolonged exercise in hypoxia. *J Appl Physiol.* **76**:1346-1349.

Kaijser, L., J. Pernow, B. Berglund, and J.M. Lundberg. 1990. Neuropeptide Y is released together with noradrenaline from the human heart during exercise and hypoxia. *Clin Physiol.* **10**:179-188.

Kibbey, M.C., D.S. Grant, and H.K. Kleinman. 1992. Role of the SIKVAV site of laminin in promotion of angiogenesis and tumor growth: An in vivo Matrigel model. *J. Nalt. Cancer Inst.* **84**:1633-1637.

Li, X., Y. Wu, R. North, and M. Forte. 1992. Cloning, functional expression, and developmental regulation of a neuropeptide Y receptor from Drosphila melanogaster. *Journal of Biological Chemistry.* **267**:9-12.

Mentlein, R., P. Dahms, D. Grandt, and R. Kruger. 1993. Proteolytic processing of neuropeptide Y and peptide YY by dipeptidyl peptidase IV. *Regulatory Peptides.* **49**:133-144.

Okazaki, M., Z.-W. Hu, M. Fujinaga, and B. Hoffman. 1994. Alpha-1 adrenergic receptor-induced c-fos gene expression in rat aorta and cultured vascular smooth muscle cells. *Journal of Clinical Investigation.* **94**:210-214.

Pineiro-Sanchez, M.L., L.A. Goldstein, J. Dodt, L. Howard, Y. Yeh, and W.T. Chen. 1997. Identification of the 170-kDa melanoma membrane-bound gelatinase (seprase) as a serine integral membrane protease. *J Biol Chem.* **272**:7595-7601.

Rohovsky, S., M. Kearney, A. Pieczek, K. Rosenfield, R. Schainfeld, P.A. D'Amore, and J.M. Isner. 1996. Elevated levels of basic fibroblast growth factor in patients with limb ischemia. *Am Heart J.* **132**:1015-1019.

Takeshita, S., L. Weir, D. Chen, L.P. Zheng, R. Riessen, C. Bauters, J.F. Symes, N. Ferrara, and J.M. Isner. 1996. Therapeutic angiogenesis following arterial gene transfer of vascular endothelial growth factor in a rabbit model of hindlimb ischemia. *Biochem Biophys Res Commun.* **227**:628-635.

Wesely, L., H. Ji, E. Grouzmann, and Z. Zukowska-Grojec. 1998. Inducible dipeptidyl peptidase IV in vascular smooth muscle cells: role in neuropeptide Y-induced mitogenesis. *FASEB J.* **12**:1912.

White, F.C., C.M. Bloor, M.D. McKirnan, and S.M. Carroll. 1998. Exercise training in swine promotes growth of arteriolar bed and capillary angiogenesis in heart [In Process Citation]. *J Appl Physiol.* **85**:1160-1168.

Zukowska-Grojec, Z. 1997. Neuropeptide Y: Implications in vascular remodeling and novel therapeutics. *Drug News and Perspectives.* **10**:587-595.

Zukowska-Grojec, Z., E. Karwatowska-Prokopczuk, W. rose, J. Rone, S. Movafagh, H. Ji, Y. Yeh, W.-T. Chen, H.K. Kleinman, E. Grouzmann, and D.S. Grant. 1998. Neuropeptide Y a novel angiogenic factor from the sympathetic nerves and endothelium. *Circ. Res.* **83**:187-195.

Zukowska-Grojec, Z., E. Karwatowska-Prokopczuk, Y. Yeh, W.T. Chen, W. Rose, and D.S. Grant. 1995. Endothelial Neuropeptide Y (NPY) system and angiogenesis. *Circulation.* **92**:I-714.

Zukowska-Grojec, Z., P. Pruszczyk, C. Colton, J. Yao, G.H. Shen, A.K. Myers, and C. Wahlestedt. 1993. Mitogenic effect of neuropeptide Y in rat vascular smooth muscle cells. *Peptides.* **14**:263-268.

Zukowska-Grojec, Z., and C. Wahlestedt. 1993. Origin and actions of neuropeptide Y in the cardiovascular system. *In* The Biology of Neuropeptide Y and Related Peptides. W.F. Colmers and C. Wahlestedt, editors. Humana Press, Totowa, NJ. 315-388.

CELLULAR EFFECTS AND SIGNALLING PATHWAYS ACTIVATED BY THE ANTI-COAGULANT FACTOR, PROTEIN S, IN VASCULAR CELLS
PROTEIN S CELLULAR EFFECTS

[1]Chryso Kanthou and [2]Omar Benzakour
[1]*Tumour Microcirculation Group, Gray Laboratory Cancer Research Trust, PO Box 100, Mount Vernon Hospital, Northwood, Middlesex HA6 2JR, UK.*
[2]*Molecular Cell Biology Laboratory, Thrombosis Resesarch Institute, Emmanuel Kaye Building, Manresa Road, London SW3 6LR, UK.*

The anticoagulant factor protein S is a secreted vitamin K-dependent γ-carboxylated protein that is mainly synthesised in the liver but is also made by endothelial cells and megakaryocytes in culture. In previous studies we have shown that protein S acts as a mitogen for cultured human vascular smooth muscle cells. The synthesis and secretion of protein S by endothelial cells suggests that in addition to its role in the coagulation cascade, protein S may be an important autocrine factor implicated in the pathophysiology of the vascular system. The effects of protein S on hVSMC proliferation, migration and survival are discussed. The activation of the components of the MAP kinase pathway, ERK1/2, JNK/SAPK and p38 is also summarised. Binding and chemical cross-linking experiments provided evidence for the existence of a cell surface protein S receptor(s). By virtue of its many cellular effects, it is suggested here that the anticoagulant factor protein S plays an important role in the pathophysiology of the vasculature.

1. INTRODUCTION

The proliferation and migration of vascular smooth muscle cells (VSMCs) and endothelial cells are important events in both the formation of atherosclerotic lesions and the process of restenosis after angioplasty as well as in angiogenesis. Two distinct physiological processes lead to the development of new blood vessels: vasculogenesis and angiogenesis.[1]

Angiogenesis: From the Molecular to Integrative Pharmacology
Edited by Maragoudakis, Kluwer Academic / Plenum Publishers, New York, 2000

Whereas vasculogenesis is restricted to embryonic development and consists in the differentiation of mesodermal precursor cells into endothelial cells followed by their organisation into the capillary plexus, angiogenesis is the formation of new capillaries from pre-existing blood vessels and takes place both during prenatal and adult life. Angiogenesis can occur under both physiological and pathological conditions such as during wound healing and solid tumour development. Moreover, intra-arterial angiogenesis is evident in atherosclerotic plaques and in recanalised thrombi.[2] The cellular and molecular events that lead to angiogenesis are not as yet fully elucidated but are known to include (i) breakdown of the extracellular matrix and of basement membrane of pre-existing blood vessels; (ii) the migration and proliferation of endothelial cells; (iii) production of extracellular matrix allowing the reconstitution of the basement membrane; (iv) recruitment of pericytes and vascular smooth muscle cells (VSMCs) to the newly formed blood vessel.

Numerous growth factors with the capacity to initiate and sustain vascular cell proliferation are secreted by platelets, macrophages, as well as endothelial cells and VSMCs themselves.[3] In addition to regulating haemostasis, some components of the fibrinolytic and coagulation systems activate cellular processes such as migration, proliferation and adhesion that are key events in response to injury. This is exemplified by numerous reports demonstrating that thrombin,[4] protein C,[5] factors Xa,[6] XII[7] and VIIa[8], uPA[9] and tPA[10] bind to cell surface receptors, activate specific signalling pathways and trigger a variety of cellular responses (including that of angiogenesis). An important role of coagulation/ fibrinolytic factors in regulating vascular cell growth is further suggested by the presence of potential inhibitors of angiogenesis within proteins such as plasminogen,[11] platelet factor 4,[12] prothrombin[13] and antithrombin-III.[14] Protein S, a plasma glycoprotein involved in the inactivation of factors Va and VIIIa was also demonstrated to exert mitogenic activities in VSMCs and thus may also be classified as yet another coagulation cascade protein with a dual role in the vascular system. In the present review, we will summarise some general notions about protein S, describe some of the cellular events and signalling pathways activated by protein S and speculate on the potential roles of protein S in the vasculature.

2. PROTEIN S

Protein S is a 69 kDa single chain plasma glycoprotein that acts as a cofactor for activated protein C in the inactivation of coagulation factors Va and VIIIa.[15] It requires no activation by proteolytic cleavage and circulates at a concentration of about 270 nM in both a free active form (40%) and in an inactive form (60%) bound to C4b-binding protein. Protein S belongs to the family of vitamin K-dependent proteins, that includes a number of zymogens and co-factors of the coagulation cascade, in which some glutamyl residues are post-translationally modified to γ-carboxyglutamic acid residues (Gla residues).[16] Gla residues allow Ca^{2+} dependent protein-phospholipid complex formation and facilitate conformational changes which are essential for zymogen activation or for cofactor activity.[17] Following the Gla region is a domain unique to protein S, which consists of a disulphide loop that is highly susceptible to proteolysis by serine proteinases such as thrombin.[18] This is followed by four epidermal growth factor-like (EGF-like) domains. The C-terminus of protein S is unrelated to other coagulation factors and is homologous to the plasma sex hormone binding globulin (SHBG) and the basement membrane proteins laminin A and merosin.[19] The latter two proteins have been shown to play a role in cell proliferation, migration and differentiation.[19]

The product of a gene specifically expressed during serum starvation in fibroblasts, named growth arrest specific gene 6 (Gas6) represents a new member of the vitamin K-dependent protein family that is homologous to protein S.[20] Apart from its lack of a thrombin-cleavage site, Gas6 protein exhibits strong homology with all the other structural domains of protein S. Gas6 is secreted by cultured VSMCs and endothelial cells and potentiates thrombin-induced proliferation in these cells.[21] Gas6 was also shown to prevent fibroblast, VSMC and endothelial cell death by apoptosis and hence is postulated to be a cell survival factor.[22,23] It is of particular interest to point out here that in a recent report, Melaragno *et al*,[24] showed in a rat model that both Gas6 and its tyrosine kinase receptor *Axl* are strongly upregulated in injured blood vessels suggesting an important role for this homologue of protein S in vascular injury and repair. The major producer of protein S and other factors of the coagulation system is the liver. However, protein S is also synthesised at extrahepatic sites including the brain and spleen[25] as well as endothelial cells,[26] megakaryocytes,[27] osteoblasts[28] and cells of the nervous system.[29] A significant proportion of circulating protein S is thought to be endothelial-derived. The production of protein S by some fully differentiated cells suggests a cell type-specific expression, which may be related to local anticoagulation or to other as yet unknown functions of this protein.

2.1 Cellular effects of protein S

Gasic *et al*[30] and our laboratory[31] have previously demonstrated that protein S induces the proliferation of cultured rat and human VSMCs (hVSMCs) respectively. In DNA synthesis experiments measuring ^{3}H-thymidine incorporation, growth arrested hVSMCs (maintained in 0.5% serum for 72 h) responded mitogenically to protein S in a dose dependent manner. A concentration of 10 nM protein S gave a clear stimulation, which reached a plateau at 200 nM. The stimulation observed with 200 nM protein S was equivalent to that induced by 100 nM thrombin which is a known potent VSMC mitogen. The mitogenic response to protein S was drastically reduced by pre-incubating it with an anti-human protein S monoclonal antibody prior to its addition to the cells indicating the specificity of its effects. Cell counts performed 3 and 6 days after the addition of protein S to growth arrested hVSMCs confirmed that the DNA synthesis assay gave a fair index of cell division. After 6 days in culture, cells in the presence of 200 nM protein S underwent at least 2 cycles of division whereas control cells declined in numbers. In addition to the mitogenic effect of protein S we have investigated possible roles for this protein in hVSMC migration. Cell migration assays were conducted in modified Boyden chambers. Protein S induced directional cell migration. A dose response experiment showed that 10 nM protein S induced the optimal level of directional migration. This represents a major difference between the chemotactic and mitogenic effects of protein S, the latter being optimal within a concentration range 100-200 nM, which may imply that, the main cellular function of protein S lies in promoting cell migration. Alternatively, protein S may activate different receptors that mediate these two processes. In comparison, PDGF-BB, the most potent known chemoattractant for VSMCs,[3] induced a 4-fold greater increase in directional migration than protein S. Protein S induced chemotaxis but not chemokinesis as when the same concentration of protein S was included in both the upper and lower wells of the chambers, migration of these cells was not significantly greater than in control chambers. Similarly Gas6, the structural homologue of protein S, although it has marginal mitogenic activity, was shown to induce the directional migration of rat and to a lesser extent hVSMCs. Maximal chemotactic activity was demonstrated using approximately 3 nM Gas6.

An increasing body of evidence suggests that VSMC apoptosis is involved in the pathogenesis of atherosclerosis and restenosis and during vascular remodelling.[32] In culture, normal hVSMCs exhibit a low level of detectable apoptosis, which has been postulated to be due, at least in part, to the ability of non-apoptotic VSMCs to phagocytose neighbouring apoptotic cells.[33] Human VSMCs derived from atherosclerotic plaques, however, were reported to exhibit a much higher level of spontaneous apoptosis.[33] In the light of the finding that protein S induces VSMC proliferation and migration, we sought to investigate whether it played also a role in vascular cell survival. The structural homologue of protein S, Gas6, protects both cultured VSMCs and endothelial cells from apoptosis in response to growth factor withdrawal.[22,23] Using the lactate dehydrogenase assay to determine total cell death and propidium iodide staining followed by FACS analysis which measures apoptosis we have investigated the putative survival effects of protein S using two known potent VSMC apoptotic agents, sodium nitroprusside (SNP)[34] and H_2O_2.[35] Incubation of hVSMCs with SNP or H_2O_2 in serum free conditions led to a significant rise in both total cell death and apoptosis. Pre-treatment of cells with protein S prior to the addition of either SNP or H_2O_2 resulted in a significant reduction in both total cell death and apoptosis. The fact that both PDGF-BB and thrombin, which are potent hVSMC mitogens, were less effective in reducing cell death than protein S suggest that the mechanisms involved in the reduction of apoptosis by protein S are independent from its mitogenic properties.

2.2 Analysis of mechanisms involved in protein S cellular effects

Using tritiated thymidine incorporation and cell count assays we have determined that the anti-coagulant factor protein S stimulates VSMC proliferation.[31] Further experiments were undertaken to determine whether protein S acts as a direct mitogen or through the production of autocrine mitogenic factors such PDGF-AA as reported for TGF-β,[36] interleukin-1[37] and thrombin.[38] Time course experiments of protein S-induced DNA synthesis, mitogen activated protein kinase ($p42/p44^{MAPK}$) phosphorylation as well as the immediate early gene (*c-fos* and *c-myc)* mRNA expression in confluent growth arrested hVSMC cultures were performed. The ability of protein S to induce a peak of DNA synthesis 20-24h post-stimulation, to activate $p42/p44^{MAPK}$ dual phosphorylation as well as transient expression of the immediate early genes within minutes demonstrate clearly that protein S stimulates hVSMC mitogenesis by rapidly acting direct mechanisms. There were no significant changes in either intracellular levels of c-AMP or inositol phosphate turnover upon the treatment of hVSMCs with protein S, and PTX failed to alter protein S-induced mitogenesis, hence suggesting that neither Gi nor Gs proteins or PLC are activated by the putative protein S receptor (s).

The activation of the mitogen activated protein kinase (MAPK) isoforms p42/p44^{MAPK} and their translocation to the nucleus is thought to be a pre-requisite for the re-entry of quiescent cells into the cell cycle.[39] p42/p44^{MAPK} undergoes rapid tyrosine and threonine phosphorylation in response to growth factors, thereby leading to the stimulation of its intrinsic kinase activity. We have, therefore, assessed the effects of protein S on both the phosphorylation/activation of p42/p44^{MAPK} and its translocation to the nucleus. Western blot analysis with an antibody specific for the dually phosphorylated/activated p42/p44^{MAPK} shows that when growth arrested hVSMCs were treated with protein S, p42/p44^{MAPK} phosphorylation occurred within minutes and was sustained for at least 4 h post-protein S addition. The extent of p42/p44^{MAPK} phosphorylation by protein S was much lower than that induced by serum which is consistent with the more potent mitogenic activity of serum. Immunofluorescence staining of hVSMC's demonstrated that protein S induced the translocation of phosphorylated p42/p44^{MAPK} to the nucleus. The sustained activation of p42/p44^{MAPK} by protein S and its translocation to the nucleus suggest that this pathway is directly involved in protein S proliferative effects. The requirement for p42/p44^{MAPK} activation/phosphorylation for protein S-induced mitogenesis was further assessed in cells that were pre-treated with the specific p42/p44^{MAPK} kinase inhibitor, PD98059. PD98059 inhibited the activation/phosphorylation of p42/p44^{MAPK} that was induced by protein S in hVSMCs and also drastically reduced the mitogenic effect of protein S demonstrating that the MAPK pathway activation is required for protein S-mediated mitogenesis.

Stress stimuli activate two groups of MAPK-related proteins, the c-jun amino terminal protein kinase (JNK) and p38 kinase which are also thought to be activated by apoptotic agents.[39] The activation of the p42/p44^{MAPK} pathway is postulated to mediate both cell proliferation and confer resistance to apoptotic stimuli. Reactive oxygen species and nitric oxide are generated in the vasculature and have been shown to induce VSMC apoptosis in vitro.[34,35] Having observed that hVSMC death induced by either H_2O_2 or SNP was substantially reduced in the presence of protein S, we then assessed the effects of H_2O_2, SNP and protein S on the activation of JNK, p38 as well as p42/p44^{MAPK} using antibodies that recognise the phosphorylated/activated forms of these kinases. In quiescent cells, JNK1 p46 protein is constitutively phosphorylated. Both H_2O_2 and SNP, stimulated the phosphorylation of JNK2 p54. Treatment of cells with protein S did not lead to any changes in either the basic pattern of JNK activation or that induced by H_2O_2 and SNP. In parallel experiments, no detectable p38 phosphorylation was evident in either quiescent hVSMC or those treated with SNP, H_2O_2 protein S or a combination of these. The inability to detect p38 activation cannot be attributed to the antibody used as in parallel experiments, sorbitol, a known p38 activator, induced a significant increase in p38 phosphorylation indicating that this pathway is functional in hVSMCs. We have also assessed whether H_2O_2 and SNP interfered with the activation of p42/p44^{MAPK} by protein S. The results obtained were negative as the extent of activation of p42/p44^{MAPK} was unaffected by the above cell death inducing agents. Altogether, the activation of the JNK2 p54 but not p38 appears to be linked with SNP or H_2O_2-mediated hVSMC death. The observed protective effect of protein S may be linked with the activation of the p42/p44^{MAPK} pathway.

Although cellular effects of protein S have been reported by several laboratories,[30,31] the identity of the cell surface receptor which is activated by protein S and the molecular mechanisms involved in protein S-induced mitogenesis are as yet unknown. Indeed, the search for receptors for Gas6 and protein S has resulted in several controversies. Gas6 and protein S were initially postulated to be ligands for the orphan tyrosine kinase receptors *Axl* and *Tyro3* respectively.[29] However, Godowski *et al* [40] demonstrated that human Gas6 and bovine protein S but not human protein S, can act as ligands for the human *Tyro3* receptor. Hence, the identity of the receptor for human protein S is still undetermined.

In our previous studies, binding and chemical cross-linking experiments provided evidence for the existence of a cell surface receptor(s) for protein S on hVSMCs.[31] The binding characteristics of ^{125}I-protein S to hVSMCs were studied and found to be saturable, reversible and cooperative. Indeed, the analysis of equilibrium binding could best be described by application of the Hill equation from which a Kd of 0.33 nM and a Hill coefficient of 1.74 were obtained. Since the Hill coefficient is a minimum estimate of the number of binding sites per receptor, it is likely that a minimum of two molecules of protein S associate with each receptor molecule. From the value of the saturation levels of bound protein S, an estimate of the number of receptors per cell was obtained as $3x10^6$. This number may, however, constitute an overestimate of cell receptor number per cell by a factor of two (or more) because the Hill coefficient indicated at least two binding sites per receptor. Autoradiographic analysis of detergent extracts of cross-linked hVSMCs separated on SDS-PAGE gel under reducing conditions showed that the radioactivity was associated with two bands migrating at 220 and 230 kDa. These bands represent complexes of ^{125}I-protein S linked to its putative receptor(s). The presence of two bands of different size suggests that the binding of protein S to hVSMC may occur either via a multimeric complex or to two distinct receptors. This observation is consistent with the deduction from the Hill equation of the presence of at least two different binding sites for protein S on hVSMCs.

3. CONCLUDING REMARKS

Gas6, the structural homologue of protein S, was discovered as a protein of unknown function that is encoded by a gene whose expression is upregulated upon serum withdrawal.[20] Thus, many studies have focused on defining its cellular roles.[20-23] These studies have demonstrated that, depending on the cellular system, Gas6 stimulates cell proliferation, migration, adhesion, potentiates thrombin's proliferative effects and acts as a cell survival factor. Although protein S was shown to be a mitogen for VSMCs for some years now[30,31], further analysis of this mitogenic or any other cellular activities was not pursued probably because its "professional" function was still considered to be purely that of a cofactor for activated protein C. In the present review, we discussed new potential roles of protein S in inducing mitogenesis, cell migration and survival of hVSMC and the signalling pathways activated by protein S. All of these cellular effects of protein S are observed within its circulating levels and thus may be of major importance for the physiopathology of the vasculature.

Calcification is commonly associated with atherosclerosis and is characterised by deposition of hydroxyapatite, a mineral found in bone, and accumulation of several bone related proteins which are thought to originate from differentiated pericytes.[41] The presence of Gla proteins was demonstrated in calcified atherosclerotic plaques. Matrix Gla protein (MGP) a component of bone matrix, is synthesised by VSMCs and plays a role in vascular calcification.[42,43] Since protein S is known to be a major component of bone matrix, it is tempting to speculate that it is also involved in vascular tissue calcification, a process occurring at the advanced stages of atherogenesis. It is of particular interest to point out here that protein S deficiency was reported to be associated with osteopenia, osteonecrosis and vascular calcification.[44,45] This raises the intriguing possibility that the anticoagulant factor protein S may also play a crucial role in both bone metabolism and vascular calcification.

In DNA synthesis experiments measuring ^{3}H-thymidine incorporation, we obsreved also that protein S substantially potentiate thrombin's mitogenic effect. This is of particular interest as upon injury to the vessel wall, there is immediate activation of the coagulation cascade and thrombotic material is often found associated with atherosclerotic lesions which during lysis can become the source of enzymatically active thrombin. This potential availability of thrombin to VSMCs has placed particular emphasis on its post-clotting role in promoting vascular cell proliferation. As well as being present in the circulation, significant quantities of protein S are also produced by endothelial cells.[26] Therefore, we suggest here that protein S contributes towards driving vascular cell migration, initiating and sustaining their proliferation by potentiating the effects of thrombin and possibly other growth factors.

ACKNOWLEDGMENTS

The authors are grateful to Professor Fedor Bachmann for critical reading of the manuscript. This work was supported by the British Heart Foundation (PG95/138) and the Thrombosis Research Trust. OB's salary was supported in part by the Gary Weston Foundation.

REFERENCES

1. Risau, W. 1997. Mechanisms of angiogenesis. *Nature* 386: 671-674.

2. Eisenstein, R, 1991. Angiogenesis in arteries. *Pharmacol. Ther.* 49: 1-19.

3. Ross R. 1993 The pathogenesis of atherosclerosis: a perspective for the 1990s. *Nature* 362: 801-809.

4. Kanthou C, Parry G, Wijelath E, Kakkar VV, Demoliou-Mason C. 1992. Thrombin-induced proliferation and expression of platelet-derived growth factor-A chain gene in human vascular smooth muscle cells. *FEBS Lett.* 314: 143-148.

5. Fukudome K, Esmon CT. 1994. Identification, cloning and regulation of a novel endothelial cell protein C/activated protein C receptor. *J Biol Chem* 269: 26486-26491.

6. Ko FN, Yang YC, Huang SC, Ou JT. 1996. Coagulation factor Xa stimulates platelet-derived growth factor release and mitogenesis in cultured vascular smooth muscle cells in rats. *J Clin Invest.* 98: 1493-1501.

7. Gordon E, Venkatesan N, Salazar R, Tang H, Schmeidler-Sapiro K, Buckley S, Warburton D, Hall F. 1996. Factor XII-induced mitogenesis is mediated via a distinct signal transduction pathway that activates a mitogen-activated protein kinase. *Proc. Natl. Acad. Sci. USA* 93: 2174-2179.
8. Rottingen J-A, Enden T, Camerer E, Iversen J-G, Prydz H. 1995. Binding of human factor VIIa to tissue factor induces cytosolic Ca^{2+} signals in J82 cells, transfected COS-1 cells, Madin-Darby canine kidney cells and in human endothelial cells induced to synthesise tissue factor. *J. Biol. Chem.* 270: 4650-4660.
9. Kanse SM, Benzakour O, Kanthou C, Kost C, Lijnen HR, Preissner KT. 1997. Induction of vascular smooth muscle cell proliferation by urokinse indicates a novel mechanism of action in vasoproliferative disorders. *Arterioscler. Throm. Vasc. Biol.* 17: 2848-2854.
10. Herbert JM, Lamarche I, Prabonnaud V, Dol F, Gauthier T. 1994. Tissue type plasminogen activator is a potent mitogen for humn aortic smooth muscle cells. *J. Biol. Chem.* 269: 3076-3080.
11. O'Reilly MS, Holmgren L, Shing Y, Chen C, Rosenthal RA, Cao Y, Moses M, Lane WS, Sage EH, Folkman J. 1994. Angiostatin: a circulating endothelial cell inhibitor that suppresses angiogenesis and tumor growth. *Cold Spring Harb Symp Quant Biol* 59:471-82.
12. Maione TE, Gray GS, Petro J, Hunt AJ, Donner AL, Bauer SI, Carson HF, Sharpe RJ. 1990. Inhibition of angiogenesis by recombinant human platelet factor-4 and related peptides. *Science* 247:77-79.
13. Lee TH, Rhim T, Kim SS. 1998. Prothrombin kringle-2 domain has a growth inhibitory activity against basic fibroblast growth factor-stimulated capillary endothelial cells. *J. Biol. Chem.* 273:28805-12
31. Benzakour O, Formstone C, Rahman S, Kanthou C, Dennehy U, Scully MF, Kakkar VV, Cooper DN. 1995. Evidence for a protein S receptor(s) on human vascular smooth muscle cells. Analysis of the binding characteristics and mitogenic properties of protein S on human vascular smooth muscle cells. *Biochem. J.* 308:481-485.
32. Newby AC, George SJ. 1996. Proliferation, migration, matrix turnover and death of smooth muscle cells in native coronary and vein graft atherosclerosis. *Curr. Opin. in Cardiol.* 11: 574-582.
33. Bennett MR, Evan GI, Schwartz SM. 1995. Apoptosis of human vascular smooth muscle cells derived from normal vessels and coronary atherosclerotic plaques. *J. Clin. Invest.* 95: 2266-2274.
34. Fukul K, Hata S, Suhara T, Nakahashi T, Shinto Y, Tsujimoto Y, Morimoto S, Ogihara T. 1996. Nitric oxide induces upregulation of fas and apoptosis in vascular smooth muscle. *Hypertension* 27: 823-826.
35. Li PF, Dietz R, von Harsdorf R. 1997. Differential effect of hydrogen peroxide and superoxide anion on apoptosis and proliferation of vascular smooth muscle cells. *Circulation* 96:3602-3609.
36. Battegay EJ, Raines EW, Seifert RA, Bowen-Pope DF, Ross R. 1990. TGF-beta induces bimodal proliferation of connective tissue cells via complex control of an autocrine PDGF loop. *Cell* 63: 515-524.
37. Raines EW, Dower SK, Ross R. 1989. Interleukin-1 mitogenic activity for fibroblasts and smooth muscle cells is due to PDGF-AA. *Science* 243:393-396.

38. Kanthou C, Benzakour O, Patel G, Deadman J, Kakkar VV, Lupu F. 1995. Thrombin receptor activating peptide (TRAP) stimulates mitogenesis, *c-fos* and PDGF-A gene expression in human vascular smooth muscle cells. *Thromb. Haemost.* 74: 1340-1347

39. Mii S, Khalil R, Morgan K, Ware J, Kent K. 1996. Mitogen activated protein kinase and proliferation of human vascular smooth muscle cells. *Am. J. Physiol.* 270: 142-150.

40. Godowski PJ, Mark MR, Chen J, Sadick MD, Raab H, Hammonds G. 1995. Reevaluation of the roles of protein S and gas6 as ligands for the receptor tyrosine kinase Rse/ *Tyro 3*. *Cell* 82: 355-358

41. Demer LL, Watson KE, Bostrom K. 1994. Mechanisms of calcification in atherosclerosis. *Trends Cardiovasc Med* 4: 45-52.

42. Luo G, Ducy P, McKee MD, Pinero GJ, Loyer E, Behringer RR, Karsenty G. 1997. Spontaneous calcification of arteries and cartilage in mice lacking matrix Gla protein. *Nature* 386:78-81.

43. Shanahan CM, Proudfoot D, Farzaneh-Far A, Weissberg P. 1998. The role of Gla proteins in vascular calcification. *Crit. Rev. Eukaryot. Gene Expr.* 8: 2357-2371.

44. Pierre-Jacques H, Glueck CJ, Mont MA, Hungerford DS. 1997. Familial heterozygous protein-S deficiency in a patient who had mutifocal osteonecrosis. *J. Bone Joint. Surg.* 79: 1079-1081.

45. Perez-Mijares R, Payan-Lopez J, Guzman-Zamudio JL, Sanchez-Angulo JI, Gomez-Fernandez P, Ramos-Diaz M, Alcala-Rueda M, Silgado-Rodriguez G, Hermosin-Ramos L, Almaraz-Jimenez M. 1996. Free protein S deficiency in hemodialysis patients due to vascular calcifications? *Nephron* 74:356-361.

14. Maione TE, Gray GS, Petro J, Hunt AJ, Donner AL, Bauer SI, Carson HF, Sharpe RJ O'Reilly MS, Pirie-Shepherd S, Lane WS, Folkman J. 1999. Antiangiogenic activity of the cleaved conformation of the serpin antithrombin. *Science* 285: 1926-1928.

15. Dahlbäck B. 1997. Factor V and protein S as cofactors to activated protein C. *Haematologic.* 82: 91-95.

16. Furie B, Bouchrd BA, Furie BC. 1999. Vitamin K-dependent biosynthesis of γ-carboxyglutamic acid. *Blood* 93: 798- 808.

17. Mann KG. 1984. Membrane-bound enzyme complexes in blood coagulation. *Prog. Hemost. Thromb.* 7: 1-23.

18. Dahlbäck B, Hildebrand B, Malm J. 1990. Chracterisation of functionally important domains in human vitamin K-dependent protein S using monoclonal antibodies. *J. Biol. Chem.* 265: 8127-8135.

19. Joseph DR. 1997. Sequence and functional relationships between androgen-binding protein/sex hormone-binding globulin and its homologs protein S, Gas6, laminin, and agrin. *Steroids* 62: 578-588.

20. Manfioletti G, Brancolini C, Avanzi G, Schneider C. 1993. The protein encoded by a growth arrest-specific gene (gas6) is a new member of the vitamin K-dependent proteins related to protein S, a negative coregulator in the blood coagulation cascade. *Mol. Cell. Biol.* 13: 4976-4985.

21. Nakano T, Higashino K, Kikuchi N, Kishino J, Nomura K, Fujita H, Ohara O, Arita H. 1995. Vascular smooth muscle cell-derived, Gla-containing growth potentiating factor for Ca^{2+} mobilising growth factors. *J Biol Chem.* 270: 5702-5705.

22. Goruppi S, Ruaro E, Schneider C. 1996. Gas6, the ligand of Axl tyrosine kinase receptor, has mitogenic and survival activities for serum starved NIH3T3 fibroblasts. *Oncogene* 12:471-480.
23. Nakano T, Kawamoto K, Higashino K, Arita H. 1996. Prevention of growth arrest-induced cell death of vascular smooth muscle cells by a product of growth arrest-specific gene, gas6. *FEBS Lett.* 387: 78-80.
24. Melaragno MG, Wurhrich DA, Poppa V, Gill D, Lindner V, Berk BC, Corson MA. 1998. Increased expression of Axl tyrosine kinase after vascular injury and regulation by G protein-coupled receptor agonists in rats. *Circ. Res.* 83: 697- 704.
25. Jamison CS, McDowell SA, Marlar RA, Degen SJ: 1995. Developmental expression of protein C and protein S in the rat. *Thromb. Res.* 78:407-412.
26. Fair DS, Marlar RA, Levin EG. 1986. Human endothelial cells synthesize protein S. *Blood* 67:1168-1172.
27. Schwarz HP, Heep MJ, Wencel-Drake JS, Griffin JH. 1985. Identification and quantification of protein S in human platelets. *Blood* 66: 1452-1458.
28. Maillard C, Berruyer M, Serre CM, Dechavanne M, Delmas PD. 1992. Protein S, a vitamin K-dependent protein, is a bone matrix component synthesized and secreted by osteoblasts. *Endocrinology* 130: 1599-1607.
29. Stitt TN, Conn G, Gore M, Lai C, Bruno J, Radziejewski C, Mattsson K, Fisher J, Gies DR, Jones PF, Masiakowski P, Ryan TE, Tobkes NJ, Chen DH, DiStefano PS, Long GL, Basilico C, Goldfarb MP, Lemke G, Glass DJ, Yancopoulos GD. 1995. The anticoagulation factor protein S and its relative, Gas6, are ligands for the Tyro3/Axl family of receptor tyrosine kinases. *Cell* 80:661-671.
30. Gasic GP, Arenas CP, Gasic TB, Gasic GJ. 1992. Coagulation factors X, Xa and protein S, as potent mitogens of cultured aortic smooth muscle cells. *Proc. Natl. Acad. Sci. USA* 89: 2317-2324.

Role of Extracellular Matrix and Adhesion Molecules in Angiogenesis

TARGETING INTEGRINS $\alpha_v\beta_3$ AND $\alpha_v\beta_5$ FOR BLOCKING TUMOR-INDUCED ANGIOGENESIS

Chandra Kumar, C., Armstrong, L., Yin, Z., Malkowski, M., Maxwell, E., Ling He., Yaremko, B., Liu, M., Varner, J.*, Smith, E. M., Neustadt, B and Nechuta, T.

Schering-Plough Research Institute, 2015, Galloping Hill Road, Kenilworth, NJ 07033, U.S.A
**University of California at San Diego, La Jolla, CA 92093, USA.*

1. INTRODUCTION

The concept that tumor growth is dependent on the formation of new blood vessels as a source of nutrients and oxygen has formed the basis for a novel therapeutic strategy against cancer (Folkman, 1992). The formation of new blood vessels permits rapid expansion of the tumor and increases the risk of metastatic spread. The process of angiogenesis involves three distinct steps as outlined in Figure 1. Angiogenesis is initiated by the secretion of cytokines by the tumor cells. These diffuse into the surrounding tissue containing preexisting blood vessels and activate normally quiescent vascular endothelial cells to proliferate and migrate towards the tumor. The migration of vascular endothelial cells is facilitated by the secretion of proteolytic enzymes such as collagenases and matrix metallo proteases which degrade the extracellular matrix and facilitate the invasion process. Finally, the vascular endothelial cells form a lumen and stop dividing. A crucial role for extracellular matrix (ECM) in the development of vasculature in physiological and pathological conditions has been demonstrated. The invasion, migration and proliferation of vascular endothelial cells during angiogenesis is regulated by a class of molecules known as integrins.

2. INTEGRIN RECEPTORS

Integrins are a class of cell surface receptors that mediate the interaction of cells with the ECM and to other cells (Hynes, 1992). Integrins are composed of noncovalently associated α and β chains which combine to give a wide array of heterodimers with distinct cellular and adhesive specificities (Ruoslahti, 1996). Adhesion to ECM is known to be important

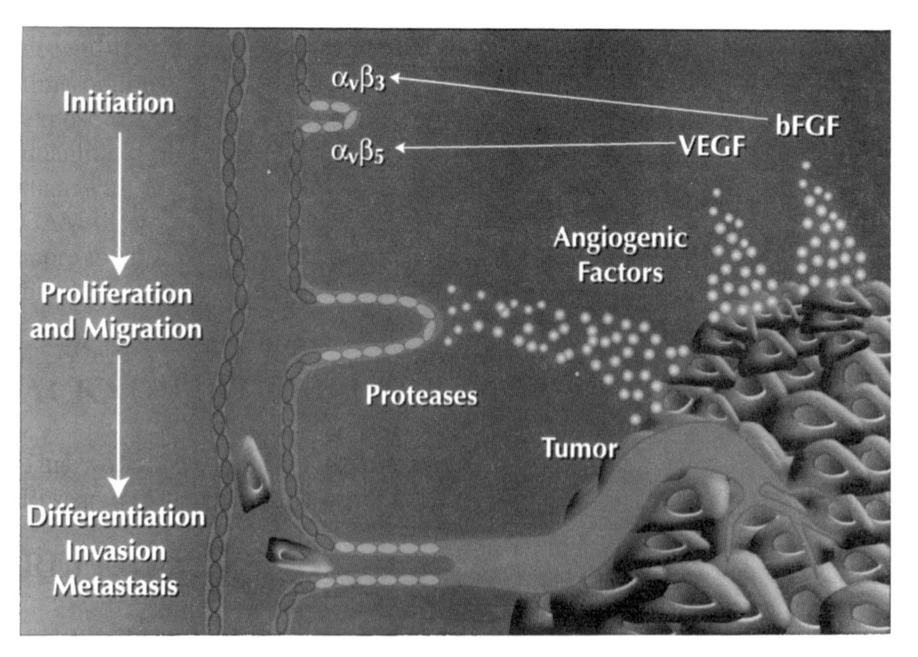

Figure 1. Tumor-induced angiogenesis

for survival and proliferation of various cell types (Ruoslahti and Reed, 1994). A number of classical signaling pathways are known to be activated by interactions of cells with matrix proteins via integrins (Kumar, 1998). The ECM is composed of structural and regulatory molecules some of which include laminin, collagen, vitronectin and fibronectin. Many of the integrins recognize the RGD (Arg-Gly-Asp) sequence present in their matrix ligands. Nevertheless, they are capable of distinguishing different RGD-containing proteins such that some bind primarily to fibronectin and others to vitronectin. A variety of physiologically important processes depend on the ability of cells to recognize and in turn respond to their immediate environment. These include angiogenesis, wound healing, bone resorption and inflammation.

2.1 Integrin $\alpha_v\beta_3$

One particular integrin, $\alpha_v\beta_3$ has been found to play a very significant role in the process of angiogenesis. Integrin $\alpha_v\beta_3$ is one of the most promiscuous members of the integrin family as it is capable of interacting with a number of ECM proteins including vitronectin, fibrinogen, fibronectin and thrombospondin (Varner et al., 1995). A characteristic feature of $\alpha_v\beta_3$ that makes it an attractive target for therapeutic intervention is its relatively limited cellular distribution. It is not generally expressed on epithelial cells and is expressed only at low levels on a subset of B cells, some cells of macrophage lineage, smooth muscle cells and activated endothelial cells (Shattil, 1995), (Cheresh, 1991)). Integrin $\alpha_v\beta_3$ is also expressed on certain invasive tumors including metastatic melanoma (Albelda, 1990; Petitclerc, 1999) and late stage glioblastoma (Gladson and Cheresh, 1991).

2.2 Role of integrins $\alpha_v\beta_3$ and $\alpha_v\beta_5$ in angiogenesis

Integrin $\alpha_v\beta_3$ is minimally, if at all, expressed on resting, or normal blood vessels, but is significantly upregulated on vascular cells within human tumors or in response to certain growth factors in vitro. For example, basic fibroblast growth factor markedly increases β_3 mRNA and surface expression in cultured human dermal micro vascular endothelial cells. Basic fibroblast growth factor and tumor necrosis factor alpha stimulate $\alpha_v\beta_3$

expression on developing blood vessels in the chick chorioallantoic membrane (CAM) (Brooks et al., 1994) and on the rabbit cornea (Friedlander, 1995). Up regulation of $\alpha_v\beta_3$ expression is also induced by human tumors cultured on the CAM, by human tumors grown in human skin explants grafted onto SCID mice and on the rabbit cornea (Brooks et al., 1995). Furthermore, antagonists of $\alpha_v\beta_3$ including cyclic RGD peptides and monoclonal antibodies significantly inhibited angiogenesis induced by cytokines and solid tumor fragments (Brooks et al., 1994). Importantly recent findings suggest that these antiangiogenic effects may be due to the ability of these antagonists to induce apoptosis in the proliferating blood vessels (Stromblad, 1996). Remarkably, $\alpha_v\beta_3$ antagonists had very little effect on preexisting blood vessels indicating the usefulness of targeting this receptor for therapeutic benefit without adverse side effects. Recent studies have provided evidence that $\alpha_v\beta_3$ receptor associates with tyrosine phosphorylated Vascular endothelial growth factor (VEGF) receptor-2 and a monoclonal antibody to β_3 subunit blocks signaling events elicited by VEGF (Soldi, 1999). These results indicate a new role for $\alpha_v\beta_3$ integrin in the activation of an angiogenic program in endothelial cells. Besides being an important survival system for nascent vessels by regulating cell adhesion to matrix, $\alpha_v\beta_3$ integrin plays a critical role in the activation of VEGF receptor-2.

Recent studies have implicated a related integrin, namely integrin $\alpha_v\beta_5$ in angiogenesis under certain conditions. For example, Friedlander et al (1995) have shown that antibody antagonists of $\alpha_v\beta_3$ inhibit bFGF stimulated angiogenesis and antagonists of integrin $\alpha_v\beta_5$ inhibit VEGF-stimulated angiogenesis in the corneal and chorioallontoic membrane models. However, angiogenesis induced by a combination of these growth factors was inhibited by either antibody. These results suggest that there are two angiogenic pathways activated by bFGF and VEGF and mediated by integrin $\alpha_v\beta_3$ and $\alpha_v\beta_5$ respectively. Therefore dual antagonists of $\alpha_v\beta_3$ and $\alpha_v\beta_5$ may be useful in blocking tumor-induced angiogenesis.

2.3 Role of integrin $\alpha_v\beta_3$ in tumor cell invasion and metastasis

Antagonists of $\alpha_v\beta_3$ have been shown to block tumor cell invasion and metastasis in many model systems in vitro and in vivo (Cheresh, 1991,; Petitclerc, 1999). Evidence suggests that $\alpha_v\beta_3$ can help modulate several proteolytic mechanisms. For example, it has been shown that ligation of

$\alpha_v\beta_3$ can result in increased expression of matrix metalloproteinase-2 (MMP-2) in certain melanoma cell lines (Seftor et. al., 1992), (Petitclerc, 1999). Furthermore, it has been shown that $\alpha_v\beta_3$ binds to MMP-2 and localizes the enzyme to the surface of invasive vascular and tumor cells, thus increasing focalized proteolysis of distinct cell-substrate contact sites (Brooks, 1996). In addition, $\alpha_v\beta_3$ receptor is found to regulate serine proteinase receptor uPAR (urokinase plasminogen activator receptor) which plays a critical role in the metastatic cascade (Nip, 111995). Thus inhibition of $\alpha_v\beta_3$ receptor may have impact on a number of proteolytic mechanisms involved in tumor progression in addition to blocking tumor-induced angiogenesis.

3. SCREENING PARADIGM

In order to discover and develop potent and selective antagonists of $\alpha_v\beta_3$ and $\alpha_v\beta_5$, we have developed a series of high-throughput receptor binding assays and cell-based secondary assays. We have used purified integrin receptors $\alpha_v\beta_3$, $\alpha_{IIb}\beta_3$ and $\alpha_v\beta_5$ to develop solid phase receptor binding assays. In addition to potency, integrin selectivity is an important factor to consider because the RGD-binding integrin receptors are widely expressed and participate in various physiological and pathological processes. Of particular concern is the fibrinogen receptor $\alpha_{IIb}\beta_3$ which has the identical β subunit and a highly homologous α subunit. $\alpha_{IIb}\beta_3$ plays a critical role in hemostasis by mediating platelet aggregation and inhibition of this receptor may cause unwanted bleeding.

3.1 Receptor binding assays

We have purified integrins $\alpha_v\beta_3$ and $\alpha_v\beta_5$ from human placenta using monoclonal antibody affinity chromatography. Echistatin, a snake venom derived peptide was used as a ligand for these three receptors. Echistatin is a 49 amino acid peptide containing the RGD sequence. We have carried out detailed biochemical characterization of the binding of echistatin to integrins $\alpha_v\beta_3$ and $\alpha_v\beta_5$. We have shown that both purified and membrane bound forms of $\alpha_v\beta_3$ and $\alpha_v\beta_5$ bind to radiolabeled echistatin with high affinity which can be competed efficiently by linear and cyclic RGD peptides (Kumar et al., 1997).

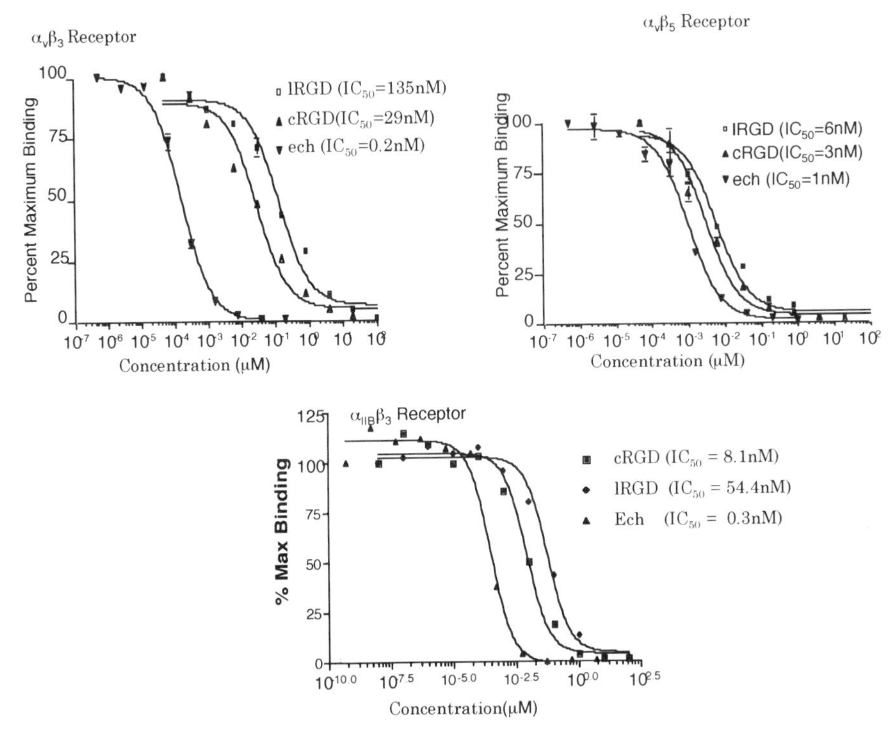

Figure 2. Competition curves for RGD peptides comparing ^{125}I-echistatin binding to $\alpha_v\beta_3$, $\alpha_v\beta_5$ and $\alpha_{IIB}\beta_3$ receptors

Our studies indicate that native echistatin binds to integrin $\alpha_v\beta_3$ and $\alpha_v\beta_5$ receptors in a non-dissociable manner similar to vitronectin (Kumar et al., 1997). In contrast, radiolabeled echistatin binds to the two related integrins $\alpha_v\beta_3$ and $\alpha_v\beta_5$ in a strikingly different manner depending on the method of labeling. Figure 2 shows the results of echistatin binding to the three integrin receptors and competition by different RGD peptides. As shown in Fig.2 the IC50 values for different RGD peptides differ significantly for different receptors.

3.2 Secondary cell-based assays

To evaluate the ability of antagonists to 1) block the binding of ligand to the receptors expressed on cell surface and to 2) inhibit adhesion of cells to

ECM proteins, we have developed stable cell lines expressing $\alpha_v\beta_3$ and $\alpha_v\beta_5$ receptors. We chose the human embryonic kidney 293 cells because they lack the $\alpha_v\beta_3$ and $\alpha_v\beta_5$ receptors (Bodary and McLean, 1990); (Bodary et al., 1989). These cells express endogenous $\alpha_v\beta_1$ receptor. HEK- 293 cells were transfected with pCDNA3 expression vectors carrying human α_v, β_3 and β_5 cDNAs. Stable transfectants obtained after selection with G418 were analyzed by FACS analysis using LM 609 and P1F6 monoclonal antibodies that specifically recognize $\alpha_v\beta_3$ and $\alpha_v\beta_5$ integrin receptors respectively. The integrin expression profile of the wild type and $\alpha_v\beta_3$ and $\alpha_v\beta_5$ transfected 293 cells was compared by FACS analysis.

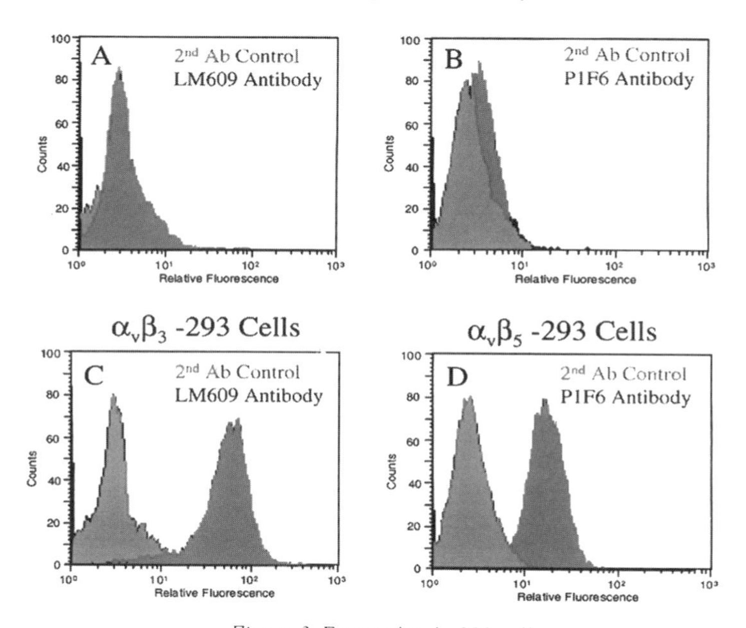

Figure 3. Expression in 293 cells

As shown in figure 3, these studies confirm that 293 cells fail to express $\alpha_v\beta_3$ (panel A) and $\alpha_v\beta_5$ (Panel B) receptors since they lack the β_3 and β_5 subunits. However they do express the α_v subunit, which associates with the β_1 subunit to form $\alpha_v\beta_1$ receptor (data not shown). Following transfection with the α_v and β_3 or β_5 cDNAs, the stable cells designated 293-$\alpha_v\beta_3$ (panel C) and 293-$\alpha_v\beta_5$ (panel D) express the $\alpha_v\beta_3$ and $\alpha_v\beta_5$ heterodimers respectively, on the cell surface.

3.3 Cell-binding and cell-adhesion assays

To measure the binding of echistatin to $\alpha_v\beta_3$ expressed on the cell surface, 293- $\alpha_v\beta_3$ cells were harvested from tissue culture flasks, placed in suspension and incubated with ^{125}I-echistatin for 2hrs. G418 resistant line (A4) that is negative for $\alpha_v\beta_3$ expression was used as a control in these experiments. The 293 - $\alpha_v\beta_3$ cell line binds 5 to 10 fold higher amount of radiolabeled echistatin compared to the $\alpha_v\beta_3$ negative A4 clone (Fig. 4a and 4b). The binding of ^{125}I-echistatin to 293-$\alpha_v\beta_3$ clone is competed by linear and cyclic RGD peptides, but not by the peptide containing RGE sequence (Fig. 4a). Addition of increasing concentrations of radioligand to the cells resulted in a linear increase in the total amount of radioligand bound to the 293-$\alpha_v\beta_3$ cells (Fig. 4b). However, there was no appreciable binding to the $\alpha_v\beta_3$ (-ve) parental cells, suggesting that the endogenous $\alpha_v\beta_1$ receptor binds poorly to echistatin.

A cell adhesion assay was established as an additional way to examine the ability of compounds to block the attachment of cells to vitronectin. The transfected 293 cells adhere very efficiently to vitronectin compared to parental 293 cells. The adhesion of 293 - $\alpha_v\beta_3$ cells can be competed by ehistatin, RGD peptides and LM609 antibodies that specifically recognize $\alpha_v\beta_3$ receptor.

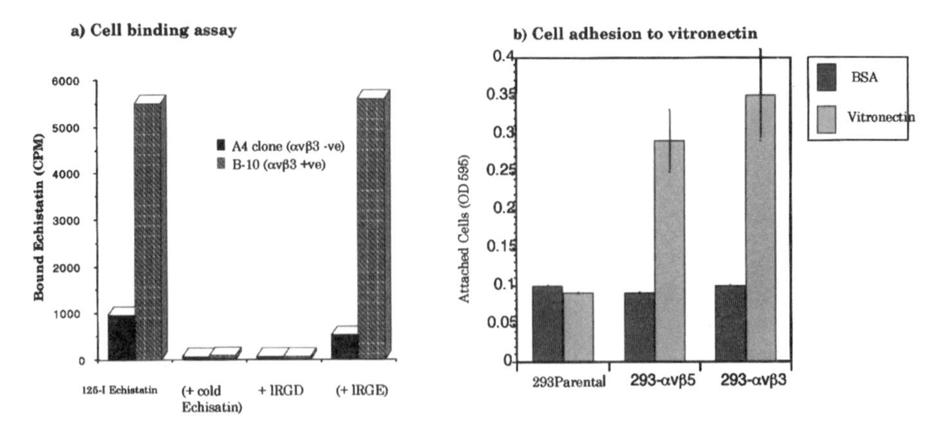

Figure 4. Secondary cell binding and cell adhesion assays

4. $\alpha_V\beta_3$ ANTAGONISTS

Many small molecule peptides containing the RGD sequence have been synthesized and their activity and specificity towards $\alpha_{IIb}\beta_3$ and $\alpha_v\beta_3$ receptor have been evaluated (Cook, 1994). These studies have shown that linear RGD peptides show very little selectivity among the integrin receptors. This non-selectivity of linear RGD peptides seems to be a result of the conformational flexibility of their structure. This is also supported by the results of solution structure analysis of disintegrins derived from snake venom, in which the RGD sequence is located at the tip of a flexible hair pin loop extruded from the globular molecule (Chen, 1991). This implies that the RGD site can flexibly associate with different types of integrin receptors by changing its conformation resulting in non-selective binding. Conformationally constrained cyclic RGD peptides have been synthesized and some of them showed relatively high specificity to the different integrin receptors (Cheng, 1994).

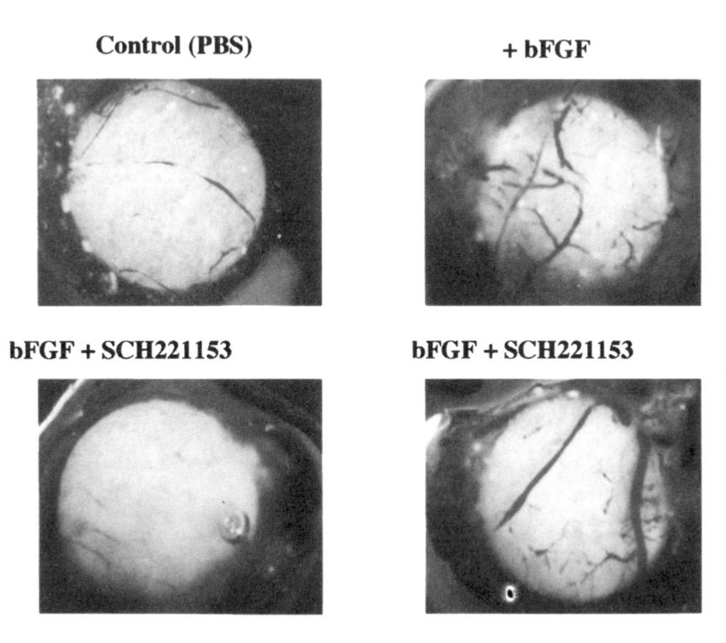

Figure 5. Inhibition of angiogenesis in the CAM assay by SCH221153

Table 1.

SCH	Dose, mpk	% Inhibition
221153	3.2	31
	8	60
	20	71
	50	67

In order to explore a diverse array of structures, combinatorial chemistry was used for the design and production of target compounds. These libraries of compounds were screened using $\alpha_v\beta_3$ receptor binding assay. A series of compounds represented by SCH 221153 were found to be potent and selective antagonist of $\alpha_v\beta_3$ receptor. This compound was found to inhibit the binding of radiolabeled echistatin to 293-$\alpha_v\beta_3$ cells and also inhibit the attachment of these cells to vitronectin matrix.

SCH 221153 was tested in the Chick chorioallantoic membrane (CAM) using basic Fibroblast growth factor (bFGF) to induce angiogenesis. In this assay, 10 day old chick embryos were used to induce angiogenesis in the CAM using filter disks soaked with bFGF and the drug was applied topically one day later. After 48 hrs, the CAMs were dissected out and the representative areas were photographed. As shown in Fig.5 SCH 221153 was effective in blocking bFGF-induced angiogenesis.

SCH 221153 was also tested to in the SCID mouse model to see if it can inhibit tumor growth and angiogenesis. In this model, LOX cells were injected intradermally on day 0 and the drug was administered twice a day by intraperitoneal route for 14 days. LOX cells were chósen because they express low levels of integrin $\alpha_v\beta_3$ receptor. As shown in Table 1 $\alpha_v\beta_3$ antagonist SCH 221153 was effective in inhibiting tumor growth in a dose dependent manner in this model.

These results demonstrate that $\alpha_v\beta_3$ antagonists will be useful in blocking tumor growth and may have therapeutic utility in the treatment of cancer.

REFERENCES

Albelda, S., M., Mette, S.A., Elder, D.E., Steward, R., Danjanovich, L., Herlyn, M and Buck, C.A. (1990). Integrin distribution in malignant melanoma : Association of the β3 subunit with tumor progression. *Cancer Res.* **50**, 6757-6764.

Bodary, S. C., and McLean, J. W. (1990). The integrin β1 associates with the vitronectin receptor α_v subunit to form a novel vitronectin receptor in a human embryonic kidney cell line. *J. Biol. Chem.* **265**, 5938-5941.

Bodary, S. C., Napier, M. A., and McLean, J. W. (1989). Expression of recombinant platelet glycoprotein IIbIIIa results in a functional fibrinogen-binding complex. *J. Biol. Chem.* **264**, 18859-18862.

Brooks, P. C., Stromblad, S., Sanders, L.C., von Schalscha, T.L., Aimes, R.T., Stetlet-Stevenson, W.G., Quigley, J.P and Cheresh, D.A. (1996). Localization of matrix metalloproteinase MMP-2 to the surface of invasive cells by interaction with integrin $\alpha_v\beta_3$. *Cell* **85**, 683-693.

Brooks, P. C., Clark, R. A. F., and Cheresh, D. A. (1994). Requirement of vascular integrin $\alpha_v\beta_3$ for angiogenesis. *Science* **264**, 569-571.

Brooks, P. C., Montgomery, A. M. P., Rosenfeld, M., Reisfeld, R. A., Hiu, T., Klier, G., and Cheresh, D. A. (1994). Integrin $\alpha_v\beta_3$ antagonists promote tumor regression by inducing apoptosis of angiogenic blood vessels. *Cell* **79**, 1157-1164.

Brooks, P. C., Stromblad, S., Klemke, R., Visscher, D., Sarkar, F., and Cheresh, D. (1995). Antiintegrin $\alpha_v\beta_3$ blocks human breast cancer growth and angiogenesis in human skin. *J Clin Invest* **96**, 1815-1822.

Chen, Y., Pitzenberger, S.M., Garsky, V.M., Lumma, P.K., Sanyal, G., Baum, J. (1991). Proton NMR assignments and secondary structure of the snake venom protein echistatin. *Biochemistry* **30**, 11625-11636.

Cheng, S., Craig, W.S., Mullen, D., Tschopp, J.F., Dixon, D., Pierschabacher, M.D. (1994). Design and the synthesis of novel cyclic RGD containing peptides as highly potent and selective integrin $\alpha_{iib}\beta_3$ antagonists. *J.Med. Chem.* **37**, 1-8.

Cheresh, D. A. (1991). Structure, function and biological properties of integrin $\alpha_v\beta_3$ on human melanoma cells. *Cancer Meta. rev.* **10**, 3-10.

Cook, N. S., Kottirsch, G., Zerwes, H.G. (1994). Platelet glycoprotein IIb/IIIa antagonists. *Drugs Fut* **19**, 135-159

Folkman, J. (1992). The role of angiogenesis in tumor growth. *Seminars in Cancer Biology* **3**, 65-71.

Friedlander, M., Theesfeld, C.T., Sugita, M., Fruttiger, M., Thomas, M.A ., Chang, S and Cheresh, D.A. (1995). Involvement of integrins $\alpha_v\beta_3$ and $\alpha_v\beta_5$ in ocular neovascular diseases. *Proc. Natl. Acad. Sci. USA* **93**, 9764-9769.

Hynes, R. O. (1992). Integrins: Versatility, modulation and signaling in cell adhesion. *Cell* **69**, 11-25.

Kumar, C. C. (1998). Signaling by integrin receptors. *Oncogene* **17**, 1365-1373.

Kumar, C. C., Nie, H., Rogers, C. P., Malkowski, M., Maxwell, E., Catino, J. J., and Armstrong, L. (1997). Biochemical characterization of the binding of echistatin to integrin $\alpha_v\beta_3$ receptor. *J Pharmacol Exp Ther* **283**, 843-853.

Nip, J., Rabbani, S.A., Shibata, H.R and Brodt, P. (111995). Coordinated expresssion of the vitronectin receptor and the urokinase-type plasminogen activator receptor in metastatic melanoma. *J.Clin.Invest.* **95**, 2096-2103.

Petitclerc, E., Stromblad, S., von Schalsck, T.L., Mitjans, F., Piulats, J., Montgomery, A.M.P., Cheresh, D.A and Brooks, P.A. (1999). Integrin $\alpha_v\beta_3$ promotes M21 melanoma growth in human skin by regulationg tumor cell survival. *Cancer Res.* **59**, 2724-2730.

Ruoslahti, E. (1996). RGD and other recognition sequences for integrins. *Ann Rev Cell Dev Biol* **12**, 697-715.
Ruoslahti, E. A., and Reed, J. (1994). Anchorage dependence, integrins, and apoptosis. *Cell* **77**, 477-478.
Seftor, R. E. B., Seftor, E. A., Gehlsen, K. R., Stetler-Stevenson, W. G., Brown, P. D., Ruoslahti, E., and Hendrix, M. J. C. (1992). Role of the $\alpha_v\beta_3$ integrin in human melanoma cell invasion. *Proc Natl Acad Sci* **89**, 1557-1561.
Shattil, S. J. (1995). Function and regulation of the β3 integrins in hemostasis and vascular biology. *Thromb. Haemostat* **10**, 3-10.
Soldi, R., Mitola, S., Strasly, M., Defilippi, P., Tarone, G and Bussolino, F. (1999). Role of $\alpha_v\beta_3$ integrin in the activation of vascular endothelial growth factor receptor-2. *EMBO J.* **18**, 882-892.
Stromblad, S., Becker, J.C., Yebra, M., Brooks, P.C. and Cheresh, D.A. (1996). Suppresion of p53 activity and p21WAF1/CIP1 expression by vascular cell integrin $\alpha_v\beta_3$ during angiogenesis. *J.Clin. Invest.* **98**, 426-433.
Varner, J. A., Brooks, P.C.,and Cheresh, D. A. (1995). REVIEW: the integrin $\alpha_v\beta_3$: angiogenesis and apoptosis. *Cell Adhes Commun* **3**, 367-374.

SUPPRESSION OF HUMAN MICROVASCULAR ENDOTHELIAL CELL INVASION AND MORPHOGENESIS WITH SYNTHETIC MATRIXIN INHIBITORS

Targeting Angiogenesis with MMP Inhibitors

Meng-Chun Jia, Martin A. Schwartz, and QingXiang Amy Sang*
Department of Chemistry and Institute of Molecular Biophysics, Florida State University, Tallahassee, Florida 32306-4390

ABSTRACT Matrix metalloproteinases (MMPs, matrixins) are a family of zinc proteinases that digest extracellular matrix and play a very important role in normal development and pathological conditions such as cardiovascular diseases and cancer metastasis. Type IV collagenases (gelatinase A/MMP-2 and gelatinase B/MMP-9) may be critical in the early steps of angiogenesis, the digestion of basement membrane and the migration of endothelial cells from the existing blood vessels. Human dermal microvascular endothelial cells were cultured on type I collagen, type IV collagen, and reconstituted basement membrane Matrigel and differentiation was examined in the presence of potent synthetic inhibitors of MMPs. The thiol inhibitor MAG-283 had IC50 values of 480 nM and 3 nM against human interstitial collagenase (MMP-1) and MMP-2, respectively, and KI value of 2.2 nM against MMP-9. The sulfodiimine inhibitor YLL-224 had IC50 values of 180 nM, 63 nM, and 44 nM against MMP-1, -2, and -9, respectively. These inhibitors at very low micromolar concentrations inhibited cell-mediated type I collagen degradation and partially blocked cell invasion through type IV collagen. These inhibitors also suppressed endothelial differentiation, i.e., formation of capillary-like tubes on Matrigel and on type I collagen. These results suggest that collagen-degrading MMPs play an important role during the initiation of angiogenesis.

Angiogenesis: From the Molecular to Integrative Pharmacology
Edited by Maragoudakis, Kluwer Academic / Plenum Publishers, New York, 2000

1. INTRODUCTION

Matrix metalloproteinases (MMPs, matrixins) are a family of highly homologous zinc endopeptidases that cleave peptide bonds of extracellular matrix (ECM) proteins, such as collagens, elastin, laminins, proteoglycans, and fibronectin[1,2]. Many types of cells under certain physiological or pathological conditions produce one or more members of the MMP family and the expression of the MMPs is regulated by microenviromental conditions such as growth factors and the integrity of the ECM. This family of proteinases plays a very important role in normal connective tissue turnover during development, morphogenesis, wound healing, reproduction, and angiogenesis[3,4]. Their proteolytic activities are controlled by a family of proteins called tissue inhibitors of metalloproteinases (TIMPs)[5]. The imbalance of the MMPs and TIMPs may be involved in many pathological processes such as cardiovascular diseases, arthritis, inflammation, corneal ulceration, and cancer metastasis.

The process of new capillary blood vessel formation from pre-existing blood vessels is called angiogenesis or neovascularization[6]. Neovascularization consists of a sequence of events that include dissolution of the basement membrane underlying the endothelial layer, migration and proliferation of endothelial cells, formation of new vascular capillary, and synthesis of a new basement membrane. Type IV collagenase activity may be important in the early steps of endothelial morphogenesis, the cleavage of basement membrane and the migration of endothelial cells out of the existing blood vessels[3,7]. At the initial stage of new blood vessel formation, basement membrane, the specialized type IV collagen and laminin-rich connective tissue matrix underlying endothelial cell layers, may be fragmented by the two type IV collagenases (gelatinase A/MMP-2 and gelatinase B/MMP-9).

A role of MMP inhibitors in suppression of prostate cancer growth and metastasis was tested [8]. Using potent and selective synthetic MMP inhibitors [9,10], this study investigated the role of collagen-degrading MMPs in the initiation of angiogenesis *in vitro*. Both MAG-283 and YLL-224 were very potent inhibitors of the two type IV collagenases (MMP-2 and MMP-9) and were less potent against the interstitial collagenase (MMP-1). These synthetic inhibitors at very low micromolar concentrations suppressed the cell-mediated type I collagen degradation of the human dermal microvascular endothelial cells (HDMVECs) stimulated by acidic fibroblast growth factor (aFGF), vascular endothelial growth factor (VEGF), and phorbol myristate acetate (PMA). They partially blocked cell invasion on type IV collagen, and prevented endothelial cell tube and cord formation on type I collagen and *Matrigel*. Our data showed that collagen-degrading MMPs may facilitate the initiation of angiogenesis.

2. SYNTHETIC MATRIXIN INHIBITORS

The thiol (sulfhydryl, mercaptosulfide) inhibitor MAG-283 and the sulfodiimine inhibitor YLL-224 were synthesized as described[9,10]. For kinetic measurements, the assay used was the fluorimetric assay based on hydrolysis of the fluorogenic matrix metalloproteinase substrate (7-methoxycoumarin-4-yl)acetyl-Pro-Leu-Gly-Leu-(*N*3-(2,4-dinitrophenyl)-L-2,3-diaminopropionyl)-Ala-Arg-NH_2 (McaPLGLDpaAR)[11]. For each assay, the enzyme and inhibitor were incubated in the assay buffer at 24 °C for 30-60 min, then the substrate was added and the rate of hydrolysis was measured by monitoring the increase in fluorescence with a Perkin-Elmer Model LS-5 Spectrofluorometer. IC_{50} values (the inhibitor concentrations at which 50% of the enzyme activities are inhibited) were determined from plots of % activity versus negative log of inhibitor concentration. The K_I

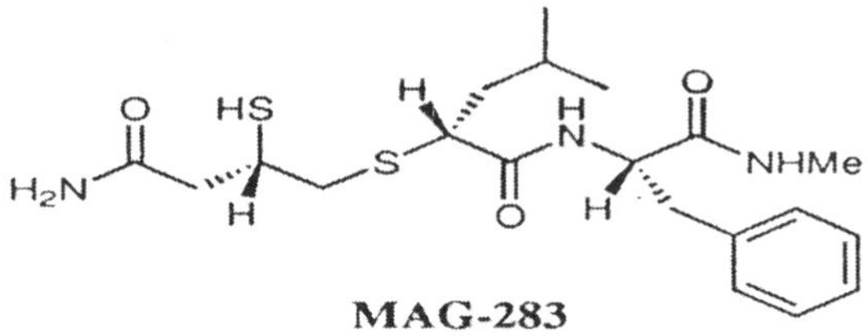

MAG-283
(MW 425.6)

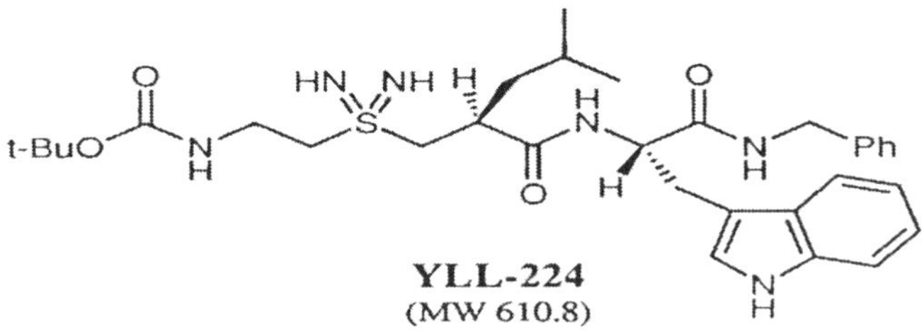

YLL-224
(MW 610.8)

Figure 1. Structures of two potent synthetic matrix metalloproteinase inhibitors. MAG-283, a thiol (sulfhydryl, mercaptosulfide) compound and YLL-224, a sulfodiimine compound based on their Zn(II)-binding functionality.

values (the inhibition constants) reported were determined from Henderson plots[12].

The potency and selectivity of the MMP inhibitors, MAG-283 and YLL-22 are shown in Figure 1. MAG-283 is a thiol (sulfhydryl, mercaptosulfide) compound and YLL-224 is a sulfodiimine compound based on their Zn(II)-binding functionality. MAG-283 was highly potent and relatively selective for MMP-2, MMP-8, and MMP-9 *vs.*MMP-1, MMP-3, and MMP-7 (Table 1). The inhibition constants (K_I values) for MMP-8 and MMP-9 were 0.47 nM and 2.2 nM, respectively, and the inhibitor concentration at which 50% of the enzyme activity was inhibited (IC_{50} value) for MMP-2 was 3.0 nM. MAG-283 was less potent against MMP-1, MMP-3, and MMP-7, with IC_{50} values at 480 nM, 280 nM, and 14 nM, respectively. YLL-224 was also a potent inhibitor of all the MMPs tested, except for stromelysin (MMP-3). It was relatively potent against the two gelatinases/type IV collagenases (MMP-2 and MMP-9); the IC_{50} values for MMP-2, MMP-8, and MMP-9 are 63 nM, 5.9 nM, and 44 nM, respectively. In addition, MAG-283 was less stable than YLL-224 because its thiol group might be oxidized easily in the enzyme assay buffer solutions and in the cell culture media. N-acetyl-L-cysteine was selected as a non-specific thiol control compound that was a very weak MMP inhibitor with greater than 9 mM IC_{50} values for MMP-2 and MMP-9.

Table 1. Potency and Selectivity of Inhibition of Six Matrix Metalloproteinases*

	IC_{50} (K_I), n*M*					
Inhibitor	HFC	HFG	HFS	MLN	HNC	HNG
MAG-28	480	3.0	280	14	(0.47)	(2.2)
YLL-224	180	63	4500	210	5.9	44

*A summary of the kinetic data, IC_{50} or (K_I) of the two potent and selective inhibitors, MAG-283 and YLL-224. HFC, human fibroblast collagenase (MMP-1); HFG, 72 kDa human fibroblast gelatinase A/type IV collagenase (MMP-2); HFS, human fibroblast stromelysin-1 (MMP-3); MLN, human matrilysin (MMP-7); HNC, human neutrophil collagenase (MMP-8); HNG, 92-98 kDa human neutrophil gelatinase B/type IV collagenase (MMP-9). The control compound, N-acetyl-L-cysteine had greater than 9 mM of IC_{50} values for MMP-2 and MMP-9.

3. INHIBITION OF ENDOTHELIAL CELL-MEDIATED COLLAGEN DEGRADATION AND ENDOTHELIAL TUBE FORMATION ON COLLAGEN

Human dermal microvascular endothelial cells (HDMVEC) were normal primary endothelial cells and were purchased from Cell Systems (Kirkland, WA). [^{3}H]-labeled type I collagen (0.06 mCi/ml) was diluted in 1 part 10x F12K media and 9 parts of 2 mg/ml non-radioactive type I collagen and the pH adjusted to 7.4 by adding several drops of 6 N NaOH. The [^{3}H] collagen mixture was pipetted into 96-well plates (60 ul/well about 5000 cpm) and allowed to gel for 1 hr at 37°C. The wells were rinsed with sterile phosphate buffered-saline (PBS) and then incubated in PBS containing 20 μg/ml trypsin overnight at 37°C, followed by two PBS rinses to remove trypsin and nonspecific radioactivity. HDMVEC were seeded into the wells at 2.5 x 10^4

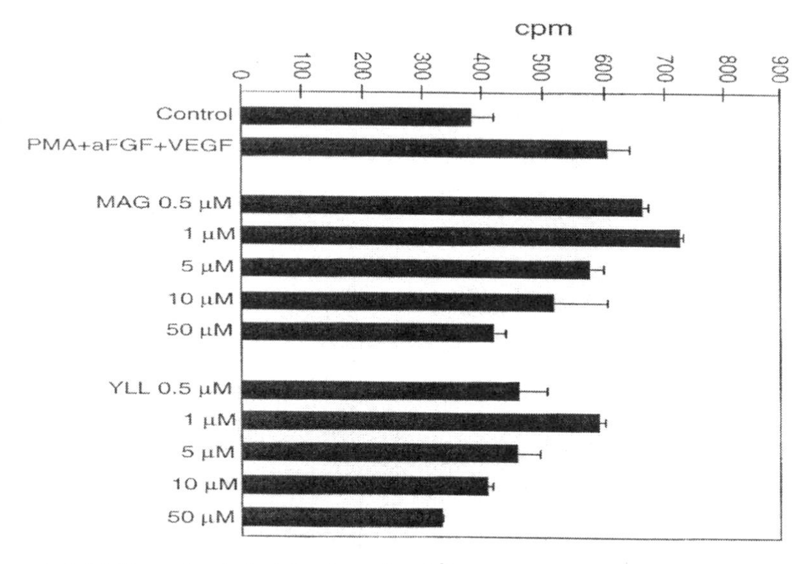

Figure 2. Inhibition of endothelial cell-mediated ^{3}[H]-type I collagen degradation by the two potent MMP inhibitors, MAG-283 and YLL-224, during a 14-day culture period. The cumulative radioactivity released in the cell culture media by the digestion of collagen by collagenases for 14 days was calculated for each treatment. Because only one tenth of the collagen substrate was radioactively labeled, this figure only represented one tenth of the actual collagen degradation by the cells. The standard deviation was calculated and shown as an error bar.

cells/well with or without different concentrations of the MMP inhibitors and 10 ng/ml aFGF, 50 ng/ml $VEGF_{165}$, and 50 nM PMA in 5% fetal bovine serum (FBS). The culture media was collected about every 3 days to measure the radioactivity released from the collagenase-digested collagen fragments. The cell culture wells were replenished with the corresponding fresh media with or without the inhibitors. The cells were cultured for 14 days. Duplicate or triplicate samples were employed for each variable.

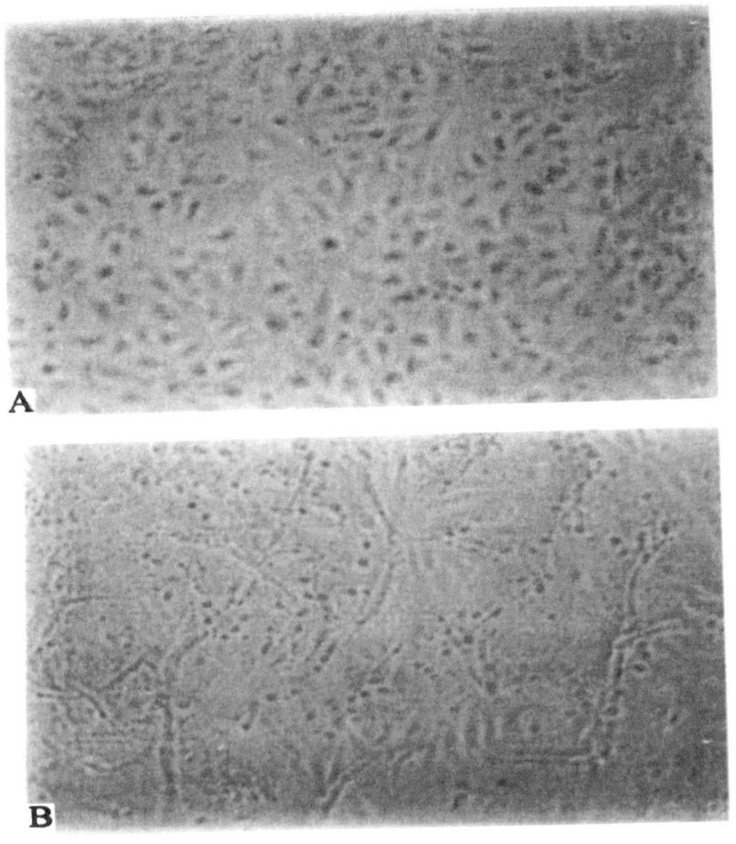

Figure 3. Endothelial cell growth on type I collagen gel for three days in the absence (A) or presence (B) of 10 ng/ml aFGF, 50 ng/ml VEGF, and 50 nM PMA in 5% FBS. The photographs were taken under a microscope (100x). In the presence of aFGF, VEGF, and PMA, endothelial tube-like structures were formed on the collagen gel. The tubes were not formed on the collagen gel in the 5% FBS control.

Endothelial cell-mediated ^{3}[H]-type I collagen degradation stimulated by aFGF, VEGF, and PMA was dose-dependently inhibited by MAG-283 and YLL-224 during a 14-day culture period (Figure 2). YLL-224 exhibited more potency in blocking cell-mediated collagen degradation than MAG-283. At inhibitor concentrations between 5μM and 50 μM, MAG-283 inhibited 14%-83% and YLL-224 inhibited 64%-over 100% of endothelial cell-mediated collagen degradation by type I collagen-degrading MMPs stimulated by the combination of PMA and growth factors, respectively. PMA and growth factors was added to the cell culture media to induce MMP-1, MMP-2, MMP-9, and MT1-MMP production and endothelial cell differentiation into tube-like structure.

The morphologies of endothelial cells grown for 3 days on type I collagen matrix in the absence or presence of 10 ng/ml aFGF, 50 ng/ml VEGF, and 50 nM PMA in 5% FBS were different (Figure 3). Growth factors were added to stimulate endothelial cell growth and PMA was added to promote endothelial cell differentiation into capillary tube-like structures. In the presence of aFGF, VEGF, and PMA, endothelial tube-like structures were formed on the collagen matrix, however, the tubes were not formed on the collagen matrix in the 5% FBS control (data not shown). By day 14, most of the endothelial tubes invaded into the collagen matrix and very few tubes were left on the surface of the collagen matrix. The metalloproteinase inhibitors partially or completely blocked the growth factor and PMA-induced tube formation. The endothelial cells were still viable on the collagen matrix at day 14 in the presence of 0.1 μM-50 μM of the inhibitors (data not shown). In the presence of 100 μM of the inhibitors, most cells were dead. Collagenase inhibitors blocked endothelial cell-mediated collagen degradation and inhibited PMA and growth factor induced endothelial tube formation on collagen matrix demonstrating that collagen-degradation was required for morphological differentiation.

4. SUPPRESSION OF PMA AND GROWTH FACTOR-STIMULATED INVASION OF ENDOTHELIAL CELLS THROUGH TYPE IV COLLAGEN

The invasion chambers containing polycarbonate filters (8 μm pore size, modified Boyden chambers, Collaborative Biomedical Products, Becton Dickinson, Bedford, MA) were coated with 200 μl of 0.5 mg/ml type

IV collagen. The Boyden chamber was dried in a laminar flow hood under UV overnight. Fifty thousand HDMVECs in 300 μl 5% FBS F12K media containing 20 nM PMA and different concentrations of one of the MMP inhibitors were cultured in each upper compartment of the chamber. Five hundred microliters of culture medium containing 20 nM PMA, 10 ng/ml aFGF, and 50 ng/ml VEGF was added into the lower compartments of the chambers. After 2 days culture, the filter of the chamber was fixed by 10% buffered formalin and stained by 0.1% Crystal Violet solution. The cells inside the chamber were removed with a cotton swab. When the inserts were dry, the filter was removed from the chamber with a sharp razor blade and mounted on a microscope slide. The cells on the bottom of the chamber were counted in two high power fields (400x) of each of the duplicate samples under a microscope.

The effects of MMP inhibitors on PMA and growth factor-stimulated human endothelial cell invasion on type IV collagen-coated filters of the modified Boyden chambers were investigated. As shown in Figure 4, at the inhibitor concentrations of 5 μM and 10 μM, MAG-283 blocked about 26% and 48% of endothelial invasion, respectively, and YLL-224 inhibited 32% and 52% invasion, respectively. Thus, type IV collagenase activities are

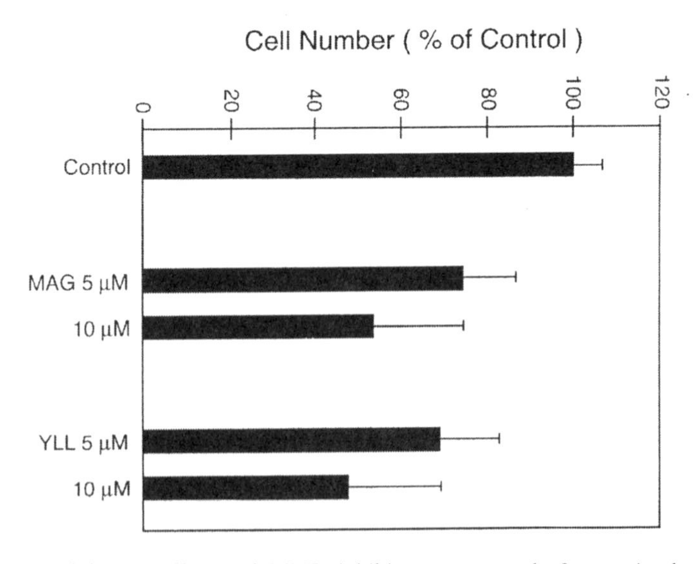

Figure 4. Inhibitory effects of MMP inhibitors on growth factor-stimulated human endothelial cell invasion on type IV collagen-coated filters of the modified Boyden chambers.

required for the PMA and growth factor-induced endothelial cell invasion through type IV collagen matrix.

5. INHIBITION OF HUMAN ENDOTHELIAL CELL TUBE AND CORD FORMATION ON *MATRIGEL*

Endothelial cell tube formation on *Matrigel* (laminin and type IV collagen-rich reconstituted basement membrane) were investigated. Fifty microliters of either *Matrigel* (thick gel) or 3-fold diluted *Matrigel* (thin gel) were coated on each well of 96-well culture plate at 37°C for more than 1 hour. Then 2.5 x 10^4 HDMVECs were plated in each well in 200 μl of CS-C serum free endothelial growth medium (Cell Systems). Different concentrations of MMP inhibitors were added into the cultures. After 3-hr or several days of culture, photographs were taken under the microscope (10

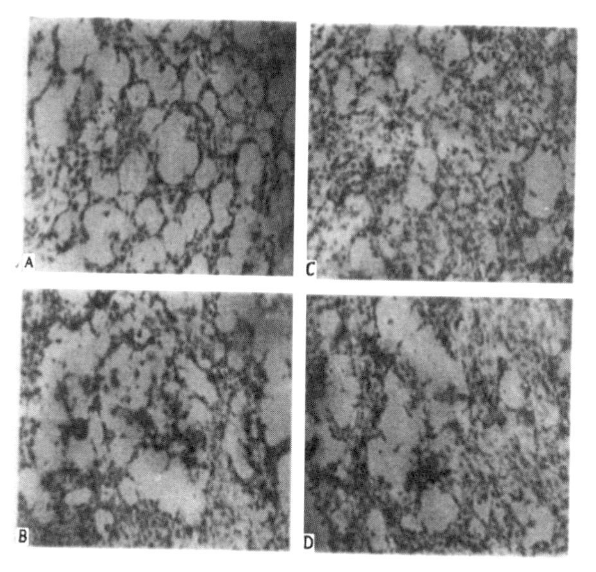

Figure 5. Inhibition of human endothelial tube formation on thin *Matrigel* by MMP inhibitors. The human endothelial cells were grown in CS-C serum-free endothelial growth medium on *Matrigel* in the absence of the MMP inhibitors (A) and presence of 10 μM N-acetyl-L-cysteine (B), MAG-283 (C), and YLL-224 (D). The cells were stained with crystal violet. The photographs were taken under a microscope (100x) at the 3.5-hr time point after the cells were seeded onto the *Matrigel*.

x 10 or 10 x 4 magnifications). The endothelial tubes were fixed with 10% buffered formalin and stained with 0.1% Crystal Violet (Sigma) Solution. Endothelial tube formation was observed both on the thick and the thin *Matrigel* after two hours and the endothelial cord formation was observed on the thick gel after 2 days. The photographs were taken under an inverted microscope with phase contrast (100x magnification, Olympus CK2-TRC 2).

We next tested the effects of the MMP inhibitors on *Matrigel* induced tube and cord formation. Human endothelial cell tube formation on the thin *Matrigel* (Figure 5) and endothelial cord formation on the thick *Matrigel* (Figure 6) was inhibited by 10 μM concentrations of MAG-283 and YLL-224. MMP inhibitors slowed endothelial tube formation and completely blocked endothelial cord formation. N-acetyl-L-cysteine at the same concentration did not prevent endothelial tube formation. These results demonstrate that potent MMP inhibitors can block the process of endothelial cell differentiation to form capillary tubes. Therefore, collagenase activity is required for human endothelial cell tube formation on *Matrigel*.

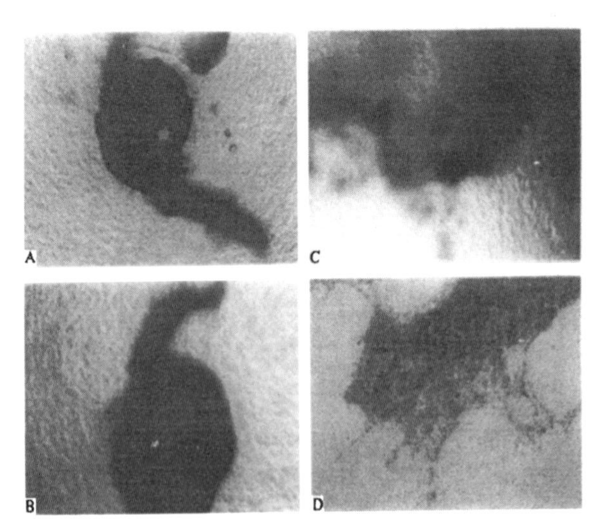

Figure 6. Inhibition of human endothelial cord formation on thick *Matrigel* by MMP inhibitors. The human endothelial cells were grown in CS-C serum free endothelial growth medium on thick *Matrigel* in the absence of the MMP inhibitors (A) and presence of 10 μM N-acetyl-L-cysteine (B), MAG-283 (C), and YLL-224 (D). The photographs were taken under a microscope (100x) at the 5-day time point after the cells were seeded onto the *Matrigel*.

6. CONCLUSIONS

We have investigated the role of collagen-degrading MMPs in the initiation of angiogenic activities *in vitro* using potent synthetic MMP inhibitors. The thiol (sulfhydryl, mercaptosulfide) inhibitor MAG-283 was about 160 and 236 fold more active against MMP-2 (IC_{50}, 3.0 nM) and MMP-9 (K_I, 2.2 nM) relative to MMP-1 (IC_{50}, 480 nM), respectively. The sulfodiimine compound YLL-224 was about 3-fold and 4-fold more potent against MMP-2 (IC_{50}, 63 nM) and against MMP-9 (IC_{50}, 44 nM) relative to MMP-1 (IC_{50}, 180 nM), respectively. Human dermal microvascular endothelial cells in culture produced MMP-1, MMP-2, MMP-9, and MT1-MMP as reported by others[13,14] and detected by collagen, gelatin zymograms and/or Western blots using antibodies specific to each MMP (data not shown). In addition, MMP-2 and MMP-9 were also present in the fetal bovine serum used for cell culture.

Although MAG-283 was a more potent inhibitor against purified MMPs than YLL-224, it had less inhibitory effect in blocking cell-mediated type I collagen degradation and cell invasion through type IV collagen matrix. This discrepancy may be due to the reduced stability of the thiol inhibitor relative to the sulfodiimine inhibitor because the thiol groups were readily oxidized in buffer or in cell culture media. At inhibitor concentrations between 5μM and 50 μM, MAG-283 inhibited 14%-83% and YLL-224 inhibited 64%-over 100% of endothelial cell-mediated collagen degradation by type I collagen-degrading MMPs stimulated by the combination of PMA and growth factors, respectively. PMA was added to the cell culture media to stimulate latent MMP-1, active MMP-9, active MT1-MMP, and active MMP-2 levels and induce endothelial cell differentiation into capillary tube-like structure[13-17]. MMP-1, MMP-2, and MT1-MMP all can digest type I collagen[1,18]. At concentrations of 5 μM and 10 μM, MAG-283 blocked about 26% and 48% of endothelial invasion through type IV collagen matrix, respectively, and YLL-224 inhibited 32% and 52% invasion, respectively. Therefore, type IV collagen-degrading MMPs, such as MMP-2, MMP-9, and MT1-MMP, are required for the endothelial cell migration and invasion induced by PMA and growth factors. These inhibitors at 10 μM prevented the endothelial differentiation into tubes on *Matrigel*; they also blocked endothelial cell tube formation on type I collagen matrix in the presence of aFGF, VEGF, and PMA. Because these compounds were much less potent against MMP-1 *vs.* other MMPs and PMA and growth factors stimulated MMP-2 and MMP-9 were active and

MMP-1 was inactive[13,15,17], the data demonstrated that in addition to MMP-1, other collagen-degrading MMPs likely contribute to the initiation of angiogenic activity: basement membrane degradation and migration and invasion of endothelial cells.

Tumor growth and metastasis are angiogenesis dependent[6]. Angiogenesis is a rate-limiting step for providing oxygen and nutrients for solid tumor growth and for generating a gateway for cancer cells to spread to other parts of the body, such as lungs, liver, brain, or bones, through blood circulation system. Synthetic MMP inhibitors batimastat (BB-94) and marimastat (BB-2516) at low micromolar concentrations were cytostatic without cytotoxicity against human gliomas and reduce cell proliferation and invasion[19]. Our inhibitors at low micromolar concentrations (<50 μM) also did not show cytotoxicity against human microvascular endothelial cells. A new MMP inhibitor KB-R7785 may control both primary and secondary tumor growth by limiting the expansion of endothelial cells and cancer cells, thus, inhibiting tumor angiogenesis and metastasis in two tumor models in mice[20].

At least three synthetic hydroxamate MMP inhibitors (marimastat, Ag3340, Bay12-9566) are in phase III human clinical trials for treatment of cancer patients as anti-angiogenesis and anti-tumor growth drugs[21]. Marimastat is a broad spectrum MMP inhibitor with IC_{50} values of 5 nM, 6 nM, and 3 nM for MMP-1, -2 and -9, respectively. Experiments in models of breast cancer have shown that MMP inhibitors can significantly reduce the growth rate of both primary and secondary tumors and can block the process of metastasis[22]. Ag3340 is 100-fold more potent against MMP-2 than against MMP-1 with K_I values of 8 nM and 0.08 nM for MMP-1 and MMP-2, respectively. Bay12-9566 is also a gelatinase selective MMP inhibitor with K_I values of >5000 nM, 11 nM, and 301 nM for MMP-1, -2, and -9, respectively. Marimastat, Ag3340, and Bay12-9566 inhibited human tumor growth. All of them showed anti-angiogenic activity when tested in the *Matrigel* implant model *in vivo* and in human cancer patients[21].

Angiogenesis inhibitors may become one type of the most promising new drugs for starving tumor cells and preventing cancer cell growth and metastasis as well as treating cardiovascular patients. Our inhibitors have different zinc-binding functionality with different functional groups of thiol and sulfodiimine from the hydroxamate inhibitors reported[21]. Both MAG-283 and YLL-224 are gelatinase selective MMP inhibitors. Because the broad spectrum inhibitor marimastat had some undesirable side effects on patients, the gelatinase selective inhibitors may reduce or eliminate some of the undesirable side effects on patients. The design, synthesis, and test of new classes and new generations of highly selective and potent MMP inhibitors are ongoing projects. Some of these inhibitors may be added to the growing list of synthetic MMP inhibitors for clinical studies as anti-angiogenesis and anti-tumor growth drugs.

ACKNOWLEDGMENTS

This work was supported in part by a Grant-in-Aid AHA 9601457 from the American Heart Association, Florida Affiliate, a grant CA78646 from the National Cancer Institute, the National Institutes of Health, and a grant from the Gustavus and Louise Pfeiffer Research Foundation (to Q.X. A. S.). The authors would like to thank Drs. Mohammad A. Ghaffari and Yi-Lin Luo for synthesizing the matrixin inhibitors, Ms. V. Lakshmi D. Badisa and Ms. Meiqin Chen for excellent technical assistance, and Drs. Michael C. Jaye, Hynda K. Kleinman, Harold E. van Wart, Hideaki Nagase, Henning Birkedal-Hansen, and L. Jack Windsor for their invaluable reagents and support.

REFERENCES

1. Sang, Q.A., and Douglas, D.A., 1996, Computational Sequence Analysis of Matrix Metalloproteinases. *J. Protein. Chem.* 15:137-160.
2. Chambers, A.F., and Matrisian, L.M., 1997, Changing views of the role of matrix metalloproteinases in metastasis. *J. Natl. Cancer Inst.* 89:1260-1270.
3. Ray, J.M., and Stetler-Stevenson, W.G., 1994, The role of matrix metalloproteinases and their inhibitors in tumor invasion, metastasis and angiogenesis. *Eur. Respir.* 7:2062-2072.
4. Sang, Q.X., 1998, Complex role of matrix metalloproteinases in angiogenesis. Minireview. *Cell Res.* 8:171-177.
5. Douglas, D.A., Shi, Y.E., and Sang, Q.A., 1997, Computational sequence analysis of the tissue inhibitor of metalloproteinase family. Minireview. *J. Protein Chem.* 16:237-255.
6. Folkman, J., 1995, Angiogenesis in cancer, vascular, rheumatoid and other disease. *Nature Med.* 1:27-31.
7. chnaper, H.W., Grant, D.S., Stetler-Stevenson, W.G., Fridman, R., D'Orazi, G., Murphy, A.N., Bird, R.E., Hoythya, M., Fuerst, T.R., French, D.L., Quigley, J.P., and Kleinman, H. K., 1993, Type IV collagenase(s) and TIMPs modulate endothelial cell morphogenesis *in vitro*. *J. Cell. Physiol.* 156:235-246.
8. Sang, Q.A., Schwartz, M.A., Li, H., Chung, L.W.K., and Zhau, H.Y.E., 1999, Targeting matrix metalloproteinases in human prostate cancer. *Ann. N. Y. Acad. Sci.* 878:538-540.
9. Schwartz, M.A., and Van Wart, H.E., 1995, Mercaptosulfide metalloproteinase inhibitors. U.S. Patent 5455262.
10. Schwartz, M.A., and Van Wart H.E., 1995a, Sulfoximine and sulfodiiimine matrix metalloproteinase inhibitors. U.S. Patent 5470834.

11. Knight, C.G., Willenbrock, F., and Murphy, G., 1992, A novel coumarin-labelled peptide for sensitive continuous assays of the matrix metalloproteinases. *FEBS Letters* 296:263-266.
12. Henderson, P.J.F., 1972, A linear equation that describes the steady-state kinetics of enzymes and subcellular particles interacting with tightly bound inhibitors. *Biochem. J.* 127:321-333.
13. Fisher, C., Gilbertson-Beadling, S., Powers, E.A., Petzold, G., Poorman, R., and Mitchell, M.A., 1994, Interstitial collagenase is required for angiogenesis *in vitro*. *Dev. Biol.* 162:499-510.
14. Foda, H.D., George, S., Conner, C., Drews, M., Tompkins, D.C., and Zucker, S., 1996, Activation of human umbilical vein endothelial cell progelatinase A by phorbol myristate acetate: a protein kinase C-dependent mechanism involving a membrane-type matrix metalloproteinase. *Lab. Invest.* 74:538-545.
15. Moscatelli, D., Jaffe, E., and Rifkin, D.B., 1980, Tetradecanoyl phorbol acetate stimulates latent collagenase production by cultured human endothelial cells. *Cell* 20:343-351.
16. Montesano, R., and Orci, L., 1985, Tumor-promoting phorbol esters induce angiogenesis *in vitro*. *Cell* 42: 469-477.
17. Puyraimond, A., Weitzman, J.B., Babiole, E., and Menashi, S., 1999, Examining the relationship between the gelatinolytic balance and the invasive capacity of endothelial cells. *J. Cell Sci.* 112:1283-1290.
18. Ohuchi, E., Imai, K., Fujii, Y., Sato, H., Seiki, M., and Okada, Y., 1997, Membrane type 1 matrix metalloproteinase digests interstitial collagens and other extracellular matrix macromolecules. *J. Biol. Chem.* 272:2446-2451.
19. Tonn, J.C., Kerkau, S., Hanke, A., Bouterfa, H., Mueller, J.G., Wagner, S., Vince, G.H., and Roosen, K., 1999, Effect of synthetic matrix-metalloproteinase inhibitors on invasive capacity and proliferation of human malignant gliomas in vitro. *Int. J. Cancer* 80:764-772.
20. Lozonschi, L., Sunamura, M., Kobari, M., Egawa, S., Ding, L., and Matsuno, S., 1999, Controlling tumor angiogenesis and metastasis of C26 murine colon adenocarcinoma by a new matrix metalloproteinase inhibitor, KB-R7785, in two tumor models. *Cancer Res.* 59:1252-1258.
21. Brown, P.D., 1999, Clinical studies with matrix metalloproteinase inhibitors. *APMIS* 107:174-180.
22. Brown, P.D., 1998, Matrix metalloproteinase inhibitors. *Breast Cancer Res. Treat.* 52:125-136.

Inhibitors of Angiogenesis and their Mechanisms of Action

TUNICAMYCIN INHIBITS CAPILLARY ENDOTHELIAL CELL PROLIFERATION BY INDUCING APOPTOSIS

Targeting dolichol-pathway for generation of new anti-angiogenic therapeutics

Juan A. Martínez, Ivette Torres-Negrón, Lilla A. Amigó, Rossely A Roldán, Alba Mendéz and Dipak K. Banerjee

Department of Biochemistry, School of Medicine, University of Puerto Rico, San Juan, PR 00936-5067, USA

Abstract: Bovine adrenal medulla microvascular endothelial cells used in this study undergo cellular proliferation and differentiation upon culturing *in vitro* as observed both by light and scanning electron microscopy. Cells also respond to the growth promoting activity of serum and basic fibroblast growth factor (FGF2). Flow cytometric analysis of a synchronized culture established that cells take 68 hours to complete one cell cycle spending 36 hours in the G1 phase, 8 hours in the S phase, and 24 hours in the G2 + M phase when cultured in EMEM containing 2% heat-inactivated fetal bovine serum (FBS). At 10% serum, or in the presence of FGF2 (10 ng/ml - 100 ng/ml) length of the cell cycle is reduced to 56 hours due to shortening of the G1 phase by 12 hours. Tunicamycin (a glucosamine-containing pyrimidine nucleotide), and an inhibitor of glucosaminyl-1-phosphate (GlcNAc 1-P) transferase, the first step of $Glc_3Man_9GlcNAc_2$-PP-Dol (OSL) biosynthesis is found to inhibit the endothelial cells proliferation by inducing apoptosis as observed by flow cytometry and DNA laddering. Cell shrinkage, compaction of nuclei, membrane fragmentation, etc., typical of apoptotic response are frequently seen by light microscopy in the presence of tunicamycin. Scanning electron microscopy also exhibited a considerable amount of cell surface blebbing. Accumulation of an immunopositive cell specific asparagine-linked (N-linked) glycoprotein, Factor VIII:C in the absence of $Glc_3Man_9GlcNAc_2$-PP-Dol in tunicamycin treated cells has been proposed as an apoptotic triggering mechanism under the current experimental conditions.

1. INTRODUCTION

Our laboratory has made an observation almost a decade ago that N-glycosylation of proteins is increased nearly 3.5-fold, and the K_m for Dol-P-Man synthase is decreased by ~50% when the capillary endothelial cells (an established cell line from the microvasculature of bovine adrenal medulla) were cultured in the absence of CO_2 (1-4). Dol-P-Man synthase is an essential intermediate in the elongation of $Man_5GlcNAc_2$-PP-Dol to $Man_9GlcNAc_2$-PP-Dol (5,6), and an allosteric activator of GlcNAc-1-phosphate transferase (7). It has also been demonstrated that Dol-P-Man synthase gene carries a cAMP-dependent protein phosphorylation consensus sequence and its activity is regulated by cAMP-dependent protein kinase-mediated protein phosphorylation signal (8- 13).

We have been interested in establishing a relationship between the dolichol pathway, especially that of Dol-P-Man synthase and angiogenesis. Earlier studies with amphomycin indicated that inhibition of $Glc_3Man_9GlcNAc_2$-PP-Dol (OSL) biosynthesis inhibited the endothelial cell proliferation (14). Amphomycin is an undecapeptide from *Streptomyces canus*, whose N-terminus is blocked due to a fatty acid substitution (15,16). Its binding to Dol-P in the presence of Ca^{2+} blocks OSL assembly by interfering with the synthesis of Dol-PP-$GlcNAc_2$, Dol-P-Man, and Dol-P-Glc, respectively (17 - 20). This paper while supports the earlier observation of amphomycin, also finds that tunicamycin, an inhibitor of GlcNAc-1P transferase and a blocker of OSL assembly (21) profoundly inhibits the endothelial cell proliferation by inducing apoptosis.

2. ALTERATION OF ENDOTHELIAL CELL MORPHOLOGY

Tunicamycin, a glucosamine-containing pyrimidine nucleoside exists in 16 different homologues, differing mostly in the fatty acid side chain is synthesized by *Streptomyces lysosuperificus*. (21,22). We exposed a synchronized population of capillary endothelial cells (23) to tunicamycin from Boehringer-Mannheim with no history of protein synthesis inhibition, at various concentrations for different lengths of time in a complete medium (EMEM) containing 2% fetal bovine serum (heat-inactivated) to evaluate the morphological changes. The cells were monitored by phase-contrast light microscopy as well as by scanning electron microscopy. In controls, cellular proliferation and differentiation were mutually exclusive, but a complete differentiation was observed only after 5 days. Examination of surface morphology by scanning electron microscopy as a function of time established cellular growth, and differentiation into capillary-like structures. Cells cultured in a higher serum concentration (i.e., 10%) duplicated the morphological changes, but took longer time to differentiate. Cells cultured in the presence of tunicamycin (1μg/ml) exhibited cell shrinkage, loss of cell to contact, apparent compaction of nuclei showing condensed pyknotic appearance and membrane fragmentation when analyzed by light microscopy. Scanning electron microscopy detected considerable surface blebbing in tunicamycin-treated cells, a morphological change which had never been observed in controls even after culturing for a longer period of time (Figure 1).

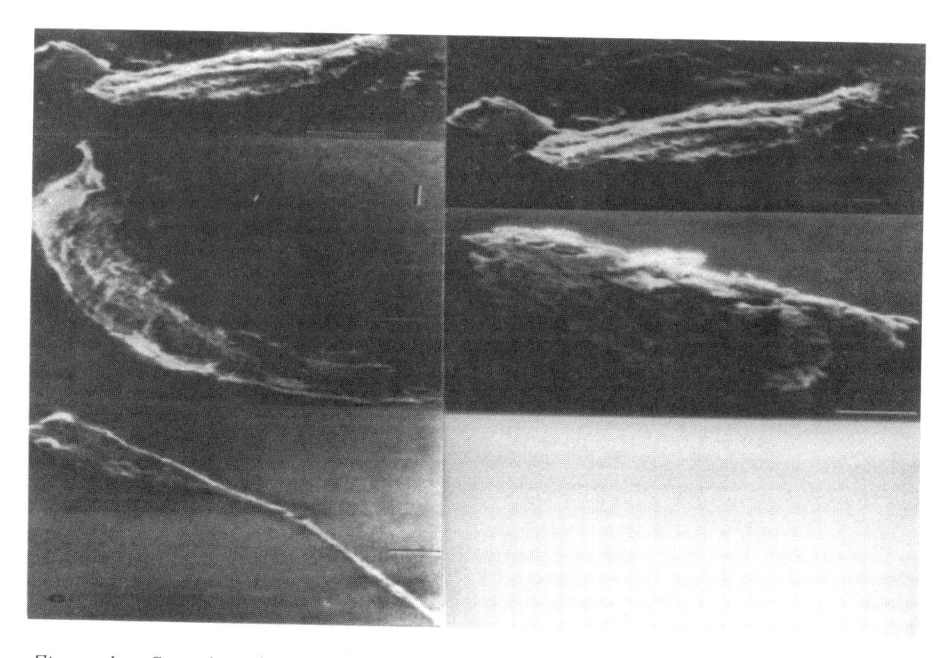

Figure 1. Scanning electron microscopy of cells before and after tunicamycin treatment. Left panels (top to bottom): cultured for 4 days in 2% serum (x 1,700; -10 μm), 4 days in 10% serum (x 700; -10μm), and 7 days in 2% serum (x 1,100; -10 μm), respectively. Right panels (top to bottom): cultured for 4 days in 2% serum (x 1,700; -10 m), and 4 days in 2% serum containing 1μg/ml tunicamycin (x 1,700; -10μm).

3. INHIBITION OF CELL GROWTH AND PROLIFERATION

Synchronized cells were cultured in complete EMEM containing fetal bovine serum (1%, 2%, or 10%), or complete EMEM with 2% fetal bovine serum containing 10 ng/ml - 100 ng/ml of FGF2, or containing 5 ng/ml, 10 ng/ml, 1 μg/ml, and 10.0 μg/ml tunicamycin. In each case, the cellular growth and viability was monitored by trypan blue exclusion, by counting the cell number, and by assessing the progression through the cell cycle by flow cytometry. Cells maintained a *status quo* for the first 32 hours whether cultured in serum alone or in the presence of FGF2. Cells proliferated normally in media containing either 2% or 10% fetal bovine serum, but a maximum cell growth was observed at 10% serum. Similarly, FGF2 at 10 ng/ml - 100 ng/ml stimulated the endothelial cell proliferation by reducing

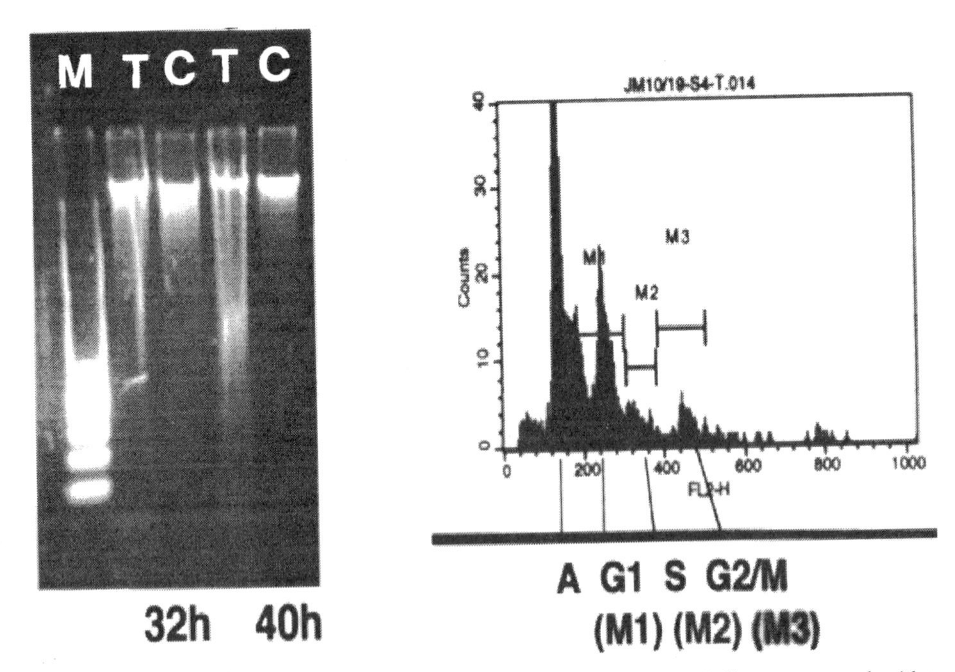

Figure 2. Effect of tunicamycin on the induction of apoptosis. Cells were treated with 1μg/ml of tunicamycin for 32 to 40 hours in EMEM containing 2% serum. DNA laddering: M = 50 bp ladder, T = tunicamycin, C = control; Flow cytometry: A = apoptosis; M1 ≈ G1, M2 ≈ S, M3 ≈ G2/M

the G1 phase. When the progression of cells through the cell-cycle was evaluated by flow cytometry, cells cultured in 2% serum needed approximately 68 hours to complete one cycle spending 36 hours in the G1 phase, 8 hours in the S phase and 24 hours in the G2 + M phase. Presence of either 10% serum, or 10 ng/ml -100 ng/ml FGF2 reduced the time to complete the cell cycle to 56 hours by decreasing the G1 phase by 12 hours.

Cells cultured in the presence of tunicamycin (1 μg/ml) maintained the same level of growth as controls for 24 hours after which the number started declining, and cells never completed the cell cycle. Flow cytometric evaluation revealed that 70% of the cell population entered into apoptosis (i.e., "programmed cell death") after an exposure to tunicamycin for 32 hours. This was also confirmed by the appearance of laddering when DNA from tunicamycin treated cells was analyzed by agarose gel electrophoresis (Figure 2.) A similar reduction in cell growth has also been observed at all tunicamycin concentrations used in this study (i.e., 5 ng/ml - 10.0 μg/ml).

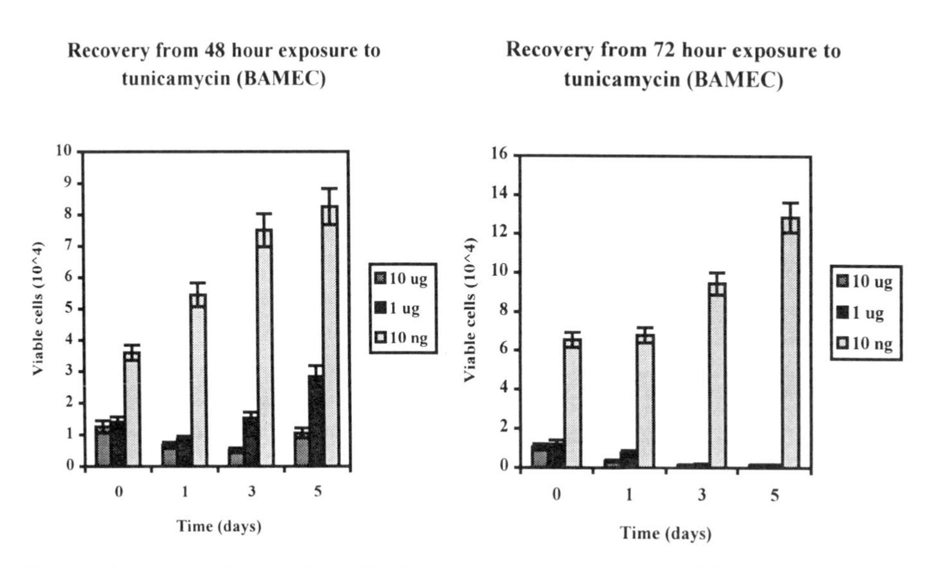

Figure 3. Recovery of endothelial cells after tunicamycin treatment for 48 hours and 72 hours.

4. ISOLATION OF A HIGHLY SENSITIVE CELL CLONE

If the behavior of tunicamycin observed here is truly anti-angiogenic and is not due to cytotoxicity then cells must recover following a tunicamycin treatment that is shorter in duration, and the concentrations are moderate. In fact, a full recovery was observed in 5 days when cells were treated with 10 ng/ml tunicamycin for 48 hours; at 1 μg/ml tunicamycin the recovery was approximately 40%, but it was only 12% when 10 μg/ml of tunicamycin was used. Cells however, never recovered when exposed to tunicamycin for 72 hours or longer at 1 μg/ml or 10 μg/ml, respectively. On the other hand, cells recovered fully from a 24 hours exposure to all tunicamycin concentrations (Figure 3).

To understand the precise molecular mechanism of tunicamycin induced apoptosis of capillary endothelial cells, we have isolated a cell clone from the parental culture that was maintained in a serum- free media for six weeks following synchronization. Morphology of these cells and their growth profile appear quite different. For example, the cell shape is more spindle-like, resembling Kaposi's Sarcoma lesions endothelial cells (24), they undergo two-population doubling in 2% fetal bovine serum, differentiate in 5 days, and complete one cell cycle in approximately 56 hours (Figure 4). Most striking however is the cells' sensitivity to tunicamycin. In 32 hours they undergo apoptosis at 10 ng/ml of tunicamycin, a concentration 100-fold less than needed for the parental cells. It has also been observed that approximately 60% of cell population recovered in 5 days from a 10 ng/ml tunicamycin exposure for 24 hours. The recovery was similar from an exposure to 1 μg/ml tunicamycin for 48 hours, but less than 20% when exposed to 10.0 μg/ml tunicamycin for the same period of time. There was no recovery after exposing the cells to 1 μg/ml – 10.0 μg/ml tunicamycin for 72 hours or more.

5. ACCUMULATION OF AN ASPARAGINE-LINKED GLYCOPROTEIN, FACTOR VIII:C

The apoptotic response induced by tunicamycin exposure during inhibition of capillary endothelial cells was due to the inhibition of protein N-glycosylation process and not due to protein synthesis inhibition.

Replacing tunicamycin with the protein synthesis inhibitor cycloheximide inhibited the cell proliferation in a time- and dose-dependent manner without inducing apoptosis (Martínez and Banerjee, unpublished observation). This argued strongly against any possibility that tunicamycin action was dependent upon the inhibition of active cellular protein synthesis. Any claim made earlier on the inhibition of protein synthesis by tunicamycin (25) therefore needs a critical examination with a correct tunicamycin homologue. This is extremely important because some homologues of tunicamycin have indeed been found to be protein synthesis inhibitors (22).

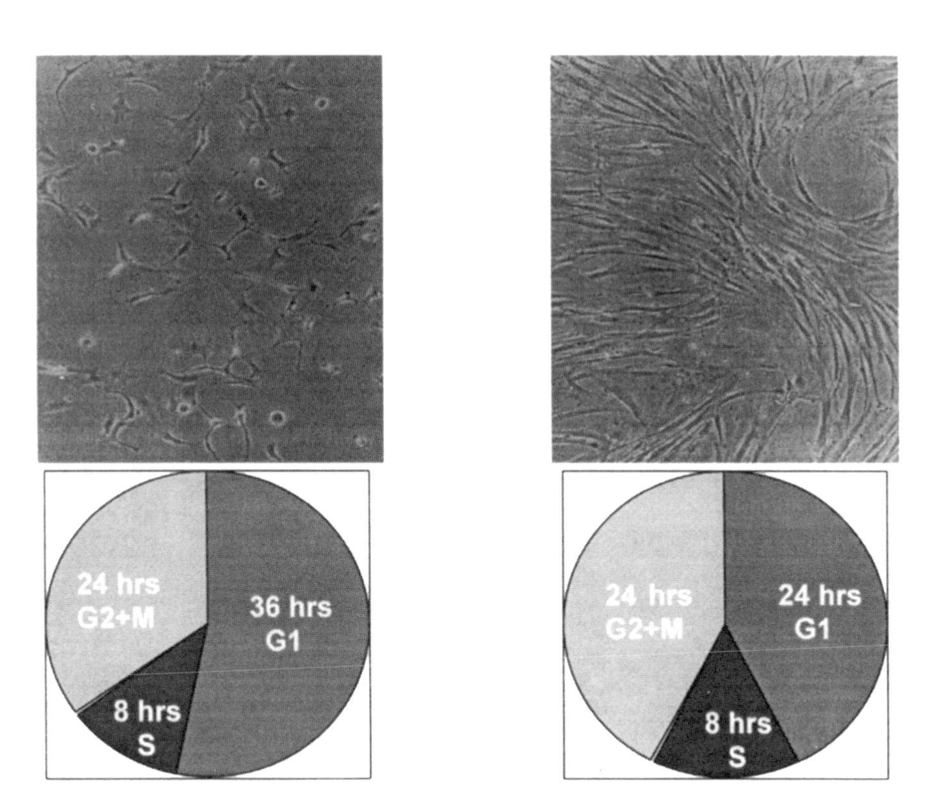

Figure 4. Comparison of cellular proliferation between the parental line (BAMEC; left panels) and the tunicamycin sensitive clone (JMEC; right panels) in EMEM containing 2% serum.

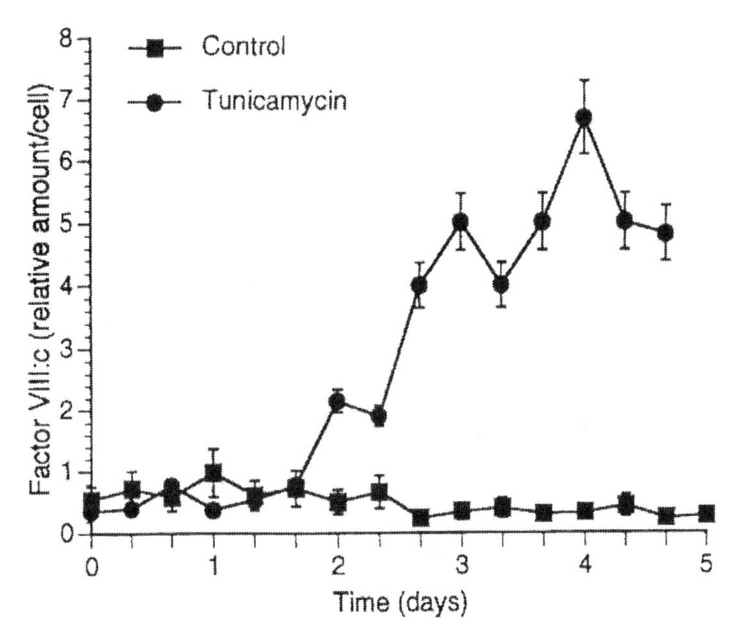

Figure 5. Factor VIII:C level in tunicamycin treated cells. Factor VIII:C was quantitatively analyzed by enzyme-linked immunoassay. Each point represents the mean ± SEM (n = 6). ■, Control; ●, tunicamycin (1 μg/ml) treated.

To support our claim that inhibition of protein N-glycosylation is a primary event in tunicamycin-mediated apoptosis, we have analyzed Factor VIII:C expression during the growth and proliferation of capillary endothelial cells. These capillary endothelial cells constitutively express a M_r 270,000 dalton N-linked glycoprotein with a biological activity of that of a Factor VIII:C of the blood coagulation cascade (26). In endothelial cells derived Factor VIII:C, the heavy-chain (M_r 215,00 dalton) and the light-chain (M_r 46,000 dalton) are joined together by disulfide bridge(s). In addition, the light-chain contains approximately 20% N-linked glycans (26), and are essential for its biological activity. Removal of N-glycan chains from the human plasma Factor VIII:C almost eliminated its ability to convert Factor X to Factor Xa (Prieto and Banerjee, unpublished observation). Apart from its participation in the blood coagulation pathway, we believe that Factor VIII:C in capillary endothelial cells is involved in activating metalloproteases during endothelial cell migration and invasion. Factor VIII:C level when analyzed in control and in tunicamycin-treated (1 μg/ml) cells by enzyme-linked immunoassay (ELISA), we have observed increased accumulation of Factor VIII:C in cells treated with tunicamycin for 32 hours and beyond (Figure 5). This indicated that either a reduced level or a complete absence of $Glc_3Man_9GlcNAc_2$-PP-Dol in the presence of tunicamycin makes under- or un-glycosylated Factor VIII:C and pushes the cells threshold to synthesize more biologically active Factor VIII:C. Thus, it is concluded that continuous accumulation of under- or un-glycosylated Factor VIII:C in tunicamycin-treated capillary endothelial cells may be responsible for the induction of apoptosis.

6. CONCLUSION AND FUTURE DIRECTION

Many asparagine-linked glycoproteins such as selectins, VEGFs, integrins and their receptors have been implicated during angiogenesis (23). Processing of the glycan chains to a "high-mannose" type or to a "complex" type has also been mentioned during endothelial cell proliferation and differentiation (27,28). All these glycoproteins, in fact, need continuous expression of $Glc_3Man_9GlcNAc_2$-PP-Dol as a pre-requisite for their structural modification. Precise involvement of the dolichol-linked glycan chain in capillary endothelial cell proliferation has been shown in our studies on environmental insult due to CO_2 depletion (4), and tying up the available dolichylmonophosphate (Dol-P) with amphomycin (17,18). We now show with a specific inhibitor of glucosaminyl-1-phosphate transferase (GlcNAc 1-P transferase), tunicamycin (21) that availability of $Glc_3Man_9GlcNAc_2$-PP-Dol is critical not only for the endothelial cell proliferation during angiogenesis but it may also act as an essential intermediate for cell proliferation and apoptosis signaling.

Accumulated evidence has suggested that $Glc_3Man_9GlcNAc_2$-PP-Dol synthesis is under genetic regulation and is also regulated by the humoral factors. Dol-P-Man synthase plays a critical role in this process. It not only provides Dol-P-Man an allosteric activator for GlcNAc 1-P transferase, but the synthase itself is regulated by a cAMP-dependent protein phosphorylation signal. The future direction of our laboratory is therefore to understand the Dol-P-Man synthase gene structure and its regulation by the phosphorylation/ dephosphorylation signal as a coupling mechanism for the angiogenic process. It has been shown that Dol-P-Man synthase in *S.*

cerevisiae is a structural gene and its mutation is lethal (8). Recently, we have observed that replacing serine 141 from the phosphorylation consensus site of the yeast Dol-P-Man synthase by alanine by site-directed mutagenesis reduces the synthase activity significantly (19), and also makes the cells grow slowly (Carrasquillo and Banerjee unpublished observation).

ACKNOWLEDGMENT

This work has been supported by a grant from the University of Puerto Rico Medical Sciences Campus. The authors would like to give thanks to Elena A. Carrasquillo and Deniliz Rodríguez for their critical comments and support.

REFERENCES

1. Banerjee, D.K., Ornberg, R.L., Youdim, M.B.H., Heldman, E., and Pollard, H.B., 1985, Endothelial cells from bovine adrenal medulla develop capillary-like growth patterns in culture. *Proc. Natl. Acad. Sci. USA* **82**, 4703-4706.
2. Banerjee, D.K., 1998, Angiogenesis: Characterization of a cellular model. *Puerto Rico Hlth. Sci. J.* **17**, 327-333.
3. Banerjee, D.K., and Martínez, J.A., 1998, Microvascular endothelial cells from bovine adrenal medulla - A model for *in vitro* angiogenesis. In *Angiogenesis: Models, Modulators and Clinical Applications* (M.E. Maragoudakis, ed), Plenum Press, New York, pp. 7-18.
4. Banerjee, D.K., 1988, Microenvironment of endothelial cell growth and regulation of protein N-glycosylation. *Ind. J. Biochem. Biophys.* **25**, 8-13.
5. Chapman, A., Trowbridge, I.S., Hyman, R., and Kornfeld, S., 1979, Structure of the lipid linked oligosaccharides that accumulates in class E *thy-1*-negative mutant lymphoma cell. *Cell (Cambridge, Mass.)* **17**, 509-515.
6. Banerjee, D.K., Scher, M.G., and Waechter, C.J., 1981, Amphomycin: Effect of the lipopeptide antibiotic on the glycosylation and extraction of dolichyl monophosphate in calf brain membranes. *Biochemistry* **20**, 1561-1568.
7. Kean, E.L., 1996, Site of stimulation by mannosyl-P-dolichol of GlcNAc-lipid formation by microsomes of embryonic chick retina. *Glycoconjugate J.* **13**, 675-680.
8. Orlean, P., Albright, C., and Robbins, P.W., 1988, Cloning and sequencing of the yeast gene for dolichol phosphate mannose synthase, an essential proteins. *J. Biol. Chem.* **263**, 17499-507.
9. Mazhari-Tabrizi, R., Eckert, V., Blank, M., Müller, R., Mumberg, D., Funk, M., and Schwarz, R.T., 1996, Cloning and functional expression of glycosyl transferase in yeast that dolichol mannose phosphate synthase from *Trypanosoma brucei*. *Biochem. J. (London)* **316**, 853-858.

10. Zimmerman, J.W., Specht, C.A., Cazares, B.X., and Robbins, P.W., 1996, The isolation of a Dol-P-Man synthase from *Ustilago maydis* that functions in *Saccharomyces cerevisiae*. *Yeast* **12**, 765-771.
11. Colussi, P.A., Taron, C.H., Mack, J.C., and Orlean, P., 1997, Human and *S. cerevisiae* dolichol phosphate mannose synthases represent two classes of the enzyme, but both function in *Schizosaccaromyces pombe*. *Proc. Natl. Acad. Sci. USA* **94**, 7873-7878.
12. Banerjee, D.K., Kousvelari, E.E., and Baum, B.J., 1987, cAMP-mediated protein phosphorylation of microsomal membranes increases mannosylphosphodolichol synthase activity. *Proc. Natl. Acad. Sci. USA* **84**, 6389-6393.
13. Banerjee, D.K., 1989, Regulation of mannosylphosphoryldolichol synthase activity by cAMP-dependent protein phosphorylation. In *Highlights of Modern Biochemistry* (A. Kotyk, J. Škoda, V. Paes and V. Kosta, eds.), VSP International Science Publishers, Zeist, pp. 379-388.
14. Banerjee, D.K., and Vendrel-Ramos, M., 1993, Is asparagine-linked protein glycosylation an obligatory requirement for angiogenesis? *Ind. J. Biochem. Biophys.* **30**, 389-394.
15. Heinemann, B., Kaplan, M.A., Muir, R.D., and Hooper, I.R., 1953, Amphomycin, a new antibiotic. *Antibiot. Chemother.* **3**, 1239-1242.
16. Bodanszky, M., Sigler, G.F., and Bodanszky, A., 1973, Structure of the peptide antibiotic amphomycin. *J. Am. Chem. Soc.* **95**, 2352-2357.
17. Banerjee, D.K., 1989, Amphomycin inhibits mannosylphosphoryldolichol synthesis by forming a complex with dolichylmonophosphate. *J. Biol. Chem.* **264**, 2024-2028.
18. Banerjee, D.K., 1994, A recent approach to the study of dolichyl monophosphate topology in the rough endoplasmic reticulum. *Acta Biochimica Polonica* **41**, 275-280.
19. Banerjee, D.K., 1987, Amphomycin: A tool to study protein N-glycosylation. *J. Biosci.* **11**, 311-319.
20. Banerjee, D.K., Diaz, A.M., Campos, T.M., Grande, C., Kozek, W.J., and Baksi, K., 1992, Monoclonal antibody to amphomycin. A tool to study the topography of dolichol monophosphate in the membrane. *Carbohyd. Res.* **236**, 301-313.
21. Elbein, A.D., 1987, Inhibitors of the biosynthesis and processing of N-linked oligosaccharide chain. *Annu. Rev. Biochem.* **56**, 497-534.
22. Duksin, D., Mahoney, W.C., 1982, Relationship of the structure and biological activity of the natural homologues of tunicamycin. *J. Biol. Chem.* **257**, 3105-3109.
23. Martínez, J.A., Torres-Negrón, I., Amigó, L.A., and Banerjee, D.K., 1999, Expression of $Glc_3Man_9GlcNAc_2PP$-Dol is a prerequisite for capillary endothelial cell proliferation. *Cellular and Molecular Biology (France)* **45**, 137-152.
24. Fiorelli, V., Gendelman, R., Samaniego, F., Markham, P.D., Ensoli, B., 1995, Cytokines from activated T cells induce normal endothelial cells to acquire the phenotypic and functional features of AIDS-Kaposi's Sarcoma spindle cells. *J. Clin. Invest.* **95**, 1723-1734.
25. Struck, D.K., and Lennarz, W.J., 1980, In *The Biochemistry of Glycoproteins and Proteoglycans* (W.J. Lennarz, ed.), Plenum Press, New York, pp. 35-83.
26. Banerjee, D.K., Tavárez, J.J., and Oliveira, C.M., 1992, Expression of bloood clotting factor VIII:C gene in capillary endothelial cells. *FEBS Letts.* **306**, 33-37.
27. Nguyen, M., Folkman, J., and Bischoff, J., 1992, 1-deoxymannojirimycin inhibits capillary tube formation *in vitro*. Analysis of N-linked oligosaccharides in bovine capillary endothelial cells. *J. Biol. Chem.* **267**, 26157-26165.

28. Pili, R., Chang, J., Partis, R.A., Mueller, R.A., Chrest, F.J., and Passaniti, A., 1995, The α-glucosidase I inhibitor castanospermine alters endothelial cell glycosylation, prevents angiogenesis, and inhibits tumor growth. *Cancer Res.* **55**, 2920-2926.
29. Carrasquillo, E.A., and Banerjee, D.K., 1998, Serine 141 is essential for Dol-P-Man synthase activity in *S. cerevisiae*. *Glycobiology* **8**, 93a.

THE CHARACTERIZATION OF ANGIOGENESIS INHIBITOR FROM SHARK CARTILAGE

Jeannie H. Liang and Kin-Ping Wong
School of Natural Sciences, California State University, Fresno, CA 93740 and Department of Medicine and Department of Biochemistry & Biophysics, University of California, San Francisco Medical School, San Francisco, CA 94143

ABSTRACT: An angiogenesis inhibitor isolated from shark cartilage, SCF2, has been characterized. SCF2 was shown to have specific angiogenesis-inhibiting activity in endothelial cell culture assays. Results of structural and functional studies indicate that the inhibitor is not a typical protein. It is a heat-stable proteoglycan, which contains keratan sulfate units and peptide. Gel filtration chromatography shows that the molecular weight of the angiogenesis inhibitor is about 10 kd.

1. INTRODUCTION

Cartilage is an interesting and unique tissue in that it is avascular and relatively tumor resistant. Early work from several laboratories had demonstrated that cartilage contains an inhibitor of angiogenesis. Langer and co-workers isolated from bovine cartilage a protein fraction, which inhibited the proliferation of blood vessels elicited in the rabbit cornea by tumor explants (Langer et al., 1976). Lee and Langer (1983) reported that shark cartilage extract incorporated into copolymer pellets inhibited angiogenesis in rabbit corneas, which results in decreased tumor size compared with controls. Using cartilage from a bovine source, a 27-kd protein was also identified that inhibited neovascularization (Moses et al., 1990). Oikawa et al (1990) isolated a molecular weight fraction between 1,000 to 10,000 kd that contained the majority of the antiangiogenic activity associated with shark cartilage. An inhibitor of neovascularization from bovine scapular chondrocyte-conditioned media was isolated and characterized (Moses et al., 1992). We have isolated an angiogenesis inhibitor from the growth media of shark chondrocyte culture (O'Hara and Wong, 1996). Recently, McGuire and his co-workers showed that the antiangiogenic effect of a fraction from shark cartilage was specific for vascular endothelium and did not affect other kinds of cells (McGuire et al., 1996). The present communication reports the characterization of an angiogenesis inhibitor from shark cartilage, SCF2.

2. MATERIAL AND METHODS

The shark cartilage used was obtained fresh from thresher sharks. It is further cleaned by trimming all the tissues attached to it, rinsed and stored at –20°C freezers until use.

The enzymes used for this study were: Chondroitinase ABC from *Proteus vulgaris*; Endo-β-Galactosidase from *Bacteroides fragilis*; Papain from *Carcica papaya*; Pepsin from *porcine stomach*; Pronase from *Streptomyces griseus* (all are obtained from Boehringer Mannheim); Enzymatic Deglycosylation Kit and Proteinase K from *Tritirachium album* are obtained from BioRad and Sigma, respectively.

All the chemicals used were Analytical Chemical Grade.

2.1 Determination of Angiogenesis-Inhibition by Endothelial Cell Culture (ECC) Assay

The assay used to determine the percentage of angiogenesis inhibition was a variation of the assay developed by Connolly et al.,(1986) for the determination of cell number by the level of acid phosphatase activity. Bovine cardiopulmonary artery endothelial cells, bovine (CPAE) acquired from American Type Tissue Culture (ATCC) were grown to 95% confluence in MEM-10E. The cells were released from the tissue culture flask (Corning) with a 0.25% trypsin solution and plated in 24-well tissue culture plates (Corning) in the same culture medium at a density of 10,000 cells/well. After the plates were cultivated for 8-12 hours at 37°C in a CO_2 incubator Model CI-44 (American Scientific Products), assay samples and controls were added. Each sample was loaded in four different wells at 100 ul/well to insure reproducibility. After incubation with the samples for 3 days, the medium was aspirated, and the number of cells was measured on the basis of the colorimetric measurement of cellular acid phosphatase.

Four other different types of cells were used to determine the specific angiogenesis inhibition of shark cartilage: Human Embryonic Lung Fibroblasts (MRC), Human Foreskin Fibroblasts (SF), Human Epidermoid Carcinoma (Hep-2), and Human Umbilical Cord Endothelial Cells (HUV). Hep-2 and HUV were from ATTC. MRC and SF were kindly provided by Dr. Dean Golden from the Valley Children's Hospital in Fresno, CA.

2.2 Analysis of SCF2 Using Chemical Methods

2.2.1 Protein Assay

The protein concentration of SCF2 was determined with the BioRad Bradford protein assay (Bradford 1976). The standard protein used was Bovine Serum Albumin (Sigma). A 800 ul sample was mixed with 200 ul of the BioRad solution and then analyzed for absorbancy at 595 nm using a Shimadzu Spectrophotometer Model UV-1601.

2.2.2 Carbohydrate Assays

Phenol-sulfuric acid reaction was used to determine the presence of neutral carbohydrates (Dubois et al 1956).

A modified carbazol assay was used to test the presence of acidic carbohydrate (Bitter and Muir 1962).

2.3 Functional Characterization of SCF2

Physical, chemical, and enzymatic methods were used to study SCF2. After each treatment, the functional angiogenesis-inhibiting activity of samples was determined by the ECC assay.

2.3.1 Heat Treatment

Samples of SCF2 were prepared with a final concentration of 0.5 mg/ml. The samples were heated in a boiling water bath for a desired time and then removed to cool down to room temperature (22-25°C). The control was a sample of SCF2 at the same concentration but without any heat treatment.

2.3.2 Acid/Base Treatment

Four samples of SCF2 were prepared with a final concentration of 0.1 mg/ml. Two were dissolved in 1N NaOH; the other two were dissolved in 1N HCl. One of the two was heated in a boiling water bath for 10 minutes. After the treatment, all the samples were neutralized and then dialyzed with Spectrum membrane (1,000 MW, Spectra 7) to remove the salt. The sample dissolved in dd H_2O with the same concentration was used as the control. The samples were sterilized by filtration through a 0.2 micron sterile Acrodisc syringe filter (Gelman Sciences) before ECC assays.

2.3.3 SDS/Mercaptoethanol Treatment

Samples of SCF2 were mixed with 1% of SDS or 5% of 2-Mercaptoethanol. The final concentration of SCF2 was 0.1 mg/ml. The solution was heated in boiling water bath for 10 minutes. After cooling, the samples were dialyzed (1,000 MW, Spectra 7) to remove the solvent. Filter-sterilization was done before the ECC assay. A sample dissolved in dd H_2O at the same concentration was used as the control.

2.3.4 Digestion of SCF2 with Proteolytic Enzymes

Twenty ul of SCF2 (0.5 mg/ml in H_2O) was mixed with reaction buffer and enzyme soluton to a final volume of 100 ul. Four different kinds of proteolytic enzymes were used: Proteinase K, Pepsin, Papain, and Pronase. The buffer for each enzyme reaction was different. For Proteinase K digestion, the buffer was 40 mM sodium phosphate, pH 7.5 (Maniatis et al 1982). For Pepsin digestion, the buffer was 0.1 mM sodium phosphate, pH 3.0. For Papain digestion, the buffer was 0.1 mM sodium phosphate, pH 6.5. For Pronase digestion, the buffer was 100 mM Tris, 10 mM EDTA, pH 7.5 (Spiro 1962). The enzyme solution was prepared by dissolving each enzyme in its reaction buffer at a concentration of 10 mg/ml. The digestions were carried out at 37°C for the desired time. And then reactions were stopped by heating in a 100°C water bath for 10 minutes. Controls for each reaction were the same amount of SCF2 and buffer, but without enzyme.

2.3.5 Digestion of SCF2 with Glycosidases

Twenty ul of SCF2 (0.5 mg/ml in H_2O) was mixed with reaction buffer and enzyme solution to a final volume of 100 ul. Four different kinds of glycosidases were used: Chondroitinase ABC, Endo-galactosidase, N-glycosidase F, and O-glycosidase. The buffers and enzyme solutions were prepared according to the protocols supplied by the manufacturers. The digestions were carried out at 37°C for the desired time, and then stopped by boiling for 10 minutes. Control for each reaction was the same amount of SCF2 and buffer, but without enzyme.

2.4 Molecular Weight Determination of Angiogenesis Inhibitor from Shark Cartilage Using Gel Filtration

Molecular weight of the angiogenesis inhibitor was estimated using gel filtration chromatography. Sephacryl S-300 HR (Sigma) was packed to a height of 30 cm in a glass column 1.5 cm in diameter. The column was equilibrated with phosphate buffered saline (PBS). The sample, 1 mg of SCF2 dissolved in 2 ml PBS, was applied to the column. Elution was achieved with the same buffer at a rate of 15 ml per hour, and 2 ml fractions were collected. Aliquot of each fraction was analyzed by the ECC assay.

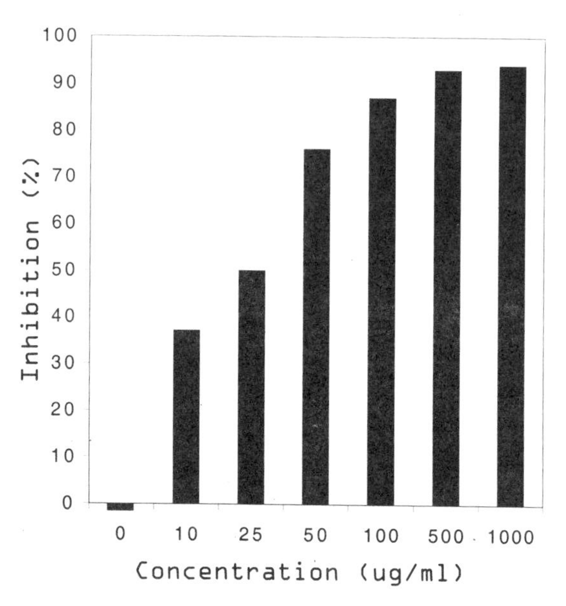

Figure 1. Concentration-dependent inhibition of SCF2. Samples of SCF2. Samples were diluted to different concentrations in dd water, and analyzed by the ECC assay.

The column was calibrated using Gel Filtration Molecular Weight Markers (Sigma).

Another Sephacryl S-300 HR profile was achieved using a different buffer. 6M Guanidine hydrochloride (GuHCl) was used to equilibrate the column, dissolve and elute the sample. The procedure was the same as above, except that the fractions collected were dialyzed (1,000 MW, Spectra 7) against dd H_2O to remove the salt before the ECC assay.

3. RESULTS AND DISCUSSIONS

3.1 Concentration Dependence of the Shark Cartilage Angiogenesis Inhibitor (SCF2)

The angiogenesis-inhibiting activity of SCF2 is concentration-dependent as shown in Figure 1. The percentage of inhibition increases with the concentration of SCF2. When the concentration of SCF2 is higher than 500 ug/ml, the inhibition reaches its maximum.

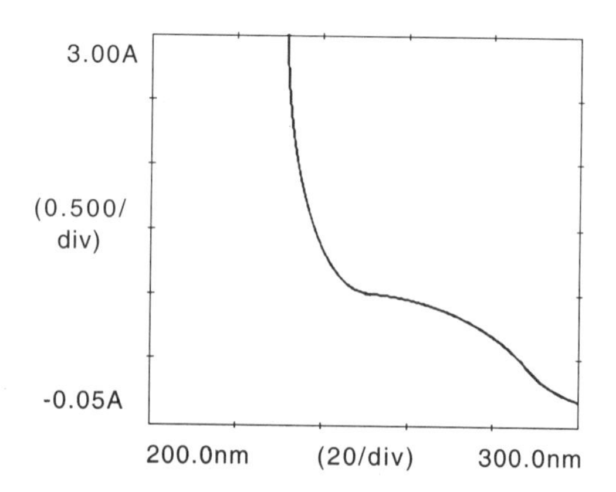

Figure 2. Ultraviolet absorption spectrum of SCF2. Notice the absence of a typical protein peak at 280 nm.

3.2 Characterization of Angiogenesis Inhibitor from SCF2

Physical, chemical, and enzymatic methods were used to characterize the angiogenesis inhibitor from shark cartilage to elucidate the structure of the SCF2.

3.2.1 Physical and Chemical Studies

A sample of SCF2 was scanned on an ultraviolet spectrophotometer (Figure 2). There was no distinct protein absorption peak with maximum at 280 nm.

Heat treatment (100°C, 2 hours) did not affect the inhibition activity as shown (Figure 3).

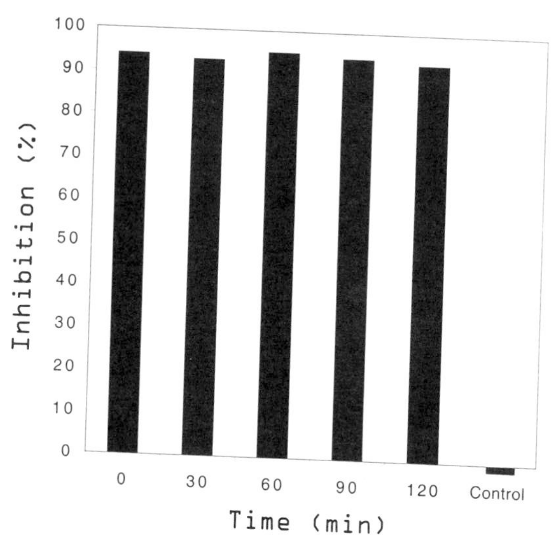

Figure 3. Inhibiting activity after heat treatment of SCF2. Samples of SCF2 were heated in boiling water bath for the time periods as indicated in the above chart. Inhibition activity of the heated samples was determined by the ECC assay. Control was dd water.

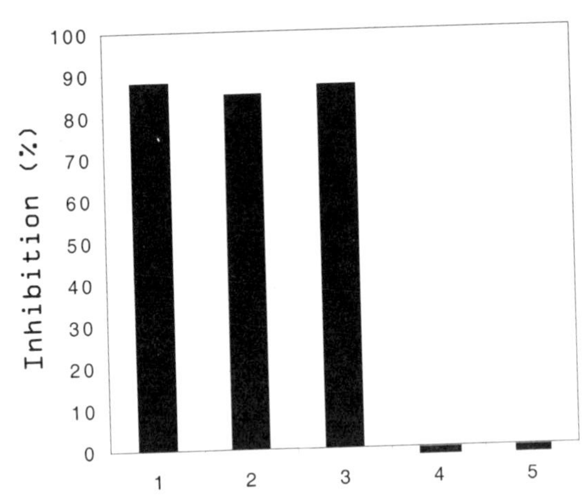

Figure 4. Inhibiting activity after SDS & Mercaptoethanol treatment of SCF2. Samples of SCF2 were mixed with SDS (column 2) or 2-Mercaptoethanol (column 3), and then heated for 10 minutes. ECC assay was performed after removing SDS and 2-Mercaptoethanol by dialysis. SCF2 dissolved in dd water with the same concentration used as the control (column 1). Column 4 and column 5 are dd water.

SCF2 retains all its angiogenesis-inhibiting activity after treated with SDS or 2-Mercaptoethanol at 100°C for 10 minutes (Figure 4).

These results suggest that the angiogenesis inhibitor is not a typical protein because under the conditions of these experiments, most proteins would be denatured and lose their biological function.

The protein and carbohydrate contents of SCF2 were determined (Table 1).

Table 1. Protein and Carbohydrate Determination

	Crude Extract	SCF2
Protein (%)	11	4.5
Neutral carbohydrate (%)	16	12
Acidic Carbohydrate (%)	29	25

Protein concentration was determined with BioRad micro-assay. Neutral carbohydrate was measured with phenol-sulfuric acid reaction. Acidic carbohydrate was determined with modified carbazol assay.

The protein concentration was low even in the Crude Extract, while the carbohydrate concentration was several times higher compared with the protein concentration. The carbohydrate concentration should be higher than that given in Table 1, because there is not a general assay method available to determine the concentration of total carbohydrates in a sample and some of the sugars in SCF2 may not be detected by the phenol-sulfuric acid reaction or the modified carbazol assay.

The effects of acid and base treatments on the inhibitory activity of SCF2 are shown in Figure 5.

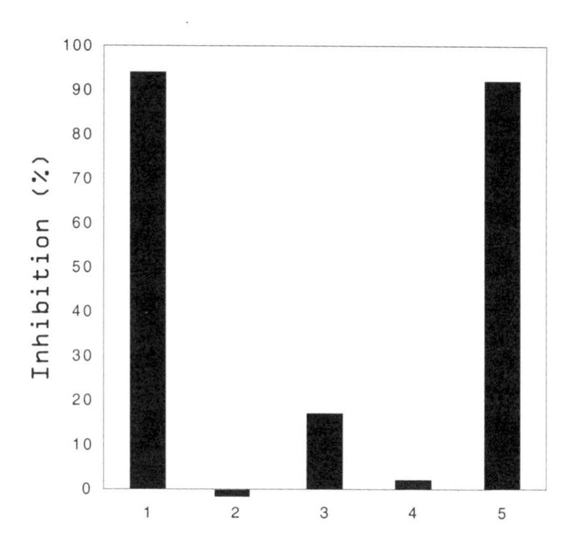

Figure 5. Inhibiting activity after acid/base treatment of SCF2. Samples of SCF2 were treated under different conditions. Column 1 (control): SCF2 dissolved in dd water. Column 2: SCF2 treated with 1N NaOH at 100°C. Column 3: SCF2 treated with 1N NaOH at room temperature. Column 4: SCF2 treated with 1N HCl at 100°C. Column 5: SCF2 treated with 1N HCl at room temperature. The inhibition was determined by the ECC assay.

At room temperature, the angiogenesis inhibiting activity dropped from 94% to 17%, after treatment with 1N NaOH; while the acid treatment did not affect the activity at all. When the treatment was performed at 100°C, both the acidic and alkaline samples lost their inhibiting activity. These results indicate that the angiogenesis inhibitor is heat stable in neutral environment but not in an extreme pH condition, and that the inhibitor is more sensitive to base than acid.

In a review, "The Structure and Function of Cartilage Proteoglycans," Carney and Muir (1988) stated that the O-glycosidic bond to core protein of the proteoglycans is alkali labile. The results presented are consistent with the hypothesis that the angiogenesis inhibitor from shark cartilage might be a glycoprotein or proteoglycan, which contains both carbohydrate and peptide. Further studies are needed to confirm this.

3.2.2 Enzymatic Digestions

We have used different enzymes to cleave specific linkages of the angiogenesis inhibitor and then analyzing the inhibition activity of digested samples by ECC assay to elucidate the specific linkages that are essential to the structure and function of the inhibitor.

Four kinds of proteolytic enzymes were used to digest SCF2 (Table 2).

Table 2. Digestion of SCF2 with Proteolytic Enzymes

Enzyme Used	Inhibitory Activity Before Digestion (%)	Inhibitory Activity After Digestion (%)
Proteinase K	90	89
Pepsin	92	90
Papain	89	90
Pronase	90	23

All the digestions were carried out at 37°C for 96 hours.

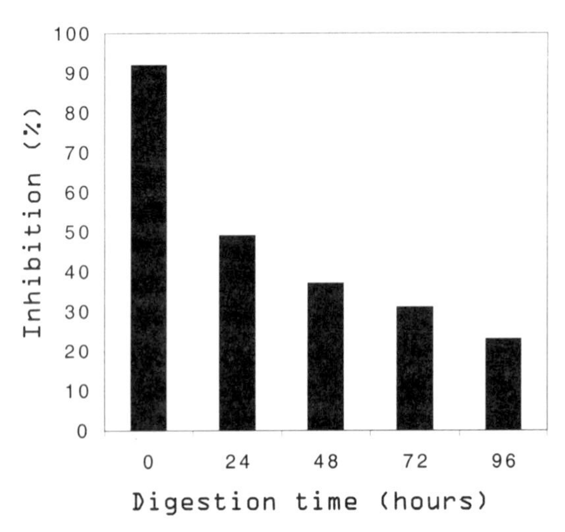

Figure 6. Inhibition after Pronase digestion of SCF2. Samples of SCF2 were digested with Pronase for several time periods as indicated in the above chart and the inhibition activity was determined by the ECC assay. Inhibition activity of SCF2 decreased after pronase digestion.

The results show that only Pronase digestion affects the activity of the inhibitor, which decreased from 92% to 23% after 4 days reaction (Figure 6).

In Spiro's study of glycoprotein (Spiro 1965), Pronase was used to digest the peptide portion of thyroglobulin to obtain the carbohydrate units of this molecules with only a few amino acid attached. The other three enzymes, Proteinase K, Pepsin, and Papain, have not been previously reported to be able to digest the peptide part of glycoproteins. These results confirm that the angiogenesis inhibitor from shark cartilage is not a typical protein, otherwise it would lose its biological activity after Proteinase K digestion. It also suggests that the inhibitor contains a peptide component, which is essential to its biological function.

Several kinds of glycosidases were used to analyze the structure of the inhibitor (Table 3).

Table 2. Digestion of SCF2 with Glycosidase

Enzyme Used	Inhibitory Activity Before Digestion (%)	Inhibitory Activity After Digestion (%)
Chondroitinase ABC (1)	90	69
Endo-β-galactosidase (2)	90	16
N-Glycosidse F (3)	92	64
O-Glycosidase (4)	89	85

(1) Catalyzes the release of chondroitin sulfate- and dermatan-sulfate-side chain from proteoglycans: GalNc-s-{chondroitin sulfate or dermatansulfate}-↓GlcA-Gal-Gal-Xyl-Ser-Core protein

(2) Hydrolyzes internal β-galactosidic linkages of the lactosamine type: GlcNAcβ1-3Galβ↓4GlcNAc

(3) Hydrolyze all types of N-glycan chains from glycopeptides and glycoproteins:

```
x-Man
     \
      Man – GlcNac-GlcNAc-↓Asn-peptide/protein
     /
x-Man
```

(4) Hydrolyze Galβ-3GalNAc from o-glycans: Galβ1-3GalNAcα1-↓(Thr or SER)-peptide/protein

The results show that the endo-β-galactosidase and chondroitinase ABC digestion affects the inhibition of SCF2. It suggests that the molecular structure of the inhibitor contains units of keratan sulphate, the substrate specific for endo-β-galactosidase, and chondroitin sulphate, the substrate specific for chondroitinase ABC (Oike et al 1980). Keratan sulphate and chondroitin sulphate are two of the major glycosaminoglycans found in cartilage (Carney and Muir 1988). The results also suggest that the keratan sulphate type linkages are more sensitive to enzyme digestion.

There are two main types of linkage between the carbohydrate and peptide in glycoprotein or proteoglycan: O-glycosidic bond and N-glycosidic bond (Dwek 1996). Table 3 shows that the inhibition decreased about one third after N-glycosidase F digestion while O-glycosidase affect the inhibition only marginally. It indicates that the structure of the inhibitor contains a N-glycosidic bond that is required for function. The experiment of acid/base treatment suggests that O-glycosidic bond may also locate in the interior of the inhibitor molecule. The reason why O-glycosidase digestion did not substantially affect the inhibition is because the linkage may either not sensitive to the enzyme or buried inside the molecule and hence protected from attack by the enzyme. The kind of O-glycosidase used here is specific to one of the O-glycan sub-classes. If the O-glycosidic bond in the inhibitor does not belong to the sub-class, it cannot be digested by the enzyme.

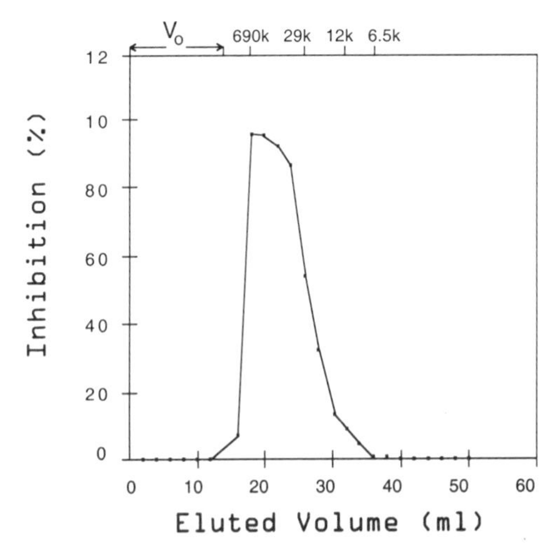

Figure 7. Gel filtration chromatography profile of SCF2. Sample of SCF2 (0.5 mg/ml) was applied to a Sephacryl S-300 HR column (1.5 x 30 cm) and eluted with PBS at a flow rate of 15 ml/hour. Fractions of 2 ml were collected and aliquot of each fraction was analyzed for angiogenesis-inhibiting activity by the ECC assay.

3.3 Molecular Weight Determination of Angiogenesis Inhibitor from Shark Cartilage Using Gel Filtration

A sample of SCF2 was applied to a Sephacryl S-300 HR column and eluted with PBS. The fractions were collected and aliquot of each fraction was analyzed for angiogenesis-inhibiting activity by the ECC assay (Figure 7). Another Sephacryl S-300 HR profile was obtained using 6M GuHCl (Figure 8).

These two profiles are different. In Figure 7, there is only one big peak and the molecular weight of the major fractions in this peak is larger than 29 kd. In Figure 8, there are two peaks: the first peak is similar to the one in Figure 7; the second peak is sharper and the molecular weight of the fraction in this peak is about 10 kd.

Conditions of the two chromatography were the same except the buffer. PBS was used in Figure 7 while 6M quanidine hydrochloride was used in Figure 8. The molecular weight of the components in SCF2 is approximately 10 kd; but in Figure 7, they came out earlier from the column than they were supposed to. One possible explanation is aggregation, which

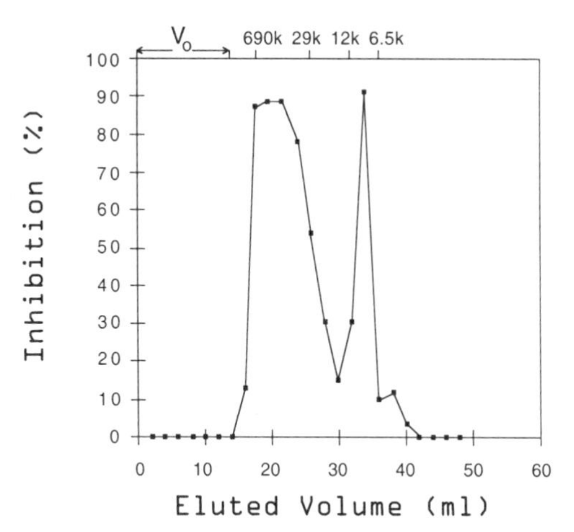

Figure 8. Gel filtration chromatography profile of SCF2 in GuHCl. Sample of SCF2 (0.5 mg/ml) was applied to a Sephacryl S-300 HR column (1.5 x 30 cm) equilibrated with 6M GuHCl. Elution was achieved with the same buffer at a flow rate of 15 ml/hour. 2 ml fractions were collected and dialyzed to remove the salt. Aliquot of each fraction was analyzed for angiogenesis-inhibiting activity by the ECC assay.

involves proteoglycan molecules interacting with one another to form large multimolecular aggregates (Carney and Muir 1996).

Guanidine hydrochloride can be used to dissociate this kind of aggregates (Hascall et al 1974). It suggests that the molecules of the angiogenesis inhibitor aggregate in PBS buffer and elute earlier from the column. In Figure 8, GuHCl dissociates some of the aggregates into monomer which are about 10 kd and elute from the column later on to form the second peak. Since 6M GuHCl could not dissociate the aggregates completely, another buffer needs to be found so that more homogeneous molecules of angiogenesis inhibitor can be obtained to perform further structural studies.

4. CONCLUSIONS

The results of physical, chemical, and enzymatic studies indicate that shark cartilage angiogenesis inhibitor SCF2 is not a typical protein. It is a heat stable proteoglycan. It is likely that keratan sulfate is the component units of the carbohydrate part of the inhibitor. N-glycosidic bond is the likely linkage between the carbohydrate and peptide. The experiment of acid/base treatment suggests that O-glycosidic bond may locate in the interior of the inhibitor.

The molecular weight of the angiogenesis inhibitor appears to be 10kd as estimated by Sephacryl S-300 HR gel filtration chromatography. However, aggregation remains a problem for a better characterization of the structure of the angiogenesis inhibitor.

REFERENCES

Bitter, T, and Muir, H., 1962, Modified Carbazol Assay. *Methods in Enzymology* 4:330.

Bradford, M., 1976, A Rapid and Sensitive Method for the Quantitation of Microgram Quantities of Protein Utilizng the Principle of Protein Dye Binding. *Anal. Biochem.* 72:248-254.

Carney, S. L. and Muir, H., 1988, The Structure and Function of Cartilage Proteoglycans. *Physiological Reviews* 68:858-898.

Connolly, D. et al, 1986, Determination of the Number of Endothelial Cells in Culture Using an Acid Phosphatase Assay. *Anal. Biochem.* 153:136-140.

Dubois, M. et al, 1956, Calorimetric Method for Determination of Sugars and Related Substances. *Anal. Chem.* 28:350.

Dwek, R. A., 1996, Glycobiology: Toward Understanding the Function of Sugars. *Chem. Rev.* 96:683-720.

Hascall, V. C. et al, 1974, Aggregation of Cartilage Proteoglycans. *J. Biol. Chem.* 249:4232-41.

Langer R. et al, 1976, Isolation of a Cartilage Factor that Inhibits Tumor Neovascularization. *Science* 193:70-72.

Lee, A. and Langer, R., 1983, Shark Cartilage Contains Inhibitors of Tumor Angiogenesis. *Science* 221:1185-7.

Maniatis, T. et al, 1982, Molecular Cloning. Cold Spring Harbor Laboratory.

McGuire, T. R. et al, 1996, Antiproliferative Activity of Shark Cartilage With and Without tumor Necrosis Factor-α in Human Umbilical Vein Endothelium. *Pharmacotherapy* 16:237-244.

Moses, M. A. et al, 1990, Identification of An Inhibitor of Neovascularization from Cartilage. *Science* 248:1408-10.

O'Hara, C. F. and Wong, K.-P., 1996, Cultured Shark Chondrocytes Secrete An Angiogenesis Inhibitor. *In Vitro: Cellular and Developmental Biology*, 32, 57A

Oikawa, T. et al, 1990, A Novel Angiogenic Inhibitor Derived from Japanese Shark Cartilage (I). *Cancer Lett.* 51:181-6.

Oike, Y. et al, 1980, Structural Analysis of Chick-Embryo Cartilage Proteoglycan by Selective Degradation with Chondroitin Lyases (Chondroitinase) and Endo-β-galactosidase (Keratanase). *Biochem. J.* 191:193-207.

Spiro, R. G., 1962, Studies on Feruin, A Glycoprotein of Fetal Serum. *J. Biol. Chem.* 237:382.

Spiro, R. G., 1965, The Carbohydrate Units of Thyroglobulin. *J. Biol. Chem.* 240:1603.

QUANTITATIVE ASSAYS FOR THE CHICK CHORIOALLANTOIC MEMBRANE

W Douglas Thompson and Allyson Reid
Department of Pathology, University of Aberdeen, Medical School, Aberdeen Royal Infirmary, Aberdeen AB25 2ZD, UK

Abstract: Focal application of angiogenic substances to the chick chorioallantoic membrane is quick and easy as a rapid screening test, but is susceptible to artefactual stimulation induced by carriers, hyperosmolarity, proteolytic activity, and indeed any cause of damage to the CAM. This can be deceptive and unanticipated. Focal application methods can be used for subsequent measurement by morphometry of the increased vascularity forming the typical spokewheel pattern of supply vessels. If test and control substances are applied in liquid form to the whole dropped CAM surface, then a much wider variety of quantitative morphometric, histological and biochemical techniques can be applied. Assessment of arterial vascularity, terminal arterial branching, supply vessels in cross-sections, and CAM haemoglobin content are direct measures of angiogenic effects, but are time-consuming. Biochemical assays of collagen, protein and DNA synthesis parallel the other assays, and these parameters can be estimated more quickly within the working week. There is inherent variability in the outbred strains of hen eggs currently commercially available. This means that all assays require substantial group numbers to achieve statistical validity, generally not less than 10 eggs per group. The biochemical assays yield interesting time-course patterns that distinguish between different types of angiogenic stimulants.

1. INTRODUCTION

This article reviews a variety of quantitative assays of morphometric, histological, and biochemical types that have been applied in Aberdeen to the chick chorioallantoic membrane model (CAM). Focal application of angiogenic substances to the chick chorioallantoic membrane is quick and easy as a rapid screening test, but is susceptible to artefactual stimulation induced by carriers, hyperosmolarity, proteolytic activity, and indeed any cause of damage to the CAM, although this can be reduced (Peek *et al* 1988). This can be deceptive and unanticipated. For example, focal slow release application of a proteolytic enzyme such as trypsin appears to result in a localised increase in angiogenesis. However application in liquid form to the whole exposed surface of the dropped CAM, in a range of dosage up to sublethal, resulted in no change in DNA synthesis or vessel content in contrast to genuine angiogenic agents, and despite an early transient influx of inflammatory pseudoeosinophils (Thompson and Kazmi 1989, Lucas and Jamroz 1961).

Quantitation of the focal spokewheel appearance can be achieved by various simple morphometric techniques such as counting the intersections of supply vessels with a three ring grid (Maragoudakis *et al* 1995). Application of an angiogenic substance in liquid form leads to increased vascularity throughout the dropped CAM (Thompson *et al* 1985), and this can be quantified by similar morphometry and also by fractal analysis of the branching supply vasculature (Parsons-Wingerter *et al* 1998, Kirchner *et al* 1996). We have shown that the volume of the CAM vascular bed can be estimated simply by assay of haemoglobin content of excised dropped CAMs (Thompson and Brown 1987). There is a progressive increase in length and number of arterial branches in the CAM after application of a stimulatory substance. The length increase is attributable to the tortuosity characteristic of accelerated angiogenesis. The morphometry was performed by tracing the vasculature using permanent preparations made by osmium staining of the lipid of red cells trapped in the arterial branches of the formalin fixed CAM (Thompson and Brown 1987) (Fig 1A and Fig1B).

These methods are accurate but slow to perform. A simpler approach is to count the number of supply vessels within the middle mesenchymal layer seen in histological cross-sections of CAM (Fig 2). The method is still tedious and there is the delay in obtaining the H&E sections of thin flimsy tissues demanding skilful preparation.

The quantitative methods that we have found to be most rapid and objective are biochemical assays of the whole dropped CAM. Results are obtainable within the working week allowing greater flexibility in planning

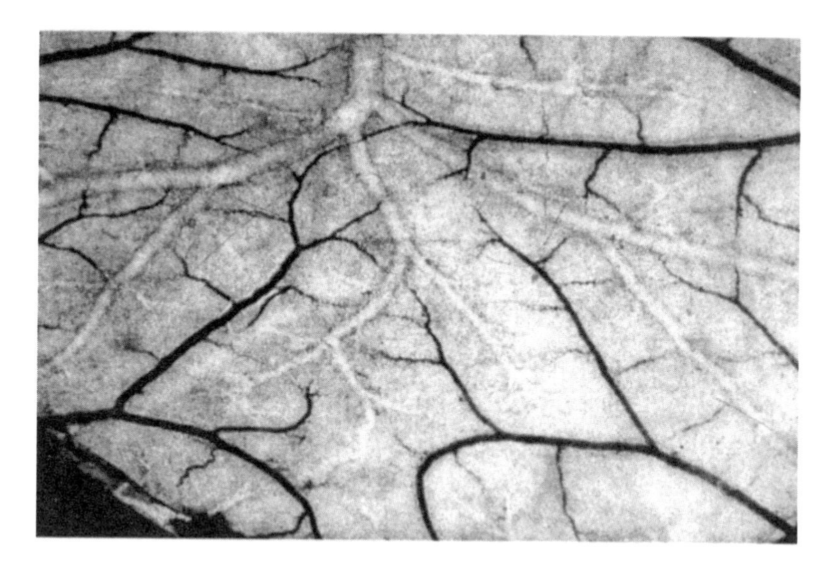

Figure 1A. Normal CAM vasculature stained with osmium tetroxide, viewed en face

Figure 1B. Stimulated CAM vasculature: there is increased vessel branching and tortuosity 3 days after application of an angiogenic agent in liquid form to the whole surface

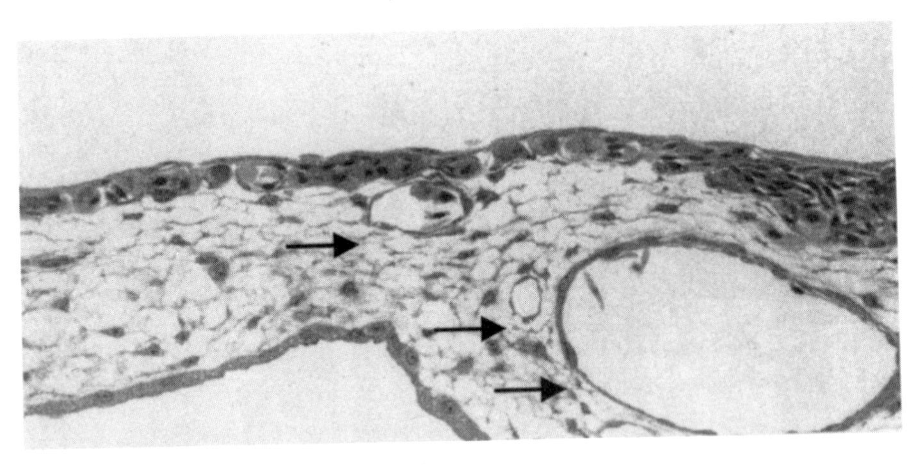

Figure 2. The chick CAM viewed in cross-section at day 14: capillaries are seen throughout the upper ectodermal layer, and supply vessels in the central mesenchymal layer(arrows)

and investigation. Protein and DNA synthesis, and also collagen synthesis are readily estimated with the aid of liquid scintillation counting. The assays for collagen synthesis are rational for angiogenesis but rather insensitive and require additional steps for specific collagenase digestion (Thompson *et al* 1986, Maragoudakis *et al* 1988). The easiest sensitive method is that for measuring the incorporation of methyl-[^{3}H]-thymidine into DNA, and this methodology and its rationale is described below. Not only does DNA synthesis follow the other parameters of angiogenesis, there is additional information to be gained from the different time patterns of DNA synthesis induced by different angiogenic substances (Thompson and Kazmi 1989, Thompson and Brown 1987, Clinton *et al* 1988, Duncan *et al* 1992,) (Figs 3, 4).

It is also suitable for measurement of stimulation of angiogenesis by extracts of pathological tissues such as atherosclerotic plaque extracts, and wound and tumour extracts (Thompson *et al* 1987, Thompson *et al* 1990, Stirk *et al* 1993, Thompson *et al* 1991, Thompson *et al* 1996, Thompson *et al* 1993), and in pursuit of active angiogenic components such as fibrin fragment E by using blocking antibodies for affinity column removal or more simply by admixture (Thompson *et al* 1992). The CAM is not stimulated by application of serum, including that of animal species used for antiserum production, in contrast to plasma which clots and lyses releasing stimulatory fibrin degradation products (Stirk and Thompson 1990).

2. METHODOLOGY

On day zero, the fertile hens' eggs, preferably from an egg production rather than a broiler strain, and which should have been brought into room temperature about the day before, are placed in an fan assisted egg incubator at 99.5°F (37.5°C) preferably fitted with automatically tilting shelves (Anderson Brown 1982). The eggs are positioned with pointed ends facing downwards. Excessive humidity allows surface organisms to penetrate the shell, but trays of clean water can be placed beneath the egg trays which should ideally have wire mesh bottoms. Addition of strong disinfectants to the water is generally too toxic.

On day 3, the eggs are removed from the incubator and placed in a 37°C oven close to a sterile hood or flow cabinet. Each egg is swabbed with 70% alcohol. Using a 2 ml syringe with a 21G needle, 2 ml of albumen is removed and the eggs returned to the incubator.

On day 4, the eggs are taken to the 37°C oven and swabbed with 70% alcohol as before. Each egg is punctured by a needle at the blunt end into the air space. Using a diamond tipped glass slide marker, a 1 cm square (size depends on further procedures) is scored in the shell. A miniature electric drill with a small cutting disk can be used, but this is slower and generates much dust even with extraction using an adjacent nozzle connected to a vacuum cleaner. The square of shell is removed using sharp pointed forceps. This causes the egg contents to drop into the space left by collapse of the air space at the end of the egg and the withdrawal of albumen. The volume of the space above the dropped CAM is approximately 7 ml. Non-viable eggs are disposed of, and for each of the remainder, the square opening in the shell is covered with Sellotape. The eggs then have to be kept level in the incubator without tilting, or transferred to a hot room at 37.5°C in racks sitting above trays of clean water.

On day 10, the surviving eggs are organised into groups of preferably no less than 10. Test and control samples are prepared and each egg receives 0.3 ml of sample injected through the Sellotape onto the upper surface of the CAM. This surface is approximately 3 cm diameter and 0.3 ml is sufficient to cover it. Application of volumes beyond 0.5ml results in increased mortality producing embryos that, to the pathologist's eye, appear to have died of congestive cardiac failure.

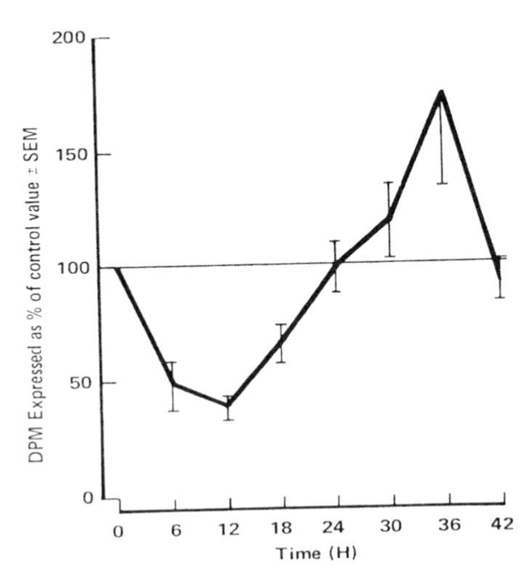

Figure 3. Pattern over time of DNA synthesis changes induced by histamine application (Thompson and Brown 1987)

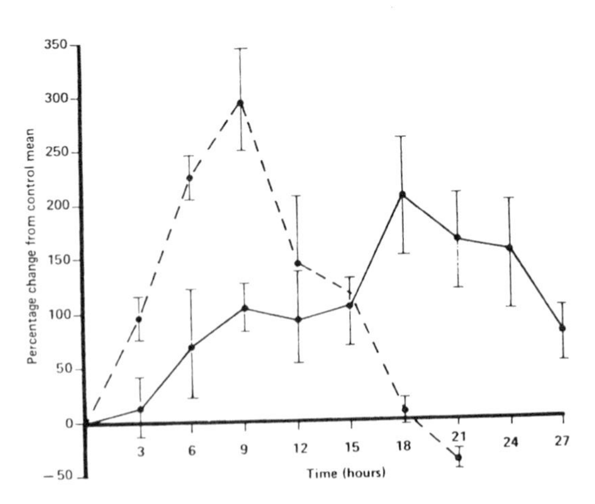

Figure 4. Pattern of protein (---) and (—) DNA synthesis stimulated by fibrin degradation products (Thompson *et al* 1985).

On day 11 after 18 h, which is optimal for wound extracts and fibrin degradation products, the assay for measurement of DNA synthesis is performed, using standard precautions such as wearing disposable gloves, cleaning and monitoring of bench surfaces and approved radioactive materials disposal for relatively minute quantities of a soft beta emitter. Methyl-[^{3}H]thymidine is applied onto the CAM surface using 0.5 ml of 2μCi/ml normal saline per egg. Methyl-[^{3}H]thymidine is used to avoid recycling of thymidine. After 20 min the eggs are removed to a radioactive type bench and all eggs are rapidly injected through the Sellotape window with 5 ml of ice-cold normal saline to halt metabolism. For large numbers, it is best to inject eggs in sequence across the groups in order to minimise any differences in incorporation between groups.

The technique of removal of each CAM is simple but crucial to the rationale of achieving results expressed as total DPM per CAM. By 20 min, methyl-[^{3}H]thymidine is incorporated only into the CAM which has been directly exposed. Beyond 30 min, [^{3}H]thymidine becomes detectable in adjacent CAM that is applied to the shell, via uptake from the bloodstream. Wearing disposable gloves, each CAM is removed by holding the egg in the palm of one hand, Sellotape window face down, and piercing the side with pointed scissors well above the level of the dropped CAM. Scissors are used to cut round the shell, and lifting the upper half cautiously, and rotating the egg if necessary, the residual yolk and embryo tipped out into a disposal container once the umbilical vessels connecting the embryo to the CAM are cut. Frequently the dropped CAM remains inflated by air and saline above the window sealed by Sellotape. There is a red ring at the edge of the dropped CAM adherent to the shell (Fig 5). It is easy to cut around this edge, but for difficult ones where the air space deflates and the CAM slips or detaches, it is permissible to cut wider around the presumed area of dropped CAM. The extra tissue sampled will not affect the values for incorporated [^{3}H]thymidine. Each CAM is rinsed briefly in saline, blotted dry and placed in a 20 ml Sterilin plastic disposable container in 5 ml of distilled water. These may be stored at 4°C overnight if desired. Plasticware suitable for radioactive disposal by incineration is employed.

Day 12: Each CAM is homogenised briefly in the 5 ml of H_2O using for example a homogeniser with a suitable head. 5 ml of 10% trichloroacetic acid (TCA) is added to each taking precautions for corrosive usage such as

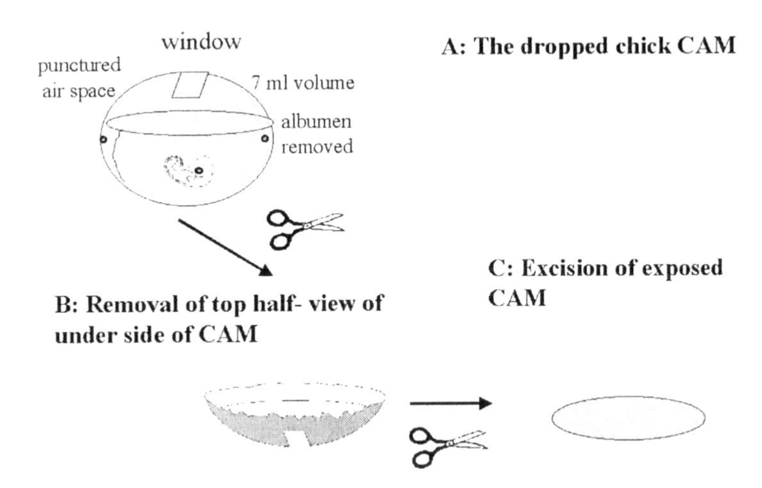

Figure 5. Excision of the area of dropped CM on day 11, previously exposed to test or control substances on day 10, and after pulsing for 20 min with [^{3}H]-thymidine

replacing each Sterilin cap before vortexing each sample. The samples are centrifuged at approximately 1,000 g for about 8 min. The supernatant is discarded and the precipitate is retained. Twice more the precipitate is resuspended in 10 ml 5% TCA, vortexed and the precipitate retained to remove free isotope. The precipitate is resuspended finally in 2.5 ml of H_2O. Scintillation vials are filled with 5 ml of Instagel Plus (Packard Biosciences). This is a thixotropic scintillation fluid. Each Sterilin is vortexed and the contents immediately poured into a vial. Each vial is vortexed until the contents become a gel. This stage is completed one Sterilin at a time. To reduce chemiluminescence, the vials are stored at 4°C overnight in the dark.

The tissue debris becomes efficiently dispersed throughout the thixotropic scintillant as a fine emulsion suitable for counting. Nevertheless there is considerable loss of efficiency due to colour, salt and protein quenching inherent to the nature of the tissue, and necessitates accurate quench correction curves created by, for example, the LKB "Hat-trick" method that can be used to check a real CAM sample. This method copes with the relatively large amount of proteinaceous material derived from each CAM. Phenol extraction of DNA (Maniatis *et al* 1982, Splawinski *et al* 1988) gives identical results but requires more handling of corrosives and entails similar quenching problems.

By applying this assay to the whole "dropped" area of CAM exposed to test or control substances, the results can be expressed per CAM, independent of changes that may be induced in weight and protein content due to, for example, oedema, and of changes in cellularity due to inflammatory cell influx.

3. CONCLUSION

Although the subject of angiogenesis research was well established by the early 1980's, there was a lack of quantitative methodologies utilising the chick chorioallantoic membrane model (CAM), even although it was by then the most popular in vivo system (Hudlická and Tyler 1986). Focal application of test and control substances first described by Folkman (1974) was, and still is, the most used method but was subject to a range of criticisms. It was argued that the CAM was too sensitive to non-specific physical carrier effects, chemical and hyperosmotic effects (Barnhill and Ryan 1983, Ryan and Barnhill 1983). The focal spokewheel response of the CAM was criticised as a local distortion of supply vessels due to fibrosis, and not a capillary increase effect. The response tended to be an all or nothing effect, and dose response data was limited to concentrations derived from serial dilutions or quantity of freeze dried protein. By now many of these doubts and problems have been overcome or at least better understood.

Considering all the quantitative assays tried by us, it appears that the greater the amplitude achievable using each method following stimulation with test substances, the greater the spread of the values, and vice versa. In other words, there is a trade-off between the ceiling of maximum stimulation and the coefficient of variation for each assay. For all assays on this living system, the standard deviation increases with the degree of stimulation of the particular parameter of angiogenesis, and this determines how statistical comparisons should be performed. Because the variation within test and control groups is unequal, the Student's t test is invalid. Statisticians favour log conversion of the data before applying any statistical test, including the t test. It should not be assumed that this will reduce the power of the test, and indeed any increase or decrease in significance is usually slight, and unpredictable in direction from viewing the raw data.

The source of considerable variability for assays of all types on the CAM is the nature of the hen. The commercial strains are all outbred stock. This is most easily illustrated by comparison of the coefficients of variation for the DNA assay. This assay has a variation of less than 5% for a pooled and aliquotted batch of CAMs. However the normal assay variation for groups of individual CAMs is around 30% to 40%, and can be worse. That this is intrinsic individual variation is further confirmed by short term organ culture of CAM samples with addition of tritiated thymidine which gives the same high assay variation. The only answer for producing high quality data is to use groups of 10 eggs minimum each to allow for a few losses.

The choice of method to use must depend on available equipment, and technical backup. Focal application methods with en face vessel assessment are quick and semi-quantifiable, economical, good for screening many novel substances using minimal amounts of scarce materials, but treacherous in inexperienced hands, with potentially significant results requiring further exploration and confirmation. The various biochemical methods provide parameters of angiogenesis and are generally quicker and completely objective, and are valuable for series of experiments exploring mechanisms and interactions, and agents that block a stimulant. The latter methods do depend however on a substantial work-up to obtain dose response data and time-optimisation information, but this in itself can be illuminating and may expose mechanisms. The one limitation of this approach concerns quantitation of interaction of antiangiogenic drugs with CAM vessels rather than with angiogenic stimulants. Here the use of sponge systems on the CAM offers a better alternative (Nguyen *et al* 1994, Ribatti *et al* 1997). The more interesting an agent becomes, the more types of visual, histological and biochemical tests need to be applied.

ACKNOWLEDGMENTS

This work has been supported by grants from the Wellcome Trust, Medical Research Council and the Scottish Home and Health Department.

REFERENCES

Anderson Brown, A.F., 1982, *The Incubation Book*. Saiga Publishing Co, Surrey, UK. pp 1-245.

Barnhill, R.L., and Ryan, T.J., 1983, Biochemical modulation of angiogenesis in the chorioallantoic membrane of the chick embryo. *J. Invest. Dermatol* 81: 485-488.

Clinton, M., Duncan, J.I., Long, W.F., Williamson, F.B., and Thompson, W.D., 1988, Effect of mast cell activator, Compound 48/80, and heparin on angiogenesis in the chick chorioallantoic membrane. *Int. J. Microcirc: Clin. Exp.* **7**: 315-326.

Duncan, J.I., Brown, F.I., McKinnon, A., Long, W.F., Williamson, F.B., and Thompson, W.D., 1992, Patterns of angiogenic response to mast cell granule constituents. *Int. J. Microcirc: Clin. Exp.* **11**: 21-33.

Folkman, J., 1974, Proceedings: Tumor angiogenesis factor. *Cancer Res.* **34**:2109-2113.

Hudlická, O., and Tyler, K.R.,1986, *Angiogenesis: the Growth of the Vascular System.* Academic Press Inc, London. pp 24-25

Kirchner, L.M., Schmidt, S.P., and Gruber, B.S., 1996, Quantitation of angiogenesis in the chick chorioallantoic membrane using fractal analysis. *Microvasc. Res.* **51**: 2-14.

Lucas, A.M., and Jamroz, C., 1961, *Atlas of Avian Hematology.* Agriculture Monograph 25. Washington: US Department of Agriculture.

Maniatis, T., Fritsch, E.F., and Sambrook, J., 1982, *Molecular Cloning – a Laboratory Manual.* Cold Spring Harbor Laboratories, Cold Spring Harbor.

Maragoudakis, M.E., Panoutsacopoulou, M., and Sarmonika, M., 1988, Rate of basement membrane biosynthesis as an index to angiogenesis. *Tissue Cell* **20**: 531-539.

Maragoudakis, M.E., Haralabopoulos, G.C., Tsopanoglou, N.E., and Pipili-Synetos, E., 1995, Validation of collagenous protein synthesis as an index for angiogenesis with the use of morphological methods. *Microvasc. Res.* **50**: 215-222.

Nguyen, M., Shing, Y., and Folkman, J., 1994, Quantitation of angiogenesis and antiangiogenesis in the chick embryo chorioallantoic membrane. *Microvasc. Res.* **47**: 31-40.

Parsons-Wingerter, P., Lwai, B., Yang, M.C., Elliot, K.E., Milaninia, A., Redlitz, A., Clark, J.I., and Sage, E.H., 1998, A novel assay of angiogenesis in the quail chorioallantoic membrane: stimulation by bFGF and inhibition by angiostatin according to fractal dimension and grid intersection. *Microvasc. Res.* **55**: 201-214.

Peek, M.J., Norman, T.M., Morgan, C., Markham, R., and Fraser, I.S., 1988, The chick chorioallantoic membrane assay; an improved technique for the study of angiogenic activity. *Exp. Pathol.* **34**: 35-40.

Ribatti, D., Gualandris, A., Bastaki, M., Vacca, A., Iurlaro, M., Roncali, L., and Presta, M., 1997, *J. Vasc. Res.* **34**: 455-462.

Ryan, T.J., and Barnhill, R.L., 1983, Physical factors in angiogenesis. In *Development of the Vascular System.* Ciba Symposium 100. Ed J Nugent, M O'Conner. Pitman Books, London. pp.80-89.

Splawinski, J., Michna, M., Palczak, R., Konturek, S., and Splawinska, B., 1988, Angiogenesis: quantitative assessment by the chick chorioallantoic membrane assay. *Methods Find. Exp. Clin. Pharmacol.* **10**: 221-226.

Stirk, C.M., and Thompson, W.D., 1990, Artificial exudate: stimulation of cell proliferation by plasma not serum is associated with fibrinolysis. *Blood Coag. Fibrinol.* **1**: 537-541.

Stirk, C.M., Kochhar, A., Smith, E.B., and Thompson, W.D., 1993 Presence of growth-stimulating fibrin degradation products containing fragment E in human atherosclerotic plaques. *Atherosclerosis* **103**: 159-169.

Thompson, W.D., Campbell, R., and Evans, A.T., 1985, Fibrin degradation and angiogenesis: quantitative analysis of the angiogenic response in the chick chorioallantoic membrane. *J. Pathol.* **145**: 27-37.

Thompson, W.D., Evans, A.T., and Campbell, R., 1986, The control of fibrogenesis: stimulation and suppression of collagen synthesis in the chick chorioallantoic membrane with fibrin degradation products, wound extracts and proteases. *J. Pathol.* **148**: 207-215.

Thompson, W.D., and Brown, F.I.,1987, Measurement of angiogenesis: mode of action of histamine in the chick chorioallantoic membrane is indirect . *Int. J. Microcirc.* **6**: 343-357.

Thompson, W.D., McGuigan, C.J., Snyder, C., Keen, G.A., and Smith, E.B., 1987, Mitogenic activity in human atherosclerotic lesions. *Atherosclerosis* **66**: 85-93.

Thompson, W.D., Kazmi, M.A., 1989, Angiogenic stimulation compared with angiogenic reaction to injury: distinction by focal and general application of trypsin to the chick chorioallantoic membrane. *Brit. J. Exp. Pathol.* **70**:627-635.

Thompson, W.D., Smith, E.B., Stirk, C.M., and Kochhar, A., 1990, Atherosclerotic plaque growth: presence of stimulatory fibrin degradation products. *Blood Coag. Fibrinol.* **1**: 489-493.

Thompson, W.D., Harvey, J.A., Kazmi, M.A., and Stout, A.J., 1991, Fibrinolysis and angiogenesis in wound healing. *J. Pathol.* **165**: 311-318.

Thompson, W.D., Smith, E.B., Stirk, C.M., Marshall, F.I., Stout, A.J., and Kocchar, A., 1992, Angiogenic activity of fibrin degradation products is located in fibrin fragment E. *J. Pathol.* **168**: 47-53.

Thompson, W.D., Smith, E.B., Stirk, C.M., and Wang, J., 1993, Fibrin degradation products in growth stimulatory extracts of pathological lesions. *Blood Coag. Fibrinol.* **4**: 113-116.

Thompson, W.D., McNally, S.J., Ganesalingam, N., McCallion, D.S.E., Stirk, C.M., and Melvin, W.T., 1996, Wound healing, fibrin and angiogenesis. In *Molecular, Cellular and Clinical Aspects of Angiogenesis.* (M. A. Maragoudakis, ed) Plenum Press, New York, pp.161-172.

Angiogenesis and Malignant Transformation

MYCN ONCOGENE AND ANGIOGENESIS: DOWN-REGULATION OF ENDOTHELIAL GROWTH INHIBITORS IN HUMAN NEUROBLASTOMA CELLS

PURIFICATION, STRUCTURAL, AND FUNCTIONAL CHARACTERIZATION

Elissavet Hatzi[1], Stephen Breit[2], Andreas Zoephel[3], Keith Ashman[4], Ulrike Tontsch[3], Horst Ahorn[3], Carol Murphy[1], Lothar Schweigerer[2] *, Theodore Fotsis[1] *

1 Lab. of Biological Chemistry, Medical School, University of Ioannina, Greece
2 Department of Hematology / Oncology and Endocrinology, Children's Hospital, University of Essen, Germany
3 Bender-Co Gmbh, Boehinger Ingelheim R+D Vienna, Austria
4 EMBL, Heidelberg, Germany
** These authors contributed equally to this work.*

Key words: angiogenesis, malignant trasnformation, oncogene, MYCN, neuroblastoma, endothelial cell, inhibitors, activn A

Abstract: Angiogenesis, the formation of new blood vessels, is seen during embryonic development and tumor progression, but the mechanisms have remained unclear. Recent data indicate that tumor angiogenesis can be induced by cellular oncogenes, leading to the enhanced activity of molecules stimulating angiogenesis. However, activated oncogenes might also facilitate angiogenesis by down-regulating endogenous inhibitors of angiogenesis. We report here that enhanced expression of the N-myc oncogene in human neuroblastoma cells down-regulates three inhibitors of endothelial cell proliferation. One of them was identified by amino acid sequencing as being identical with activin A, a developmentally-regulated protein. Down-regulation involves interaction of the N-myc protein with the activin A promoter. Work is ongoing to characterize the other two endothelial cell inhibitors. We suggest that the N-myc induced down-regulation of angiogenesis inhibitors could contribute to tumor angiogenesis.

1. INTRODUCTION

Angiogenesis, the generation of new capillaries from pre-existing vessels, is a tightly regulated and self-limiting physiological processes. However, in certain pathological conditions, angiogenesis is dramatically enhanced and loses its self-limiting capacity. Uncontrolled production of vessels is seen during the development and progression of many diseases, such as rheumatoid arthritis, psoriasis, and diabetic retinopathy. One of the most important manifestations of pathological angiogenesis is that induced by solid tumors. Well vascularized tumors expand both locally and by metastasis while avascular tumors do not grow beyond a diameter of 1 to 2 mm [1,2].

The pattern of new blood vessel growth in transgenic mouse models and human malignancies argues that the angiogenic switch is a discrete process which is an essential part of the phenotypic repertoire that characterizes the successful tumor [3]. This switch is often activated during early, preneoplastic stages in the development of a tumor and is subject to specific regulatory controls. Indeed, in situ solid malignancies may exist for months or years without neovascularization and as a consequence they remain limited to a volume of few cubic millimeters. Suddenly, a group of malignant cells (approx. 4-10%) acquire the angiogenic phenotype, induce angiogenesis and finally lead to the formation of a solid, malignant tumor [4].

Though the mechanisms responsible for the persistent angiogenesis are still unclear and certainly not the same in each disease, recent evidence indicates that they lead to an imbalance between angiogenic stimulators and inhibitors [3,5]. But what triggers these changes and activates angiogenesis in malignant cells ? Neoplastic transformation is a multistep process [6] in which sequentially accumulated activation of proto-oncogenes and / or inactivation of tumor suppressor genes [7] leads to increased loss of normal control mechanisms. It is conceivable that tumor angiogenesis is the result of such genetic changes. Indeed, normal p53 appears to regulate the expression of the angiogenesis inhibitors thrombospondin [8,9] and glioma-derived angiogenesis inhibitory factor [10]. Up-regulation of the expression of angiogenesis stimulators like basic fibroblast growth factor (bFGF) or vascular endothelial growth factor (VEGF), has been reported to be induced by activation of oncogenes [11]. Despite these initial observations, the molecular mechanisms involved in angiogenesis activation by the genetic alterations of tumors are still unclear.

An interesting candidate to study the role of oncogene activation in initiation of angiogenesis is the MYCN oncogene. Amplification of MYCN is a frequent event in advanced stages (III and IV) of human neuroblastomas [12]. MYCN amplification correlates with poor prognosis [13] and enhanced vascularization [14] of human neuroblastomas, suggesting that the MYCN oncogene could stimulate tumor angiogenesis and thereby allows neuroblastoma progression. We have previously developed an *in vitro* model for enhanced expression of MYCN in human neuroblastoma cells [15]. In this system, we have stably transfected a neuroblastoma cell line (SH-EP) with the MYCN oncogene and have increased the expression of the latter by 100-fold in the transfected cells (WAC 2) [15]. The WAC 2 cells acquired an enhanced proliferative potential and, unlike the SH-EP 007 cells (the control transfectants), were able to form colonies in soft agar and induce well vascularized tumors in nude mice [15].

2. RESULTS

2.1 MYCN overexpression down-regulates three inhibitors of endothelial cell proliferation

We have screend conditioned media from SH-EP 007 and WAC 2 cells for differentially expressed inducers and inhibitors of the *in vitro* proliferation of endothelial cells [16]. We have reasoned that differentially expressed angiomodulators would strongly suggest that these molecules are regulated by MYCN oncogene and thus would provide information regarding the molecular mechanisms linking oncogene activation in malignant transformation with initiation of tumor angiogenesis. Towards this purpose, we subjected supernatants of human neuroblastoma cells with normal (SH-EP007) or enhanced (WAC2) MYCN expression to various chromatographic purification procedures. At each stage, the fractionated supernatants were examined for their abilities to either stimulate or inhibit the proliferation of capillary endothelial cells in order to demonstrate potential stimulators or inhibitors of angiogenesis. This resulted in highly reproducible elution profiles of endothelial cell growth modulators both on the qualitative and quantitative level. The endothelial cell modulators were named according to their origin (S = SH-EP007, W = WAC2) and abilities to either inhibit (I) or stimulate (S) capillary endothelial cell proliferation. Analysis of supernatants of human neuroblastoma cells with normal MYCN expression (SH-EP 007) by cation exchange chromatography identified one stimulator (SS.1) and one inhibitor (SI.4) as a bFGF-like molecule and transforming growth factor β-1 (TGF-β1), respectively, and three additional inhibitors (SI.1, SI.2 or SI.3) of unknown identity (16). In contrast, WAC 2 supernatants contained a bFGF-like molecule and TGF-,1 at concentrations identical to those in the SH-EP 007 cells, but ehhibited only minor or undetectable quantities of SI.1, SI.2 and SI.3 [16]. This suggested that SI.1, SI.2 and SI.3 were down-regulated in WAC 2 cells by the enhanced N-myc expression.

Regarding the quantitative comparison of the inhibitors between SH-EP 007 and WAC 2 cells, we have subjected to cation exchange chromatography repeatedly identical volumes of CMs (3 to 5 liters). Throughout the collections, in which identical volumes of serum-free medium per collecting surface were used in both cell lines, WAC 2 cells achieved roughly less than double the final density and had a bit more than double protein concentration in CM comparing to SHEP-007 cells. Despite the increased number of cells and protein concnetration during collections of CM, we were unable to detect or found negligible activity of SI.1, SI.2, and SI.3 in any of the batches of WAC 2 CM. Thus, SI.1, SI.2 and SI.3 appear to be down-regulated dramatically in WAC 2 CM.

1.2 SI.3 is identified as activin A and is regulated by MYCN oncogene

The active fractions containing SI.3 were pooled and subjected to chromatofocusing chromatography (Fig. 1A) with a pH-gradient from 8.3 to 5.0. The SI.3 inhibitory activity was resolved into three further subactivities with IPs of 8.0, 7.0 and 6.5 (Fig. 1A). The same protein band was found in all active fractions, indicating a distribution of the same protein in several isoelectric points during chromatofocusing purification. The inhibitory activity with isoelectric point 7.0 was further electrophoresed using preparative SDS-PAGE gel. Coomassie staining revealed the well separated 23 kD band which was excised for amino acid sequence analysis (Fig. 1B).

The excised band was digested by chymotrypsin, and the proteolytic fragments were sequenced by Edman degradation. We obtained 7 peptides of 5 to 14 amino acid length corresponding to a total of 68 amino acids. A

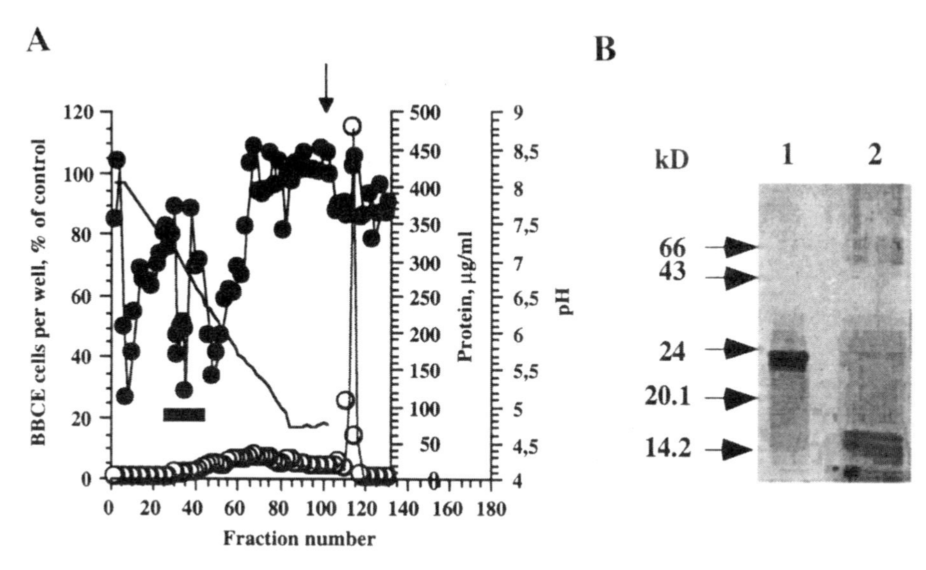

Figure 1. Chromatofocusing chromatography (A) and SDS-PAGE analysis of SI.3 (B).

comparison of their sequences with those of known protein sequences showed them to be identical with partial sequences of the human inhibin beta A, but not inhibin beta B or inhibin alpha, gene product (Fig. 2). Since, in addition to the reported 7 peptides no other sequences were obtained, we conclude that SI.3 is very closely related and probably identical to a homodimer of inhibin A subunits therefore representing human activin A.

As outlined above, human neuroblastoma cells with enhanced N-myc expression (WAC2) appeared to have lost or severely reduced their ability to synthesize endothelial cell growth inhibitors including activin A. To determine whether the correlation of enhanced N-myc expression with a reduced activin A expression was a general phenomenon, we examined N-myc and activin A expression in a series of human neuroblastoma cell lines with various geno- and phenotypes using a multiplex RT-PCR approach. Indeed, activin A transcripts were detected in all neuroblastoma cell lines with normal N-myc expression. In contrast, no activin A transcripts were detectable in cell lines with an enhanced N-myc expression, in neuroblastoma cell lines with N-myc amplification and in stable transfectants with constitutively high N-myc expression (data not shown).

INBA ---MP---------LLWLRGFLLASCWIIVRSSPT---------PGSEGHSAAPDCPSCA 39

INBB MDGLPGRALGAACLLLLAAGWLGPEAWGSPTPPPTPAAPPPPPPPGSPGGSQD-TCTSCG 59

INA ---------------------MVLHLLLFLLLTPQ----------G--GHS----CQGLE 23

INBA LAALPKDVPNSQPEMVEAVKKHILNMLHLKKRPDVTQPVPKAALLNAIRKLHVG-KVGEN 98

INBB GFRRPEELGRVDGDFLEAVKRHILSRLQMRGRPNITHAVPKAAMVTALRKLHAG-KVRED 118

INA LAR-ELVLAKVRALFLDALGPPAVTRE--GGDPGVRR-LPRRHALGGF--THRGSEPEEE 177

INBA GYVEIEDDIGRRAEMNELMEQTSEIITFAESG---TARKTLHFEISKEGSDLSVVERAEV 155

INBB GRVEIPHLDGHASPGADGQERVSEIISFAETDGLASSRVRLYFFISNEGNQNLFVVQASL 178

INA EDVSQAILFPATDASCEDKSAARGLAQEAEEG-----LFRYMFRPS-QHTRSRQVTSAQL 131

INBA WLFLK-VPKANR--TRTKVTIRLFQQQKHPQGSLDTGEEAEEVGLKGERSELLLSEKVVD 212

INBB WLYLKLLPYVLEKGSRRKVRVKVYFQE---QG-------------HGDR--WNMVEKRVD 220

INA WFHTGLDRQGTAASNSSEPLLGLLALS---PG--------------GP----VAVPMSLG 170

INBA ARKSTWHVFPVSSSIQRLLDQGKSSLDVRIACEQCQESG---ASLVLLGKKKKKEEEGEG 269

INBB LKRSGWHTFPLTEAIQALFERGERRLNLDVQCDSCQELA---VVPVFVD-------PGE- 269

INA HAPPHWAVLHLATSALSLLTHP--VLVLLLRCPLCTCSARPEATPFLVAHTRTRPPSGG- 227

INBA KKKGGGEGGAGADEEKEQSHRPFLMLQARQSEDHPHRRRRR**GLECDGKVNI**CCKKQFFVS 329

INBB -----------------ESHRPFVVVQARLG-DSRHRIRKRGLECDGRTNLCCRQQFFID 311

INA -----------------ERARRSTPLMSWPWSPSALRLLQRPPEEPAAHANCHRVALNIS 270

INBA F**KDIGWND**W**IIAPSGY**HANY**CEGECPSHIAGTSG**SSLSF**HSTVINHYRMRGHSP**FANLKS 389

INBB FRLIGWNDWIIAPTGYYGNYCEGSCPAYLAGVPGSASSFHTAVVNQYRMRGLNP-GTVNS 370

INA FQELGWERWIVYPPSFIFHYCHGGCGLHIP--PNLSLPVPGAPPTPAQPYSLLP--GAQP 326

INBA CC--VPTKLRPMSMLY**YDDG**-**QNIIKKDIQN**MIVEECGCS 426

INBB CC--IPTKLSTMSMLYFDDE-YNIVKRDVPNMIVEECGCA 407

INA CCAALPGTMRPLHVRTTSDGGYSFKYETVPNLLTQHCACI 366

Figure 2. Amino acid sequences of SI.3 peptides and comparison with inhibin beta A, inhibin beta B or inhibin alpha chains. Dashes indicate a gap; sequences of the SI.3 peptides are in bold and underlined.

Transcripts of activin beta or inhibin alpha were not detected in either of the cells (not shown).

The inverse correlation of N-myc and activin A expression suggested that N-myc could down-regulate activin A expression by repressing activin A promoter activity. We therefore transiently transfected human 293 embryonic kidney cells with a construct of the human genomic activin A gene promoter linked to the reporter gene bacterial chloramphenicol acetyl transferase (CAT) [17]. Subsequently, the activin A-CAT-construct was co-transfected with either the N-myc expression vector pNMYC which is known to induce high N-myc expression, or with the control vector pNMYC$^{d(351-387)}$ encoding a mutant N-myc protein which, due to a deletion of its DNA-binding domain, is unable to bind to DNA [18]. Co-transfection

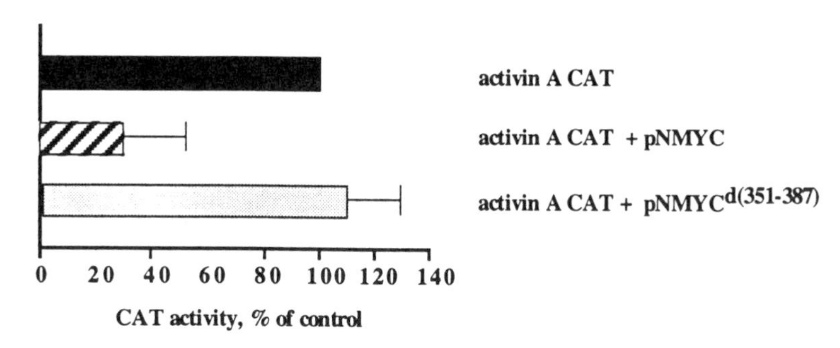

Figure 3. Effect of N-myc on expression and promoter activity of activin A

of the activin A promoter-reporter construct and the control N-myc vector pNMYC$^{d(351-387)}$ had no influence on CAT activity. In contrast, co-transfection of the activin A promoter-reporter construct and the functional N-myc vector reduced CAT activity to 29% of controls (Fig. 3) indicating that N-myc regulates activin A expression by repressing its promoter activity.

1.3 SI.1 is a glycosylated protein of approx. 45 Kda molecular size

The flow-through of the cation exchange step containing the SI.1 activity was further purified by a ConA Sepharose column (Fig. 4A). Small scale experiments had already indicated to us that this inhibitor is glycosylated. ConA Sepharose chromatography provided a well separated peak eluted in the methylpyranoside gradient (Fig. 4A). Though a considerable degree of purification had been achieved with this column (very little protein in the active fraction in Fig. 4A), SI.1 was not pure (data not shown). The active fractions from the ConA Sepharose column were subjected preparative SDS-PAGE in tubes followed by reversed phase HPLC. The SI.1 activity could be renatured following both these high resolution steps greatly facilitating the purification of this molecule (data not shown). Analytical SDS-PAGE of the active fractions revealed a broad band at Mr around 45 kD (Fig. 4B). In the near future, this band will be excised from a preparative SDS-PAGE gel for amino acid sequence analysis.

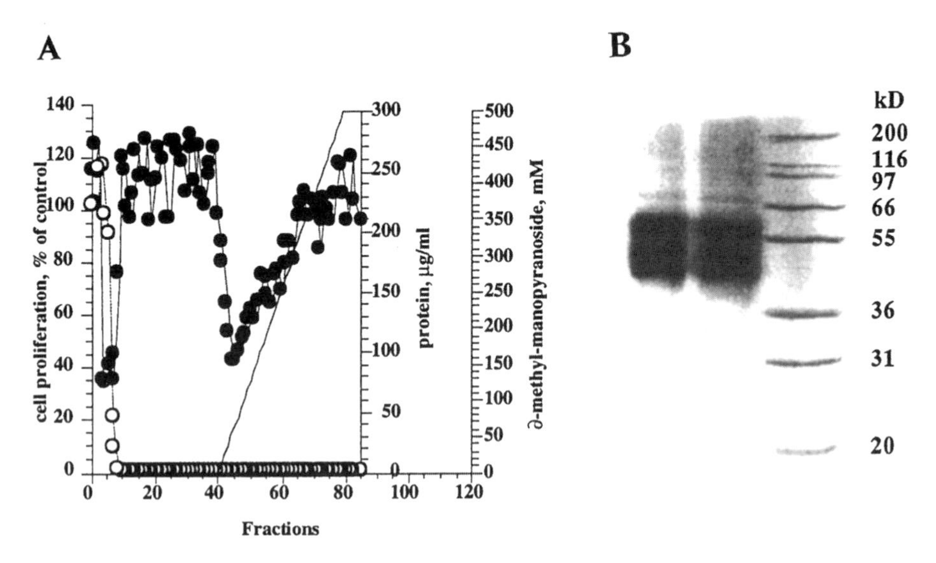

Figure 4. ConA Sepharose chromatography (A) and SDS-PAGE analysis of SI.1 (B).

2. CONCLUSIONS – FUTURE WORK

Genetic alterations of tumor cells appear to contribute to the initiation of tumor angiogenesis, either by up-regulating stimulators or down-regulating inhibitors of angiogenesis or both. In the present study, we have attempted to define the angiogenic profile induced in neuroblastomas by the enhanced expression of the MYCN oncogene. We have observed that MYCN overexpression in neuroblastoma cells resulted in an almost complete down-regulation of three inhibitors, SI.1, SI.2 and SI.3, of endothelial cell proliferation.

These results clearly demonstrate that overexpression of an oncogene can cause down-regulation of endothelial cell proliferation inhibitors. Our data corroborate the hypothesis that changes in the relative concentrations of stimulators and inhibitors of angiogenesis can activate the angiogenic switch [3]. Down-regulation of inhibitors of endothelial cell proliferation in neuroblastoma cells by enhanced MYCN expression might have important implications in the progression of this solid tumor. It may permit neovascularization and progression of neuroblastromas and may explain the previously reported correlation of MYCN application with increased neuroblastoma vascularization [14] and poor prognosis [13].

Elucidation of the molecules and mechanisms involved in the initiation of angiogenesis in solid malignancies will provide targets for therapeutic intervention and open the horizon for an anti-angiogenic approach for the treatment of solid tumors. The list of angiogenesis inhibitors is constantly increasing and there are already nine compounds in clinical trials as anti-angiogenic agents [3]. Elucidation of the identity of the SI.1, SI.2, and SI.3 molecules, which is underway, is expected to contribute in understanding the regulation of the angiogenic switch in tumors and providing further tools in anti-angiogenic therapeutic strategies.

ACKNOWLEDGMENTS

This work was supported by a grant from Bender-Co Gmbh/Boehringer Ingelheim R+D Vienna (to T.F. and L.S.). We thank L. Kyrkou, R. Frenk and G. Frank for technical assistance.

REFERENCES

[1] Klagsbrun, M. & Folkman, J. (1990) Angiogenesis. In Peptide growth factors and their receptors II (Sporn, M.B., & Roberts A.B., eds], pp. 549-586. Springer Verlag, Berlin.
[2] Folkman, J. & Cotran, R.S. (1976) Relation of vascular proliferation to tumor growth. Int. Rev. Exp. Pathol., 16, 207-248.
[3] Hanahan, D. & Folkman, J. (1996) Patterns and emerging mechanisms of the angiogenic switch during tumorigenesis. Cell, 86, 353-364.
[4] Folkman, J. (1995) Angiogenesis in cancer, vascular, rheumatoid and other disease. Nature Medicine, 1, 27-31.
[5] Liotta, L.A., Steeg, P.S. & Stetler-Stevenson, W.G. (1991) Cancer metastasis and angiogenesis: An imbalance of positive and negative regulation. Cell, 64, 327-336.
[6] Vogelstein, B. & Kinzler, K.W. (1993) The multistep nature of cancer. Trends Genet, 9, 138-141.
[7] Bishop, J.M. (1987) The molecular genetics of cancer. Science, 235, 305-311.
[8] Dameron, K.M., Volpert, O.V., Tainsky, M.A. & Bouck, N. (1994) Control of angiogenesis in fibroblasts by p53 regulation of thrombospondin-1. Science, 265, 1582-1584.
[9] Volpert, O.V., Stellmach, V. & Bouck, N. (1996) The modulation of thrompospondin and other naturally occuring inhibitors of angiogenesis during tumor progression. Breast Cancer Res. Treat., 36, 119-126.

[10] Van Meir, E.G., Polverini, P.J., Chazin, V.R., Su Huang, H.-J., De Tribolet, N. & Cavenee, W.K. (1994) Release of an inhibitor of angiogenesis upon induction of wild type p53 expression in glioblastoma cells. Nature Genet., 8, 171-176.

[11] Bouck, N., Stellmach, V. & Hsu, S.C. (1996) How tumors become angiogenic. Adv. Cancer Res., 69, 135-174.

[12] Brodeur, G.M., Seeger, R.C., Schwab, M., Varmus, H.E. & Bishop, J.M. (1984) Amplification of MYCN in untreated human neuroblastomas correlates with advanced disease stage. Science, 224, 1121-1124.

[13] Grady-Leopardi, E., Schwab, M., Ablin, A. & Rosenau, W. (1986) Detection of NMYC oncogene expression in human neuroblastoma by in situ hybridization and blot analysis: Relationship to clinical outcome. Cancer Res., 46, 3196-3199.

[14] Meitar, D., Crawford, S.E., Radmaker, A.W. & Cohn, S.L. (1996) Tumor angiogenesis correlates with metastatic disease, N-myc amplification, and poor outcome in human neuroblastoma. J. Clin. Oncol., 14, 405-414.

[15] Schweigerer, L., Breit, S., Wenzel, A., Tsunamoto, K., Ludwig, R. & Schwab, M. (1990) Augmented MYCN expression advances the malignant phenotype of human neuroblastoma cells: Evidence for induction of autocrine growth activity. Cancer Res., 50, 4411-4416.

[16] Fotsis, T., Breit, S., Lutz, W., Rössler, J., Hatzi, E., Schwab, M., & Schweigerer, L. (1999) Down-regulation of endothelial cell growth inhibitors by enhanced MYCN oncogene expression in human neuroblastoma cells. Eur. J. Biochem. (in press).

[17] Tanimoto, K., Yoshida, E., Mita, S., Nibu, Y., & Fukamizu, A. (1996) Human activin A gene. J. Biol. Chem., 271, 32760-32769.

[18] Wenzel, A., Csiepluch, C., Hamann, U., Schürmann, J., & Schwab, M. (1991) The N-myc oncoprotein is associated *in vivo* with the phosphoprotein Max (p20/22) in human neuroblastoma cells. EMBO J., 10, 3703-3712.

FROM HYPERPLASIA TO NEOPLASIA AND INVASION: ANGIOGENESIS IN THE COLORECTAL ADENOMA - CARCINOMA MODEL

Kaklamanis L[1*], Kakolyris S[2*], Koukourakis M[2+], Gatter K.C*, Harris A.L[+]
[1]Department of Pathology, Onassis Cardiac Surgery Centre, Athens, Greece, [2]Clinical Oncology and Radiotherapy, University of Heraklion, Crete, Greece, [*]Nuffield Department of Clinical Biochemistry and Cellular Science, [+]ICRF Clinical Oncology Unit, John Radcliffe Hospital, University of Oxford, OX3 9DU, UK.

1. INTRODUCTION

Tumor angiogenesis, the formation of new blood vessels from the existing vascular network, is a complex multistep process, essential in tumor growth and metastatic dissemination *[Blood and Zetter,1990]*. It involves endothelial cell migration and proliferation, remodelling of the extracellular matrix, expansion, differentiation and anastomosis of the capillary network *[Paweletz et al. 1989]*. Since the very first reports by Folkman *[Folkman 1986; Folkman 1990]* who demonstrated that neoplastic cell proliferation and further tumor growth are angiogenesis dependent, a large number of peptides has been implicated in the regulation of this process, either as molecules inducing the expansion of the vascular bed or as anti-angiogenic factors downregulating this vasculature *[Bicknell and Harris 1991]*.

One of these angiogenic factors is platelet-derived endothelial cell growth factor (PDECGF). This molecule was initially described as angiogenic and present in platelets, distinct from other endothelial cell growth factors because of its unique sequence and lack of heparin binding *[Ishikawa et al. 1989]*. It was described both as angiogenic in model systems and as mitogenic for endothelial cells. Recently PDECGF was shown to be thymidine phosphorylase (TP), an enzyme which catalyses the reversible phosphorolytic pathway of thymidine to deoxyribose-1-phosphate and thymine *[Moghaddam et al. 1995; Moghaddam and Bicknell 1992; Usuki et al. 1992; Barton et al. 1992; Finnis et al. 1993]*. It may potentially affect proliferation since high thymidine concentrations inhibit cell growth in vitro and TP can degrade thymidine, allowing cell growth *[Moghaddam and Bicknell 1992]*.

Its effect on incorporation of thymidine into HUVECs is due not to stimulation of de novo DNA synthesis but is caused by the direct effect of the enzyme on the availability of extracellular thymidine *[Moghaddam and Bicknell 1992]*. TP transfected into transformed fibroblasts in nude mice increased the tumor vasculature *[Ishikawa et al. 1989]*. TP was also found to be angiogenic in the rat sponge angiogenesis assays and in a freeze-injured skin graft model *[Moghaddam et al. 1995]*. Although the mechanism by which TP modulates angiogenesis is not known, the metabolite deoxyribose has been shown to be angiogenic *[Haraguchi et al. 1994]*.

TP levels were shown to be elevated in the plasma of cancer patients. Its overexpression in ovarian tumors was strongly correlated with ovarian malignancy and increased blood flow in these tumors *[Reynolds et al. 1994]*. High levels of TP have been described in other tumor types such as liver, genitourinary, haemopoietic and gastrointestinal malignancies *[Vertongen et al. 1984; Yoshimura et al. 1990; Kono et al. 1984]*.

In most of these studies only a few cases were analysed using tumor homogenates to detect the expression of the enzyme and there was no information regarding the tissue distribution and the cell types expressing it.

Overexpression of TP in MCF-7 breast carcinoma cells had no growth effect in vitro but in vivo tumor growth was enhanced. Recently, using recombinant TP, a reliable monoclonal antibody was produced (P-GF44C) and the tissue distribution of TP was described *[Fox et al. 1995]*. Also, using immunocytochemistry and ribonuclease protection analysis techniques, TP expression was observed in 55% of breast tumor cells along with stromal, inflammatory and endothelial elements, inversely correlated with tumor grade and size *[Fox et al. 1996]*.

As normal cells undergo a variety of genetic alterations, which lead to neoplastic transformation, they have to reduce the effects of anti-angiogenic effects and to switch to an angiogenic phenotype. Thrombospondin-1 (TSP-

1), a large, multifunctional matrix glycoprotein *[Asch and Nachman 1989]* is regarded as a potent inhibitor of neovascularization *[Rastinejad et al. 1989.]*

Recently it was shown that in cultured fibroblasts from Li-Fraumeni patients, wild type (wt) p53 inhibits the angiogenic phenotype by stimulating the production of endogenous TSP-1 *[Dameron et al. 1994]*. When these cells lost the wt allele, secreted reduced amounts of TSP-1 and switched to an angiogenic phenotype. Since mutations of the p53 gene are the most common genetic alterations in human tumors it would be of interest to analyse the effect of p53 on angiogenesis.

During the last few years an increasing number of studies have shown, by different quantitative methods, that angiogenesis is a powerful prognostic tool in a variety of human tumor types.

In patients with breast carcinoma, tumor vascularity provides independent prognostic information for relapse-free and overall survival *[Bosari et al. 1992; Horak et al. 1992]* information for stratifying patients requiring different treatments and response to chemotherapy *[Protopapa et al. 1995]* and also identifying the pre-invasive phase of neoplastic benign diseases *[Fregene et al. 1994]*.

In colorectal cancer cases, tumor vascularity detected by microvessel counting, was an early event during tumor progression and significantly higher in carcinomas than in adenomas *[Bossi et al. 1995; Vermeulen et al. 1995]*. However microvessel density did not provide any significant prognostic information or correlation with disease stage or metastases *[Vermeulen et al. 1995]*.

This study was undertaken in order to analyse in a large series of colorectal tumors:

i) the angiogenic profile during the colorectal neoplastic transformation, including solitary adenomas, adenomatous and carcinomatous areas and transitional zones from the same tumor,

ii) its relationship to clinicopathological variables,

iii) the differential expression of thymidine phosphorylase in normal mucosa, solitary adenomas and carcinomas arising from adenomas,

iv) the tissue distribution of TP expression in relation to angiogenic areas of the tumor and

v) the p53 overexpression in relationship to the angiogenic profile of the neoplasms.

2. CLINICAL MATERIALS AND METHODS

2.1 Tissue Specimens

Representative samples were taken from paraffin-embedded archival files including 16 hyperplastic polyps, 35 solitary tubular and tubulovillous adenomas (ranging from 1cm-7.5cm, median 3.2cm) and 47 cases of sporadic colorectal carcinomas arising on the basis of preexisting adenomas (25 cases were Dukes' A, 10 Dukes' B and 12 Dukes' C).

Sections of non-neoplastic colonic mucosa were also examined separately from the above cases. Samples from both the adenomatous and the carcinomatous areas from the carcinoma cases were included in the analysis. All samples were stained for TP and assessed for vascularity.

2.2 Antibodies and immunohistochemistry

This was performed on formalin-fixed parrafin-embedded 3mm thick sections cut onto coated slides. The monoclonal antibodies PG44c-recognising TP previously validated by western blotting, absorption with recombinant protein and labeling of transfectants *[Fox et al. 1995]*, JC70A against CD31 (endothelial marker-DAKO, code: M 0823) and p53 DO7 (DAKO, code: M 7001) were used. Staining was performed using the streptavidin-biotin peroxidase method with the PG44c and DO7 antibodies and the APAAP method for JC70A *[Mason et al. 1983]*. Predigestion with 12.5mg of protease type XXIV (Sigma, Poole, UK) in 100ml of phosphate-buffered saline (PBS) for 20 min at 37^0C was necessary for optimal JC70A immunostaining but no treatment was required for TP.

2.3 Assessment of Tumor Vascularity

CHALKLEY COUNT

The three most vascular areas, where the highest number of microvessels were stained, were chosen over a conference microscope by two observers. Vessels within areas of abscess formation, granulation tissue reaction or extravasation of mucin, were not included. Microvascular spaces were defined as any immunoreactive endothelial cells separate from adjacent microvessels. These areas of vascularity were detected by scanning at low power and estimated by both observers using a 25-point Chalkley point

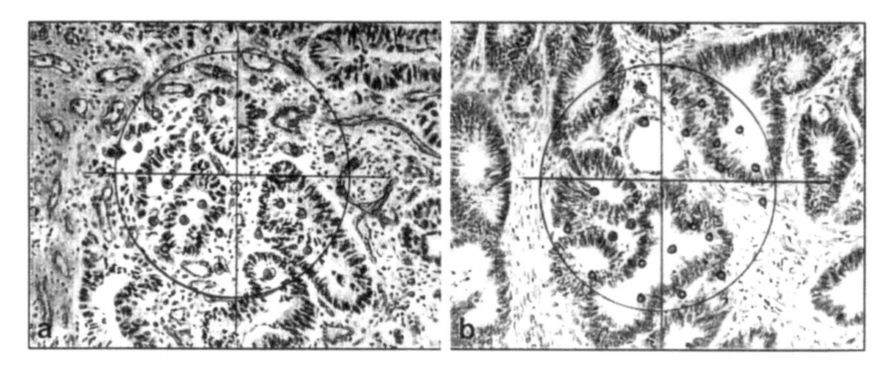

Figure 1: Rich (a) and poor (b) angiogenic tumors using the endothelial marker JC70. Application of the 25-point Chalkley count graticule.

eyepiece graticule *[Chalkley 1943]* at x250 magnification covering an area of 0.155 mm^2. The maximum number of graticule dots overlaying immunocyto-chemically detected vessels or their lumens, was counted. Vascular counts (VCs) were produced using the mean of the three graticule counts (Fig. 1.).

This method relies on the concept that the ratio of positive points (those on or within the vascular spaces) to negative will give the area fraction of the tumor vasculature occupied to the remainder of the tissue elements. This is because the ratio of the sums approaches that of the area fractions occupied by any individual component. The shape of the individual component does not influence this consideration.

2.4 Assessment of TP expression

All the samples were assessed for TP. Tumors were considered positive when more than 20% of cells were stained. The presence and distribution of TP was also documented in the stroma, the endothelial and inflammatory cells, in the transformational zone of adenomatous-carcinomatous areas and in the invading edge of the carcinomas.

2.5 Statistics

Statistical analysis and graphic presentation was performed using the Graph Pad Prism 2.01 package (USA). Chalkley score (CS), was used as a

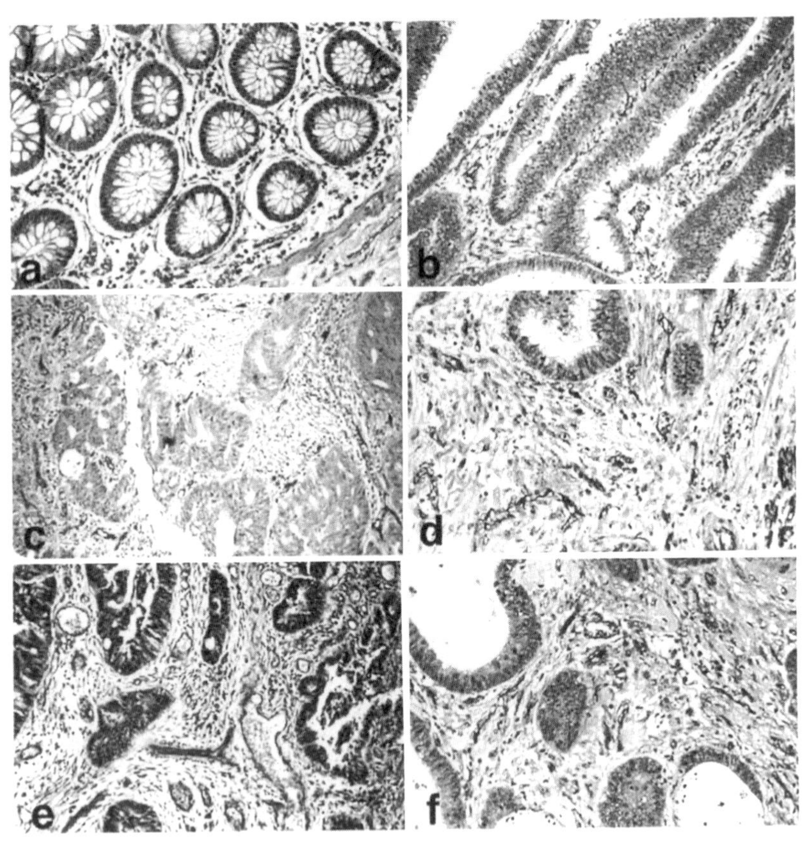

Figure 2: Angiogenic phenotypes in (a) normal mucosa, (b) adenoma, (c) carcinoma with low vascular count, (d) carcinoma with angiogenic invading edge, (e),(f) carcinomas with high vascular counts.

continous variable whilst p53 reactivity of as a categorical one. Two-tailed t-test with Welch's correction, Wilcoxon matched pairs test or x^2 with Yate's correction were used. Linear regression analysis was used to assess correlation between categorical tumor variables.

3. RESULTS

3.1 Tumor Vascularity in the colorectal adenoma carcinoma neoplastic progression

Non-neoplastic mucosa

Occasional vascular spaces are noted surrounding the colonic glands. Hot spots are not detected (Fig. 2a).

Hyperplastic polyps

Areas with high angiogenesis, the so-called "hot spots" were not identified. The Chalkley counts were performed in three randomly selected areas. The range was 1-4 (mean: 1.7, SD: 1.0).

Adenomas

Areas with increased vascular spaces were not identified. The range was 1-6 (mean: 3.1, SD: 1.5, Fig. 2b).

Carcinomas

Homogeneous staining pattern in CD31 positive endothelial cells was identified and Chalkley counts per x250 magnification showed a range from 3-14 (mean: 6.3, SD:2.8) in the carcinomatous areas (Fig. 2c-f). The adenomatous parts did not show high angiogenic areas and the Chalkley counts were mostly below 3 (range 1-8, mean: 3.7, SD: 1.9). Comparing the angiogenic profile of these two groups a statistical significant difference was observed ($p=0.0001$).

3.2 TP expression in normal mucosa, hyperplastic polyps and adenomas

The epithelial cells of the non-neoplastic mucosa were all negative and only occasional macrophages and fibroblasts showed weak cytoplasmic immunoreactivity for TP (Fig. 3a). The endothelial cells were mostly

Table 1: Clinicopathological features

	Hyperplastic polyps	Solitary. adenomas	Adenomatous area	Carcinomatous area
number	16	35	47	47
size	0.5cm	1-7.5cm (med:3.2)		2.7 - 8.9cm (med:4.5)
Dukes' stage A				25
Dukes' stage B				20
Dukes' stage C				12
Chalkley counts	mean:1.7	1-6 (mean:3.1)	1-8 (mean 3.7)	3-14 (mean: 6.3)
epithelium TP +	0	3	1	18
epithelium TP -	16	32	46	29
stromal TP +	scanty	focal	4	39
stromal TP -			43	8

negative. The same pattern was observed for the 16 hyperplastic polyps examined (Table 1.).

In the solitary adenomas there was no immunoreactivity of the neoplastic epithelial cells, apart from three cases in which weak staining was detected in few epithelial cells. Stromal macrophages, fibroblasts and occasional endothelial cells showed focal positivity with cytoplasmic and occasionally nuclear staining of variable intensity (Fig. 3b).

3.3 TP expression in carcinomas

Neoplastic cells in the carcinomatous part of the tumor were positive for TP in 18/47 (36%) cases. Both nuclear and cytoplasmic staining was detected but in a few cases only one of these was present. The staining pattern was heterogeneous, usually up-regulated at the infiltrating margin of the neoplasm, but occasionally focal. In 12 more cases scattered epithelial cells were positive and in the rest 17 (35%) none of them showed any immunoreactivity (Fig. 3 c,d,e,f).

Within the tumor and more pronounced at the infiltrating edge, tumor macrophages and stromal cells were strongly positive (Fig. 3g). Endothelial cell reactivity was also detected in 23 (48%) cases, especially in the areas described above although occasionally scattered positive endothelial cells

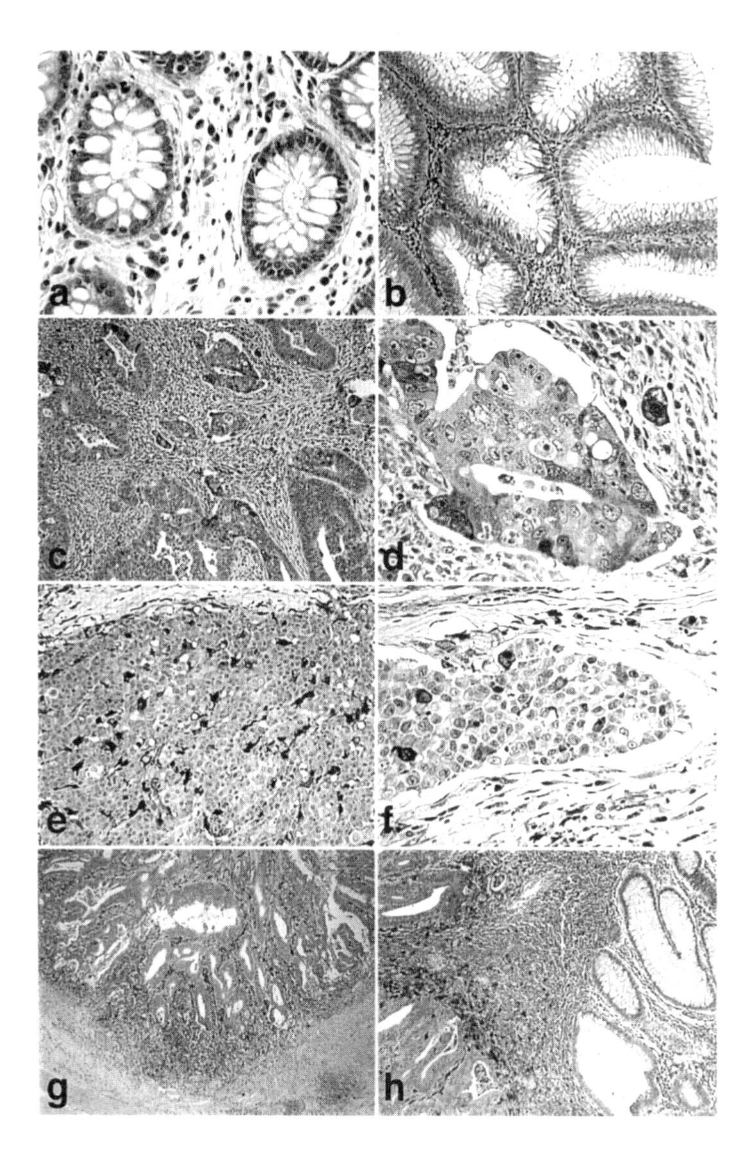

Figure 3: TP expression: a : normal mucosa with occasional positive macrophages and no labelling of the epithelial cells, b: adenoma, with unlabelled epithelium and positive macrophages, c, d : carcinoma: positive neoplastic and stromal cells, e: carcinoma: positive macrophages in between negative tumor cells, f : tumor embolus within a lymphatic with occasional positive cells, g: increased expression at the invading edge of the tumor stroma, h: The transitional area between an adenoma-scanty stromal positive cells (right) and carcinoma-increased stromal positivity.

were observed. The areas that showed neoplastic cell positivity were also heavily populated by labeled macrophages.

We did not detect any epithelial immunoreactivity in areas where the surrounding tumor stroma cells showed scanty or no labeling. On the other hand there were cases with pronounced stromal cell positivity which were not accompanied by any epithelial TP expression. In areas where there was strong inflammatory reaction and granulation tissue formation because of release of mucin, the stromal macrophages, inflammatory cells and fibroblasts showed intense and diffuse immunoreactivity (Fig. 3h).

Almost none of the neoplastic epithelial cells of the adenomatous part of the tumor showed any immunoreactivity . The staining of the stromal macrophages and fibroblasts was usually focal and much less compared to the stroma of the carcinomatous areas. In 4 cases there was multifocal, intense staining in macrophages and other stromal cells, in 39 cases occasional cells were positive and 4 cases completely negative.

Comparison of TP expression between adenomas and carcinomas

There was a highly significant difference in TP expression in neoplastic epithelial cells when adenomas were compared with carcinomas (p=0.0001). The same was true when the immunoreactivity of the stromal cells in these two neoplastic groups was compared (p=0.0001-Table 2).

Correlation of Dukes' stage with TP expression

There was no significant statistical correlation either between Dukes' stage and cases with positive neoplastic epithelial cells (p=0.65) or Dukes' stage and TP stromal expression (p=0.92).

Table 2: TP in adenomatous vs carcinomatous parts of tumors

	TP expression in neoplastic epithelial cells		
	TP positive	TP scanty	TP negative
adenomatous	0	1	46
carcinomatous	18	12	17
	p=0.0001		
	TP expression in neoplastic stromal cells		
adenomatous	4	39	4
carcinomatous	39	7	1
	p=0.0001		

3.4 Tumor vascularity and TP expression

Although areas with high angiogenesis such as those at the invading edge of the tumor showed intense epithelial, endothelial and stromal TP immunoreactivity, overall there was no significant correlation between TP expression and Chalkley counts. There were cases with no epithelial staining for TP and increased microvascular density and cases which strong stromal positivity was not accompanied by increased vascularity.

There was significant statistical correlation when TP expression in stromal cellular elements of the adenomatous parts was compared with the Chalkley count results of these areas. Most of the negative /weakly stained for TP cases showed also low vascular counts (p=0.01).

3.5 Tumor vascularity and p53 immunoreactivity

Of the 47 carcinoma cases 26 showed p53 immunoreactivity (55%). The staining was diffuse and nuclear. In these cases the tumor vascularity detected by the Chalkley counts were 7-12 with a mean of 8.6 and SD 3.8. However p53 positive cells were detected not only in areas with increased vascularity (hot spots) but also in areas where the CS was low. In 14 of these 26 cases the adenomatous part of the neoplasm was also p53 positive (30%). The range of the CS was between 3-8 (mean 4.7, SD 1.2), (Table 3, Fig. 4).

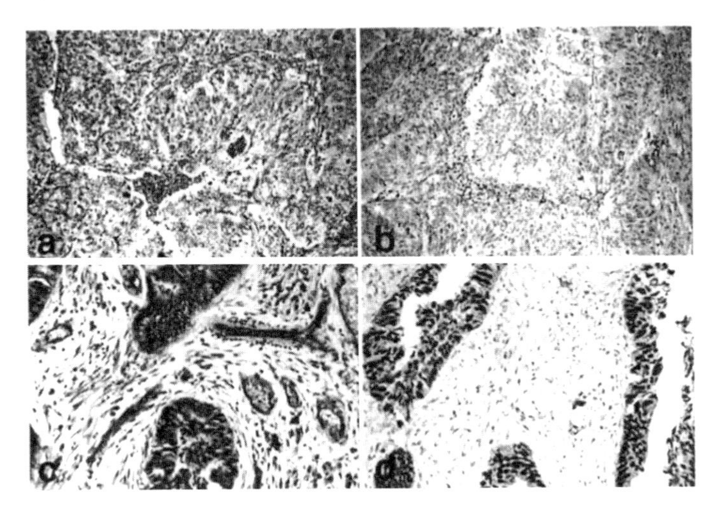

Figure 4:(a) and (b) represent a case with low vascular counts and no detection of p53; (c) and (d) is a case with high angiogenesis and intense p53 immunoreactivity.

Table 3: Tumor vascularity and p53 immunoreactivity in adenomatous and carcinomatous parts of tumors

	p53 positive carcinomas		
	CHALKLEY COUNTS	p53 POSITIVE	
adenomatous	3-8 (mean 4.7)	14/47	
carcinomatous	7-12 (mean 8.6)	26/47	
	p53 negative carcinomas		
	CHALKLEY COUNTS	p53 NEGATIVE	
adenomatous	1-8 (mean 3.2)	33/47	
carcinomatous	3-13 (mean 7.1)	21/47	

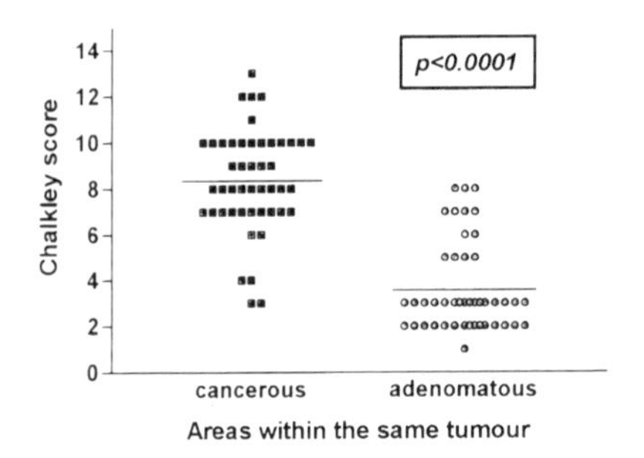

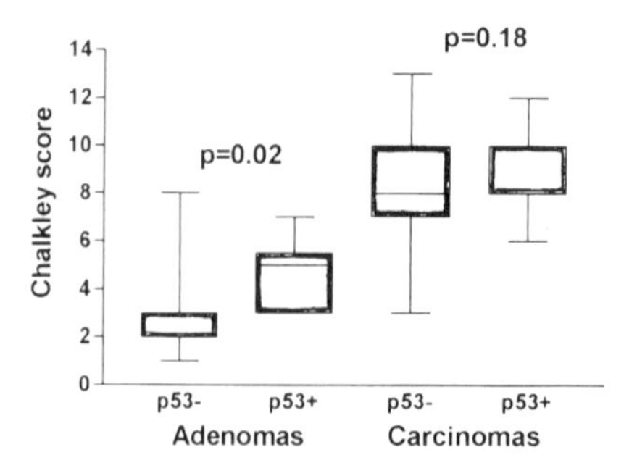

Figure 5: Graphic presentation of the CS in carcinomatous and adenomatous parts of the same tumor (top) and in relationship to p53 immunoreactivity (bottom).

There were also 21 cases which did not show p53 immunoreactivity and the Chalkley counts were ranging between 3-13 (mean: 7.1, SD: 2.4). The respective 33 negative adenomatous areas showed a range between 1-8 (mean: 3.2, SD: 1.8). The CS was higher in adenomatous areas expressing p53 (mean 4.7+/-1.2) as compared to p53 negative adenomas (mean 3.2+/-1.8, $p=0.02$). Although the p53-positive carcinoma cases showed an increased vascularity (mean 8.6+/-2.1) compared to the p53-negative cases (mean 7.1+/-2.4), statistical significant difference between the two groups was not detected ($p<0.1$, Fig. 5).

4. DISCUSSION

This is the first study which demonstrates a remarkable up-regulation of TP expression both in the stromal and neoplastic epithelial cells of colorectal carcinomas compared to the adjacent adenomatous parts from which they arose. TP expression was detected in the macrophages, fibroblasts and to a lesser degree in endothelial cells in most of the carcinomas whereas the respective cellular elements of the lamina propria of the non-neoplastic colonic mucosa showed scanty and weak immunoreactivity. The stromal elements of the adenomatous parts showed enhanced TP expression in few cases, but much less than that seen in the carcinomas.

While this study was in press another study was published on colon showing that *[Takebayashi et al. 1996]* TP expression was detected in 43% of the cases and only in 6% of the non-neoplastic mucosa samples. In another report *[Takahashi et al. 1996]* epithelial expression was only noted in 5% of the carcinoma cases. Both these studies confirmed increased TP expression in the neoplastic stromal cells. Takahashi's study showed that TP is associated with increased vascularity and that its expression is higher in cases with down-regulated VEGF expression and lower when VEGF is higher. Takebayashi's study showed that TP expression correlates with local tumor invasion, metastasis to lymph nodes, Dukes' stage and vascular invasion. TP immunoreactivity was a factor predicting poor prognosis even after adjustment for Dukes' stage and microvessel count by multivariate analysis. However we did not find any correlation between TP expression and Dukes' stage, an observation which was also made in Takahashi's study.

In our study overall there was no statistical correlation between TP expression and microvascular density detected by Chalkley counts. However in the majority of the cases TP positivity on macrophages, endothelial and epithelial cells was pronounced at the invading carcinomatous edge, an area which is also highly vascularised and angiogenesis is most active. More interestingly the low TP expression in adenomas was strongly associated with low vascular counts. These observations imply that TP could be a mediator of angiogenesis although it is not clear at which step its role is most important. It could act at a later stage, when other angiogenic factors have already been expressed, influencing and enhancing the vascular density of the invading periphery of the tumor and/or remodeling an existing vasculature.

The observation that high TP expression is detected in stromal cells and especially in macrophages, implies these cells might play an important role in the regulation of angiogenesis. This phenomenon also called "macrophage potentiation of tumor angiogenesis" was first described in a study *[Polverini and Leibovich 1984]* which showed that the induction of angiogenic activity by tumor-associated macrophages was much stronger compared to those induced by peritoneal macrophages. Endothelial cell proliferation was increased 10-fold by conditioned media taken from tumor-associated macrophages.

In tumor cell lines *[Eda et al. 1993]* it has been demonstrated that cytokines such as interferon-α interleukin 1, tumor necrosis factor a up-regulate TP. It is very likely in vivo that cytokines directly influence and regulate TP expression through autocrine and paracrine pathways. The expression of these cytokines by tumor cells or stromal cells might enhance the recruitment of TP-laden macrophages, which may in turn accentuate TP expression in the neoplastic epithelial cell population.

In fact macrophages expressed TP more uniformly than tumor cells in all three studies and in ours we only found high tumor TP when there was high stromal TP. This suggests that the macrophages are the major mediator of TP induction in epithelium. The low number of macrophages in adenomas suggests that in adenoma-carcinoma progression there is induction of macrophage infiltrates by tumor chemoattractants.

In other tumor types such as breast carcinomas *[Fox et al. 1996]* no significant correlation was observed with relapse-free or overall survival. It is quite possible that TP might influence the sensitivity of tumor cells to 5FU prodrugs, an observation proved in vitro *[Paterson et al. 1995]*. Obviously the amount of TP produced by the neoplastic cells or stroma might be predictive of the response to chemotherapy. Furthermore tumors rich in TP may also be a future target of antiangiogenic therapy. There are parallels with chronic inflammation and necrosis where macrophages play a major role and it would be of interest to study TP in such cases eg: myocardial infarction.

It has been recognised for many centuries that neoplastic tissue is more vascular than its normal counterpart, however it is only since Folkman's hypothesis on anti-angiogenesis *[Folkman 1971]* that a more quantitative method for evaluating angiogenesis in tissue sections has been introduced. In 1972 Brem et al. *[Brem et al. 1972]* developed a microscopic angiogenesis grading system (MAGS) based on tinctorially identified vessels and the degree of hyperplasia of the endothelial cells and cytology to quantify tumor angiogenesis. With the advent of endothelial markers increasing number of publications appeared examining the vasculature in tissue sections *[Mlynek et al. 1985; Svrivastava et al. 1988; Porschen et al. 1994]*. Many of these markers were not specific for endothelial cells and it is only in the last five years as improved endothelial markers have become available that quantitation studies have been performed and validated.

In 1989 Folkman et al. *[Folkman et al. 1989]*, using a model with trasgenic mice expressing an oncogene in the b-cells of the pancreatic islets, showed that hyperplastic islets become angiogenic in vitro at a time when such islets are neovascularized in vivo and at a frequency that correlates closely with subsequent tumor incidence. Our study is in keeping with these observations where pre-neoplastic lesions (hyperplastic polyps) or early neoplastic tumors (adenomas) lack an obvious neovascularized profile compared to invasive carcinomas where a switch to an angiogenic phenotype usually takes place in most of the cases.

Both adenomatous and carcinomatous parts of tumors that showed p53 immunoreactivity were also more angiogenic compared to the respective parts of the p53 negative neoplasms. In relation to carcinomas however, this observation did not reach a statistical significance. In the p53 positive cases the angiogenic areas were scattered and fields with low angiogenic counts were easily observed, implying that if mutant p53 plays a role in downregulating anti-angiogenic factors (eg TSP), the subsequent phenotypic effect has an heterogeneous localization within the tumor mass.

Neoplastic transformation and tumor formation is the result of the accumulation of genetic defects causing either inhibition of genes, repressing growth or inducing cell death or activation of genes, promoting growth and inducing cell proliferation. This abnormal cell proliferation sets the stage for additional genetic or epigenetic secondary phenomena. Induction of angiogenesis seems to be of paramount importance for tumor growth and progression.

REFERENCES

Asch AS, Nachman RL (1989). Thrombospondin: phenomenology to function. *Prog.Hemost.Thromb.*, **90**, 157-176.

Barton GJ, Ponting CP, Spraggon G, Finnis C, Sleep D (1992). Human platelet-derived endothelial cell growth factor is homologous to *Escherichia coli* thymidine phosphorylase. *Protein Science,* **1**, 688-690.

Bicknell R, Harris AL (1991). Novel growth regulatory factors and tumor angiogenesis. *Eur. J. Cancer*, **27,** 781-785.

Blood CH, Zetter BR (1990). Tumour interactions with the vasculature: angiogenesis and tumor metastasis. *Biochem. Biophys. Acta.*, **1032** , 89-118.

Bosari A, Lee AK, DeLellis RA, Wiley BD, Heatley GJ, Silverman ML (1992). Microvessel quantitation and prognosis in invasive breast carcinoma. *Hum. Pathol.*, **23**, 755-761

Cavallo T, R. Sade, J. Folkman and R. Cotran (1973). Ultrastructural autoradiographic studies of the early vasoproliferative response in tumor angiogenesis. *Am J Pathol,* **70**, 345-362.

Chalkley H (1943). Method for the quantative morphological analysis of tissues. *J Nat Cancer Inst* **4**, 47-53.

Cordell JL, Falini B, Erber FB, Ghosh AK, Abdulaziz Z, McDonald S, Pulford KAF and Mason DY (1984). Immunoenzymatic labelling of monoclonal antibodies using immune complexes of alkaline phosphatase and mono-clonal anti-alkaline phosphatase (APAAP complexes).*J Hist.Cyt.*, **32**, 219-229 .

Dameron KM., OV. Volpert, MA. Tainsky and N. Bouck (1994). Control of angiogenesis in fibroblasts by p53 regulation of thrombospondin-1. *Science* ,265(5178), 1582-1584.

Eda H, Fujimoto K, Wanatabe S (1993). Cytokines induce thymidine phosphorylase expression in tumor cells and make them more susceptible to 5'-deoxy-5-fluorouridine. *Cancer Chemother. Pharmacol*, **32**, 333-338.

Finnis C, Dodsworth N, Pollitt CE, Carr G, Sleep D (1993). Thymidine phosphorylase activity of platelet-derived endothelial cell growth factor is responsible for endothelial cell mitogenicity. *Eur J Biochem,* **212** , 201-210.

Folkman J (1990). What is the evidence that tumours are angiogenesis dependent? *J.Natl. Cancer Inst*, **82**, 4-6.

Folkman J (1986). How is blood vessel growth regulated in normal and neoplastic tissue? G.H.A. Clowes Memorial Award Lecture. *Cancer Res*, **46**: 467-473.

Folkman J, Watson K, Ingber D, Hanahan D (1989). Induction of angiogenesis during the transition from hyperplasia to neoplasia. *Nature*, **339**, 58-61.

Fox SB, Moghaddam A, Westwood M, Hickson I, Turley H, Gatter K, Harris AL (1995). Platelet-derived endothelial cell growth factor/Thymidine phosphorylase expression in normal tissues: an immunohistochemical study. *J. Path.* ,**176** , 183-190.

Fox SB, Westwood M, Moghaddam A, Hickson I, Turley H, Gatter K, Harris AL (1996). The angiogenic factor platelet-derived endothelial cell growth factor/Thymidine phosphorylase is up-regulated in breast cancer epithelium and endothelium. *Br. J. Cancer,*, **73** , 275-280.

Fregene TA, Kellog C, Pienta KJ (1994). Microvessel quantitation as a measure of angiogenic activity in benign breast tissues lesions: a marker for precancerous disease? *Int J Oncol*, **4**, 1199-1202.

Haraguchi M, Kazutaka M, Uemura K, Sumizawa T, Furukawa T, Yamada K, Akiyama S-I (1994). Angiogenic activity of enzymes. *Nature*, **368** ,198-201.

Horak ER, Leek R, Klenk N, LeJeune S, Smith K, Stuart N, Greenall M, Stepniewska K and Harris AL (1992). Angiogenesis, assessed by platelet/endothelial cell adhesion molecule antibodies, as indicator of node metastases and survival in breast cancer. *Lancet*, **340**, 1120-1124.

Ishikawa F, Miyazono K, Hellman U (1989). Identification of angiogenic activity and the cloning and expression of platelet-derived endothelial cell growth factor. *Nature*, **338**, 557-562.

Kono A, Hara Y, Sugata S, Matsusima Y, Ueda T (1984). Substrate specificity of a thymidine phosphorylase in human liver tumor. *Chem. Pharm. Bull.*, **32,** 1919-1921.

Mlynek, M., D. van Beunigen, L.-D. Leder and C. Streffer (1985). Measurement of the grade of vascularisation in histological tumour tissue sections. *Br J Cancer* 52: 945-948.

Moghaddam A, Bicknell R (1992). Expression of platelet-derived endothelial cell growth factor in *Escherichia coli* and confirmation of its thymidine phosphorylase activity. *Biochemistry*, **31**, 12141-12146.

Moghaddam A, Zhang HT, Fan TP, Bicknell R (1995). Thymidine phosphorylase is angiogenic and promotes tumor growth. *PNAS (USA)*, **92,** 998-1002.

Paweletz N, Knierim M (1989). Tumor-related angiogenesis. *Crit.Rev. Oncol. Hematol.* **9**, 197-242.

Polverini PJ, Leibovich SJ, (1984). Induction of neovascularization in vivo and endothelial proliferation in vitro by tumor-associated macrophages. *Lab. Invest.*, **51**, 635-642.

Porschen, R., S. Classen, M. Piontek and F. Borchard (1994). "Vascularization of carcinomas of the esophagus and its correlation with tumor proliferation." *Cancer Res* 54(2): 587-91.

Protopapa, E., G. S. Delides and L. Revesz (1993). "Vascular density and the response of breast carcinomas to mastectomy and adjuvant chemotherapy." *Eur J Cancer*.

Rastinejad F, PJ. Polverini and NP. Bouck (1989). Regulation of the activity of a new inhibitor of angiogenesis by a cancer suppressor gene. *Cell,* **56**(3), 345-55.

Reynolds K, Farzaneh F, Collins WP (1994). Correlation of ovarian malignancy with expression of platelet-derived endothelial cell growth factor. *J. Natl. Cancer Inst.*,**86**, 1234-1238.

Svrivastava, A., P. Laidler, R. Davies, K. Horgan and L. Hughes (1988). The prognostic significance of tumor vascularity in intermediate-thickness (0.76-4.0 thick) skin melanoma. *Am J Pathol* 133: 419-423.

Takahashi Y, Bucana CD, Liu W, (1996). Platelet-derived endothelial cell growth factor in human colon cancer angiogenesis: role of infiltrating cells. *J. Natl. Cancer Inst.*, **88**,1146-1151.

Takebayashi Y, Akiyama S, Akiba S, Yamada K, Miyadera K, Sumizawa T, Yamada Y, Murata F, Aikou T (1996). Clinicopathologic and prognostic significance of an angiogenic factor, thymidine phosphorylase in human colorectal carcinoma. *J. Natl. Cancer Inst.*, **88** , 1110-1117.

Usuki K, Saras J, Waltenberger J (1992). Platelet-derived endothelial cell growth factor has thymidine phosphorylase activity. *Biochem. Biophys. Res. Commun.* **184,** 1311-1316, .

Vertongen F, Fondu P, Van Den Heule B, Cauchie C, Mandelbaum IM (1984). Thymidine kinase and thymidine phosphorylase activities in various types of leukemia and lymphoma. *Tumor Biology* , **5**, 303-311.

Vermeulen PB, Verhoeven D, Hubens G, Van ME, Goovaerts G, Huyghe M, De Bruijn EA, Van OA, Dirix LY (1995). Midrovessel density, endothelial cell proliferation and tumour cell proliferation in human colorectal adenocarcinomas. *Ann Oncol*, **6**, 59-64.

Yoshimura A, Kuwazuru Y, Furukawa T, Yoshida H, Yamada K, Akiyama S, (1990). Purification and tissue distribution of human thymidine phosphorylase; high expression in lymphocytes, reticulocytes and tumors. *Biochim. Biophys. Acta.* **1034,** 107-113.

TUMOR ANGIOGENESIS, MACROPHAGES, AND CYTOKINES

HIROKO BANDO and MASAKAZU TOI
Tokyo Metropolitan Komagome Hospital, 3-18-22, Honkomagome, Bunkyo-ku, Tokyo 113, Japan

1. INTRODUCTION

Angiogenesis is a multistep program that involves the activation, proliferation, and migration of endothelial cells. In healthy body, except for reproductive cycle or embryogenesis, angiogenesis is strictly regulated by numerous factors to maintain homeostasis. Factors can be divided mainly in two categories, positive regulators and negative regulators. In pathological process like acute inflammation or wound repair, both positive and negative regulators are induced and activated, but quickly the response comes to the reduction. However, in oncogenesis, new vessel formation occurs repeatedly and chronically[1-2]. In the early stage of tumor progression, the alteration of the balance between positive regulators and negative regulators is limited only in the primary tumor site, but in the late stage/ clinical stage, this alteration is no longer localized in the local foci. Over expression of positive regulators and/or down regulation of negative regulators at systemic level are frequently observed. In this process, not only tumor cells but also various stromal cells such as macrophages, other leukocytes, fibroblasts and endothelial cells are responsible for the explosive induction of positive regulators. It is estimated that continuous interaction between tumor cells and stromal cells in a positive feedback loop enables a ceaseless. In this manuscript, we focused on the importance of stromal cells, particularly monocytic cells, and reviewed the role of cytokines in tumor angiogenesis.

Angiogenesis: From the Molecular to Integrative Pharmacology
Edited by Maragoudakis, Kluwer Academic / Plenum Publishers, New York, 2000

2. POSITIVE ENDOTHELIAL REGULATORS AND MICROENVIRONMENTAL CONDITIONS

Using breast cancer materials, we have examined the expression of a variety of endothelial regulators[3,4,5] Among the factors, vascular endothelial growth factor (VEGF) and thymidine phosphoryalse (TP) had a positive correlation with intratumoral microvessel density (MVD)[6] (Table 1). As well known, VEGF and its receptors, flt and KDR, are frequently overexpressed in various types of solid tumors. VEGF protein concentration was an independent prognostic indicator in 260 node-negative and no-treatment primary breast cancer patients.[7] The prognostic value of VEGF content in node-negative breast cancer was recently confirmed in a study with larger number of cases by Linderholm et al.[8] Eppenberger et al. also studied the value of VEGF content in primary breast cancer, and showed a poor prognosis of VEGF-rich tumors compared to VEGF-poor tumors.[9] In addition, VEGF and urokinase-type plasminogen activator (uPA) provided a combination effect for predicting the prognosis, which suggests that VEGF and uPA can cooperatively function. This result also suggests a close interaction between tumor cells and stromal cells, because VEGF is mainly derived from tumor cells, but uPA is mainly from stromal cells such as monocytic cells and endothelial cells in breast cancer tissuues. In other types of tumors including esophagus, lung, ovary, stomach, bladder, head and neck cancer and glioblastoma, many studies have described the positive association between VEGF expression and increment of MVD, and poor prognosis.[10-17] A recent report also showed that plasma cells in the bone marrow from patient with multiple myeloma expressed VEGF. Furthermore, both the Flt-1 and KDR high affinity VEGF receptors were elevated in the normal bone marrow myeloid and monocytic cells surrounding tumor, which

Table 1. Angiogenesis regulators and breast cancer

Association with MVD	Association with VEGF	(Irrelevant)	Combination effect for MVD
		bFGF	
VEGF	TP	HGF	VEGF & TP
TP	TNF-α	CD68 count	VEGF & MMP-2
MMP-9	IL-6	EGFR	VEGF & MT1-MMP
	MMP-2	Erb B2	VEGF & CD68
	MCP-1	P53	
		ER, PgR	

suggests a significant role of VEGF in the growth of haematopoietic neoplasm.[18] Salven et al. reported that high serum VEGF level was associated with poor outcome in 82 patients with non-Hodgkin's lymphoma.[19]

According to immunocytochemical analysis, high-density overexpression of VEGF, either in tumor cells or stromal cells, is often observed in the surrounding area of the necrosis as well as in the front of invasion site of the tumor tissue. VEGF expression level in in vivo condition sometimes differs from that in in vitro condition in the experiment with cancer cell lines. Recently, Carmeliet et al. demonstrated that loss of hypoxia-inducible factor (HIF)-1α reduced hypoxia-induced expression of VEGF and prevented formation of large vessels in ES-derived tumors.[20] Maxwell et al. also documented that tumors of HIF-1 deficient hepatoma (Hepa-1) cells had reduced vascularity and grew more slowly as compared to tumors of wild type cells in xenograft model.[21] These findings show that HIF-1 activation, occurs in hypoxic regions of tumors, can influence on VEGF expression, angiogenesis, and eventually tumor growth. Recently, it was documented that von Hippel-Lindau (VHL) tumor suppressor gene product plays a crucial role in the activation of HIF-1, which might lead a new understanding for the mechanism of hypoxia-induced neovascularization.[22] In addition, Brooks et al. showed that VEGF protein production induced by hypoxia was enhanced further by increasing pH and increasing glucose, but decreased by low pH and low glucose production in C6 glioma cells and in Muller cells.[23] The changes of the glucose concentration or pH did not result in significant induction of VEGF protein under normoxic conditions. In murine thioglycolate-induced peritoneal macrophages (MPMs) model, Xiong et al. reported that hypoxia, lactate, and IFN-γ/LPS-activated MPMs express angiogenic activity, and angiogenic activity produced by hypoxia, lactate, and by IFN/LPS was neutralized by anti-VEGF antibody.[24] According to Satake et al., glucose deprivation induced both VEGF mRNA expression and VEGF protein production in U-937 cells, and the low glucose-induced VEGF expression returned to the control level after supplementation with glucose.[25] Because oligomycin, a mitochondrial ATP synthase inhibitor, increased VEGF protein production, intracellular ATP depletion due to glucose deprivation is thought to be crucial for the up-regulation of VEGF.

Recent studies suggest that VEGF may also function as a survival factor to the endothelium.[26] Spyridipoulos reported that VEGF could support the endothelial cells to survive from exposure to the apoptosis-inducing cytokine, TNF-α[27] Nor et al. showed that human dermal microvascular endothelial cells grown on collagen type 1 gels in the presence of VEGF exhibited a significant reduction of apoptosis, and its prolonged survival was associated with increase of Bcl-2.[28] Kondo et.al showed that aortic endothelial cells over expressing Bcl-2 were protected from the apoptotic effects of bFGF withdrawal.[29] Prolonged survival of endothelial cells and the maintenance of function of newly-formed vessels induced by VEGF might cause the difference between tumor vasculature and the vasculature developed in other pathological condition.

TP is a nuclear acid metabolic enzyme and positively involved in angiogenesis through 2-deoxy-D-ribose, a degradation product of thymidine caused by TP. 2-deoxy-D-ribose is a potent chemotactic factor to the endothelium.[30-32] TP expression is selectively increased in tumor tissues as compared to adjacent normal tissues in a variety of tumor types, although only a low or faint amount of TP expression was detected in cultured cancer cell lines.[33-35] Thereby, TP expression in tumor tissues is also thought to be microenvironment-dependent (Fig. 1). In sense, several cytokines [tumor

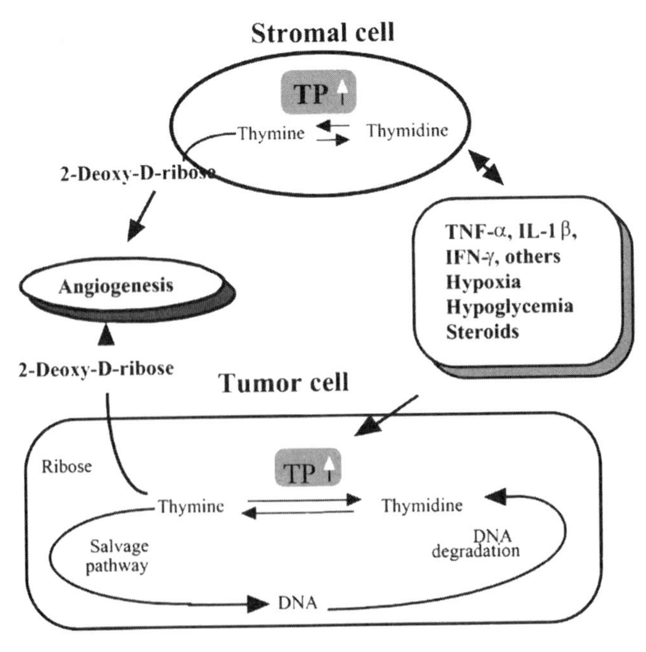

Figure 1.

necrosis factor-, interleukin (IL)-1, interferon (IFN)-γ] are known to be potent inducer of TP.[36,37] Hypoxia is also characterized as an inducer of TP, and it is known that cytokine-induced TP expression is enhanced in hypoxic condition compared to normoxia condition. As to the function of TP, Nishino recently showed an interesting result that that loss-of-function in TP caused impaired replication or maintenance of mtDNA in mitochondrial neurogastrointestinal encephalomyopathy.[38] On the other hand, Kitazono et al. reported that human KB epidermoid carcinoma cells transfected with TP cDNA were resistant to hypoxia-induced apoptosis, in vitro. TP might confer resistance to apoptosis induced by hypoxia.[39]

Importance of TP as a prognostic marker has been reported in various types of human tumors including stomach, colon, kidney, lung and pancreas tumor [40-44]. In breast cancer, TP status was not a potent prognostic indicator, although a significant correlation between TP expression and MVD was found.[1] However, in a combination with VEGF, both positive phenotype for VEGF and TP showed the worst prognosis as compared to other three categories[45]. In ovarian cancer, the level of TP expression in stage 3-4 was significantly higher than that of in lower stage, and TP expression was a significant prognostic factor.[46] In the study of non-small-cell lung cancer, cancer cell TP overexpression was also related to high vascular grade, and advanced T stage, and poor prognosis.[43]

From the point of prediction of the effect of FU/ FU derivativies containing treatment, TP seems to be a promising indicator. Fox et al. showed that TP status determined by immunocytochemical assay was a significant predictor of CMF (cyclophosphamide, methotrexate and 5FU) treatment[47]. In a recent cohort study by Gasparini and Toi, high TP concentration in primary breast tumor tissues was potently associated with the reduction of relapse rate by adjuvant CMF treatment [48]. Although it is difficult to obtain the direct evidence in clinical setting, the selective antitumor effect of CMF treatment in TP-overexpressing tumors would be due to inhibition of angiogenesis activity, at least in part.

TP localizes in both tumor cells and stromal cells such as macrophages, fibroblasts and endothelial cells. In particular, TP expression in monocytic cells is frequently detected. Recently, we compared immunocytochemical TP expression with TP protein concentration in tumor tissues in the same samples and found that both tumor-cell TP expression and monocytic TP expression were significantly correlated with TP protein concentration (Table 2). Predominant expression of TP in stromal cells is also found in gastrointestinal tumors.

Table 2. TP: Comparison between ICA and EIA

TP-ICA	No.	TP-EIA (U/mg) Average	(SD)	P-value (t-test)
Tumor cell TP status	+ 65	223.6	(158.9)	0.062
	-70	169.2	(82.6)	
Monocytic TP status	+67	243.7	(162.1)	0.016
	-68	170.4	(98.3)	

Basic FGF is one of the most studied positive endothelial regulators in human tumors, but still many remains controversial. Colomner et al, measured bFGF levels with 140 breast tumor cytosols and reported that there was no significant difference in bFGF protein level between benign and malignant breast tumors. Although, breast cancer patients with low levels of bFGF were associated with increased tumor size, stage of disease, shorter disease-free survival, and a poor prognosis compared to increased bFGF cases. [49] Basic FGF receptor concentrations were also positively associated with longer relapse-free survival and overall survival in 225 breast cancer tissues. Tumors with high bFGF level showed longer survival in ovarian cancer. [50] Elevation of serum bFGF was found to be a good prognostic factor in 68 non-small cell lung cancer patients. [51] However, circulating bFGF level measured by ELISA was noted to be significantly correlated with the size of the primary tumor in renal cell carcinoma, and in addition, bFGF, ICAM-1, IL-6, and TNF-α had a significant prognostic value, respectively. In children with high-grade gliomas, tumor bFGF expression was strongly associated with progression free survival and overall survival. High bFGF expression group showed a poor prognosis. [52] The role and implication of bFGF expression and activation seems to depend upon organ background.

3. MACROPHAGES

The accumulation of macrophages, more or less, observed in the most of tumor tissues. The tumor associated monocytic cells (TAMs) have a pleotropic function; phagocytosis of degraded cells or matrixes, differentiation to dendritic cells, antigen presentation and so on. Several studies have showed that the TAMs are closely involved in angiogenesis in human tumor tissues. TAMs are known to secrete protumor factors including epidermal growth factor (EGF), vascular endothelial growth factor (VEGF), basic fibroblast growth factor (bFGF), uPA, cathepsin D and matrix metalloproteinases (MMPs).[53-60] Particularly, a close association between TAMs content and uPA levels has been underlined.[61] From these points of view, we have measured CD68 positive TAMs count in primary breast tumors, and investigated its clinical significance.[62] Although we failed to demonstrate the prognostic value of TAMs count, we found that TP positive TAMs count can predict postoperative early relapse. A multivariate analysis confirmed that TP status of TAMs provided an independent prognostic value. Furthermore, interestingly, monocytic TP status could categorize the CD68+ patients, who had an extensive accumulation of CD68 positive TAMs, into two subgroups with contrastive prognosis: good prognostic monocytic TP- group and poor prognostic monocytic TP+ group. This finding results in a hypothesis that TAMs are consisted of antitumor and protumor ones. Many experimental results have already indicated the heterogenous subpopulation of the TAMs, but no good marker has been reported to characterize the phenotype of TAMs. TP might be a candidate to classify the property of TAMs, roughly to protumor and antitumor. In the subsequent analysis, monocytic TP status and MT1-MMP expression showed a combination effect for predicting the poor prognosis.[63] Inactivated pro-form of MMPs produced by TAMs can be converted to activated from by MT1-MMP on the surface of tumor cells, which would facilitate the tumor cells to grow.

4. CYTOKINES AND CHEMOKINES

It is well known that a variety of cytokines and chemokines are involved in the new vessel formation and its remodeling. (Fig. 2). According to our screening analysis by ELISA using the cytosols from 120 primary breast tumors, the levels of VEGF, TP, IL-4 and IL-8, among a variety of cytokines and chemokines, were markedly increased in tumor tissues rather than normal tissues. The mean value of VEGF, TP, IL-4 and IL-8 were 313 (range: 0-2957) pg/mg and 195 (range:7-1125) U/mg, 6 (range:-53) pg/mg and, 83 (-952) pg/ml, respectively. MCP-1 also showed the elevated median value of 48 (-434) pg/ml in tumor tissues. Expressions of negative regulators were fairly low, for IL-12 and IL-10, the median 0.4 (-3) pg/ml and 0.6 (-21) pg/ml, respectively.

IL-4 has been characterized as a positive endothelial regulator.[64] Mitogenic assay using capillary endothelial cells demonstrated that IL-4 is a

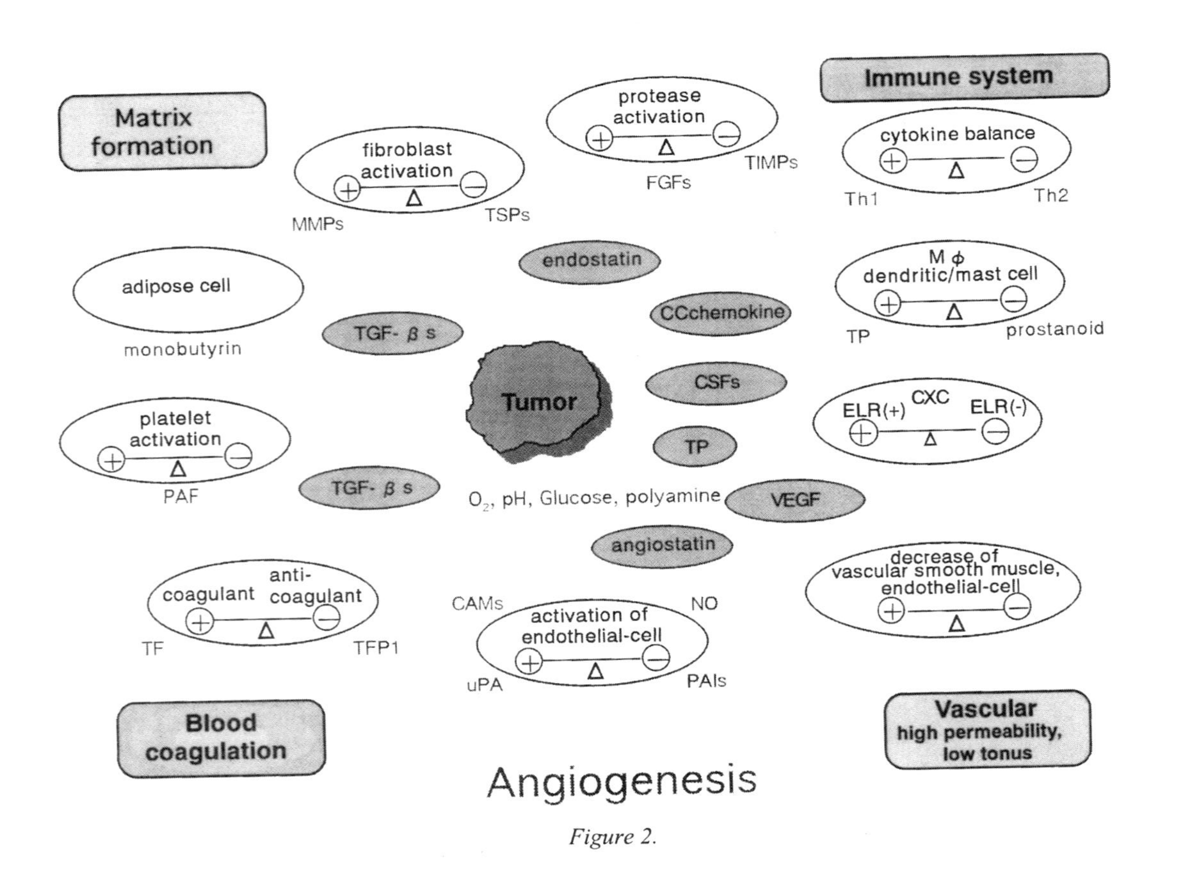
Matrix formation
fibroblast activation
MMPs
TSPs
protease activation
FGFs
TIMPs
Immune system
cytokine balance
Th1
Th2
adipose cell
monobutyrin
endostatin
TGF- β s
CCchemokine
M φ dendritic/mast cell
TP
prostanoid
Tumor
CSFs
ELR(+) CXC ELR(-)
platelet activation
PAF
TP
TGF- β s
O2, pH, Glucose, polyamine
VEGF
angiostatin
coagulant anti-coagulant
TF
TFP1
CAMs
NO
activation of endothelial-cell
uPA
PAIs
decrease of vascular smooth muscle, endothelial-cell
Blood coagulation
Vascular
high permeability, low tonus
Angiogenesis

Figure 2.

potent endothelial regulator. In addition, recently, tube-forming activity of IL-4 was also confirmed in bovine aortic endothelial cells and human microvascular endothelial cells.[65] IL-4 can up-regulate expression of VCAM-1 and uPA in endothelial cells in a synergy with IL-1.[66-68] Yao et al. reported that IL-4 induced a five- to nine fold increase in P-selectin messenger RNA (mRNA) in human umbilical vein endothelial cells (HUVEC).[69] As well known, IL-4 is a key cytokine in Th2 arm of the T cell immune response and B cells to produce IgE. [70,71] IL-4 inhibits the tumorgenicity of tumor cell lines, including those of lymphoid origin, or breast and colon carcinoma. Saleh et al. reported that rat C6 glioma cells engineered to express mouse IL-4 system failed to form tumors when the cells were implanted to athymic nude mice.[72] The expression of IL-4 by C6 cells resulted in eosinophil infiltration, as well as inhibition of tumor angiogenesis. They also illustrated that exogenous mouse IL-4 can suppress the expression of VEGFR2 in cultured mouse endothelial cells. Because IL-4 directly can inhibit tumor cell growth in a variety of carcinoma types, role of IL-4 in tumor progression seems to be double-edged.[73] From the point of immune surveilance, Th-2 dominant stage might be better than Th-1 dominant stage for tumor cells to evade.

IL-8, a member of ELR+CXC chemokine, is 8kD polypeptide with neutrophil recruitment, endothelial cell chemotactic and proliferative activity. [74,75] IL-8 is produced in macrophages, fibroblasts, endothelial cells and cancer cells. IL-1 can induce IL-8. Melanoma cell lines with higher metastatic potential expressed higher levels of IL-8 mRNA. [76] Also in xenograft model, enforced expression of IL-8, with IL-8 cDNA transfection, rendered the melanoma cells highly tumorigenic and increased their metastatic potential, and in vitro, IL-8-transfected cells displayed up-regulation in MMP-2 mRNA and collagenase activity and increased invasiveness. [77] Expression of IL-8 correlated with tumor size in non-small cell lung carcinoma. [78] Galffly et al. found that significant high levels of IL-8 in mesothelioma pleural fluids, and neutralization of IL-8 markedly decreased proliferative activity of some kinds of MM cell lines. [79] Overexpression of IL-8 was also found in human gastric cancer cell lines and in the most of primary gastric cancer tissues. IL-8 expression was significantly correlated with increased MVD. [80] In nude mice model with human ovarian carcinomas, IL-8 expression was associated with rapid growth of carcinoma cells, and inversely associated with mice survival. [81] Desbaillets et al. reported that hypoxic/anoxic status on glioblastoma cells in vitro using anaerobic chamber or within spheroids developing central necrosis induced an increase in IL-8 mRNA and protein expression. mRNA for IL-8 binding protein receptors CXCR1, CXCR2, and Duffy antigen receptor for chemokine (DARC) were found in all astrocytoma grades by reverse transcription PCR analysis. It was suggest that either directly or indirectly augmented IL-8 can promote angiogenesis by binding to DARC and by inducing leukocyte infiltration and activation by binding to CXCR1 and CXCR2. [82,83] In head and neck squamous cell carcinoma (HNSCC), Richards et al. showed that IL-8 receptors were expressed by cancer cells and microvessel endothelial cells. IL-8 may act in both autocrine and paracrine fashion to stimulate proliferation of tumor cells and endothelial cells. [84] Eisma et al. reported that in HNSCC tumor specimens both VEGF and IL-8 were concomitantly overexpressed and patients with high level of VEGF, and IL-8 had aggressive disease, manifested by higher TNM stage or recurrence rate, and shorter survival. [85]

IL-10 is known to inhibit the production of various types of cytokines in the immune response, such as IL-2, IL-3, lymphotoxin, IFN-γ, and granulocyte-macrophage colony-stimulating factor (GM-CSF) in Th1 cells; IL-4, and IL-5 in Th2 cells; IFN-γ and TNF-α in NK cells. [86-90] Huang et al. showed that in human melanoma A375P cells transfected with a murine IL-10 cDNA exhibited the suppression of the tumorigenicity and metastatic potential in nude mice. Suppression of tumor growth and metastasis was significantly associated with decrease in neovascularity. Murine IL-10 down-regulated the expression of VEGF in macrophages, in a dose-dependent manner. Other factors involved in angiogenesis such as IL-1β, TNF-α, IL-6, and MMP-9 were also inhibited in activated macrophage in vitro, suggesting that IL-10 produced by tumor cells inhibits macrophage-derived angiogenic factors. [91] Richter et al. showed that IL-10 blocked tumor growth in ovary by blocking angiogenesis and macrophage penetration. [92] Kundu et al. also showed that IL-10 transfection to murine mammary tumor cells inhibited tumor growth completely and reduced metastasis significantly in nude mice. [93] Stearns et al. demonstrated that human primary prostate cells induced human bone marrow endothelial cells to form microvessels, and IL-10 treatment blocked induction of microvessel formation, though IL-10 receptor antibodies and promoted angiogenesis. [94,95]

The study of Watanabe et al, showed that the treatment with recombinant of murine IL-12 in combination with recombinant human IL-2 suppressed the growth of human bFGF/B16F10 melanoma cells in SCID mice. The treatment revealed an inhibition of tumor growth and absence of vascularity even in the absence of a mononuclear infiltration. T and/or NK cells were not the principal mediators of the anti-tumor response, but the inhibition of neovascularization resulted in tumor regression in this case. [96] Dias et al. also described the inhibition of tumor growth by IL-12, where VEGF level and MMP-9 expression was decreased markedly, but TIMP-1 level was increased. Since VEGF expression in tumor cells was reduced by IFN-γ, downregulation of VEGF by IL-12 treatement might be through IFN-γ, which is induced by IL-12. [97]

IFN-γ inducible protein (IP)-10, an ELR- CXC chemokine, is another mediator of IL-12. IP-10 inhibits migration and proliferation of endothelial cells induced by bFGF, IL-8 and VEGF. [98] In a recent report, IP-10 and platelet factor 4 (PF-4), another antiangiogenic ELR- CXC chemokine, bind together to heparan sulfate proteoglycans binding site and inhibit the endothelial cell proliferation. It is reported that IP-10 regulates non-small cell lung carcinoma derived angiogenesis, tumor growth, and spontaneous metastasis, in a xenograft model. [99] To the hematopoietic cells, IP-10 drives chemotactic activity on normal CD-4 positive lymphocytes and inhibits the proliferation of early subsets of normal and leukemic progenitors. [100-102] Expression of IP-10 is often found in inflamed tissues (epithelium predominantly and few in subepithelial cells) where Th1-type cytokine IFN-γ is dominant.

MCP-1 is a proinflammatory CC chemokine that stimulates chemotaxis of peripheral blood monocytes specifically, but not of lymphocyte or polymorphonuclear leukocytes. [103,104] In primary breast cancer tissues, MCP-1 concentration was significantly increased compared to normal tissues and it was positively associated with CD68 TAMs accumulation. It is widely accepted that chemoattractant proteins including MCP-1 and IL-8 play major roles in inflamatory disorders. Mice lacking receptors for MCP-1 are known to be less susceptible to atherosclerosis and have fewer monocytes in vascular lesions. MCP-1 and IL-8 can cause rolling monocytes to adhere firmly onto monolayers expressing E-selectin. [105] IL-1βand TNF-α are potent inducer of IL-8 and MCP-1 secretion. The induced IL-8 and MCP-1 expression is modulated by other proinflammatory factors including IL-4 and GM-CSF in a synergistic manner. [106,107] IL-6 is also responsible for the expression of MCP-1 in peripheral blood mononuclear cells, but did not other chemokines, including RANTES, MIP-1α, MIP-1β, and IL-8. [108] VEGF is also capable of inducing MCP-1 mRNA expression in bovine retinal microvascular endothelial cells in a dose- and time-dependent manner. Binding activity of transcription factor AP-1, which is suggested to regulate induction of the MCP-1 gene together with NFkB, was also stimulated by VEGF. [109] In human ovarian cancer, in situ hybridization detected mRNA for the macrophage MCP-1 in most of tumors. In serous tumors, mRNA expression was found to localize to the epithelial areas. [110] In cervical squamous cell carcinoma cells, CD40L exposure was able to induce MCP-1 production and CD40 ligation led to a marked MCP-1 induction, which suggests a significant role of MCP-1 for Th1-type immune response against tumors.[111] On the other hand, urinary MCP-1 levels were significantly associated with tumor stage, grade and distant metastasis in bladder carcinoma. The low-grade RT4 bladder cancer cell line produced only traces of MCP-1, however, the highly malignant T24 bladder cancer cell line spontaneously secreted large amounts of MCP-1. In such cases, MCP-1 might function as pro-tumor mediator, possibly for neovascularization.[112]

Still little is known about the relationship between angiogenesis and immune network system. Recent studies showed that VEGF functions not only for new vessel formation but also for suppressing the immune reaction, which leads an intriguing hypothesis that switch-on of angiogenesis and switch-off of immune response, particularly inhibition of the Th-1 type reaction, might occur concomitantly. In this process, TAMs seem to play crucial roles. Further studies are warranted to analyze the microenvironmental conditions in tumor tissues.

REFERENCES

1) Folkman J., 1995, Angiogenesis in cancer, vascular, rheumatoid and other disease. Nat Med. 1: 27-31.
2) Folkman J. et al.,1987, Angiogenic factors. Science 235: 442-447.
3) Gasparini G . et al.,1995, Clinical importance of the determination of tumor angiogenesis in breast carcinoma: Much more than a new prognostic tool. J. Clin. Oncol. 13: 765-782.
4) Toi M. et al., 1996, Clinical significance of the determination of angiogenic factors. Eur .J. Cancer . 32A: 2513-2519.
5) Toi M. et al., 1995, Vascular endothelial growth factor and platelet-derived endothelial cell growth factor are frequently coexpressed in highly vascularized human breast cancer. Clin. Cancer Res. 1: 961-964.
6) Toi M. et al., 1995, Expression of platelet-derived endothelial cell growth factor in human breast cancer. Int. J. Cancer 64: 79-82.
7) Gasparini G. et al., 1997, Prognostic significance of vascular endothelial growth factor protein in node-negative breast carcinoma. J Natl. Cancer Inst. 89:139-47.
8) Linderholmm B. et al., 1998, Vascular endothelial growth factor is of high prognostic value in node-negative breast carcinoma. J. Clin. Oncol. 16: 3121-3128.

9) Eppenberger U. et al.,1998, Markers of tumor angiogenesis and proteolysis independently define high- and low-risk subsets of node-negative breast cancer patients. J.Oncol. 16:3129-3136.

10) Ohta Y. et al., 1997, Vascular endothelial growth factor and lymph node metastasis in primary lung cancer. Br. J. Cancer76 :1041-1045.

11) Kitadai Y. et al.,1998, Significance of vessel count and vascular endothelial growth factor in human esophageal carcinomas. Clin. Cancer Res. 4:2195-200.

12) Takanami I. et al., 1997, Vascular endothelial growth factor and its receptor correlate with angiogenesis and survival in pulmonary adenocarcinoma. Anticancer Res. 17:2811-4.

13) Paley P.J. et al., 1997, Vascular endothelial growth factor expression in early stage ovarian carcinoma. Cancer 80:98-106.

14) Maeda K. eg al.,1996, Prognostic value of vascular endothelial growth factor expression in gastric carcinoma. Cancer 77:858-863.

15) Crew J.P. et al., 1997, Vascular endothelial growth factor is a predictor of relapse and stage progression in superficial bladder cancer. Cancer Res. 57:5281-52855

16) Sauter E.R. et al., 1999, Vascular endothelial growth factor is a marker of tumor invasion and metastasis in squamous cell carcinomas of the head and neck. Clin. Cancer Res. 5:775-782.

17) Abdulrauf S.I. et al.,1998, Vascular endothelial growth factor expression and vascular density as prognostic markers of survival in patients with low-grade astrocytoma. J. Neurosurg. 88:513-520.

18) Bellamy W.T. et al., 1999, Expression of vascular endothelial growth factor and its receptors in hematopoietic malignancies. Cancer Res. 59:728-733.

19) Salven P. et al., 1997, A high pretreatment serum vascular endothelial growth factor concentration is associated with poor outcome in non-Hodgkin's lymphoma. Blood 90:3167-3172.

20) Carmeliet P. et al., 1998, Role of HIF-1alpha in hypoxia-mediated apoptosis, cell proliferation an tumour angiogenesis. Nature. 394:485-490.

21) Maxwell P.H. et al., 1997, Hypoxia-inducible factor-1 modulates gene expression in solid tumors and influences both angiogenesis and tumor growth. Proc. Natl. Acad. Sci .U S A. 94:8104-8109.

22) Maxwell P.H. et al., 1999, The tumour suppressor protein VHL targets hypoxia-inducible factors for oxygen-dependent proteolysis. Nature 399:271-275.

23) Brooks S.E. et al., 1998, Modulation of VEGF production by pH and glucose in retinal Muller cells. Curr. Eye Res. 17:875-882.

24) Xiong M. et al., 1998, Production of vascular endothelial growth factor by murine macrophages: regulation by hypoxia, lactate, and the inducible nitric oxide synthase pathway. Am .J. Pathol. 153:587-598.

25) Satake S. et al., 1998, Up-regulation of vascular endothelial growth factor in response to glucosedeprivation. Cell 90:161-168.

26) Solovey A. et al.,1999, Sickle cell anemia as a possible state of enhanced anti-apoptotic tone: survival effect of vascular endothelial growth factor on circulating and unanchored endothelial cells. Blood 93:3824-30.

27) Spyridopoulos I. et al.,1997, Vascular endothelial growth factor inhibits endothelial cell apoptosis induced by tumor necrosis factor-alpha: balance between growth and death signals. J. Mol. Cell. Cardiol. 29:1321-1330.

28) Nor J.E. et al.,1999, Vascular endothelial growth factor (VEGF)-mediated angiogenesis is associated with enhanced endothelial cell survival and induction of Bcl-2 expression. Am. J. Pathol. 154:375-384.
29) Kondo S. et al., 1994, bcl-2 gene prevents apoptosis of basic fibroblast growth factor-deprived murine aortic endothelial cells. Exp. Cell Res. 213:428-432.
30) Moghaddam K. et al., 1995, Thymidine phosphorylase is angiogenic and promotes tumor growth. Proc. Natl. Acad. Sci. 92: 998-1002.
31) Miyadera K. et al. , 1995, Role of thymidine phosphorylase activity in tthe angiogenic effect of platelet-derived endothelial cell growth factor/ thymidine phosphorylase. Cancer Res. 55: 1687-1690.
32) Haraguchi M. et al. Angiogenic activity of enzymes. Nature 368, 198, 1994
33) Moghaddam A. et al.,1995, Thymidine phosphorylase is angiogenic and promotes tumor growth. Proc. Natl. Acad. Sci. 92: 998-1002.
34) O'Brien T. et al., 1995, Differential angiogenic pathways characterize superficial and invasive bladder cancer. Cancer Res. 55: 510-513.
35) Reynolds K. et al., 1994, Association of ovarian malignancy with expression of platelet-derived endothelial cell growth factor. J. Natl. Cancer Inst. 86: 1234-1238.
36) Eda H. et al., 1993, Cytokines induce thymidine phosphorylase expression in tumor cells and make them more susceptible to 5'-deoxy-5-fluorouridine. Cancer Chem. Pharm., 32: 333-338..
37) Griffiths L. et al., 1997, The influence of oxygen tension and pH on the expression of platelet-derived endothelial cell growth factor/ thymidine phosphorylase in human breast tumor cells grown in vitro and in vivo. Cancer Res. 57: 570-572.
38) Nishino I. et al., 1999, Thymidine phosphorylase gene mutations in MNGIE, a human mitochondrial disorder. Science 283:689-692.
39) Kitazono M. et al., 1998, Prevention of hypoxia-induced apoptosis by the angiogenic factor thymidine phosphorylase. Biochem. Biophys. Res. Commun. 253:797-803.
40) Tanigawa N. et al., 1996, Tumor angiogenesis and expression of thymidine phosphorylase/platelet derived endothelial cell growth factor in human gastric carcinoma. Cancer Lett. 108:281-290.
41) Takebayashi Y. et al., 1996, Clinicopathologic and prognostic significance of an angiogenic factor, thymidine phosphorylase, in human colorectal carcinoma. J. Natl. Cancer Inst. 88:1110-1117.
42) Imazano Y. et al., 1997, Correlation between thymidine phosphorylase expression and prognosis in human renal cell carcinoma. J. Clin Oncol. 15: 2570-2578.
43) Koukourakis M.I. et al., 1998, Different patterns of stromal and cancer cell thymidine phosphorylase reactivity in non-small-cell lung cancer: impact on tumour neoangiogenesis and survival. Br. J. Cancer 77:1696-1703.
44) Takao S. et al., 1998, Expression of thymidine phosphorylase is associated with a poor prognosis in patients with ductal adenocarcinoma of the pancreas. Clin. Cancer Res. 4:1619-1624.
45) Toi M. et al., 1997, Co-ordination of the angiogenic factors thymidine phosphorylase and vascular endothelial growth factor in node-negative breast cancer: prognostic implications. Angiogenesis 1: 71-83
46) Hata K. et al., 1999, Expression of the thymidine phosphorylase gene in epithelial ovarian cancer. Br. J. Cancer 79:1848-1854.

47) Fox S.B. et al., 1997, Relationship of elevated tumour thymidine phosphorylase in node-positive breast carcinomas to the effects of adjuvant CMF. Ann. Oncol. 8: 271-275.
48) Gasparini G. et al., 1999, Clinical relevance of vascular endothelial growth factor and thymidine phosphorylase in patients with node-positive breast cancer treated with either adjuvant chemotherapy or hormone therapy. Cancer J .Sci. Am. 5:101-111.
49) Colomer R. et al., 1997, Low levels of basic fibroblast growth (bFGF) are associated with a poor prognosis in human breast carcinoma. Br. J. Cancer 76:1215-1220.
50) Thibault A. et al., 19898, A phase II study of 5-aza-2'deoxycytidine (decitabine) in hormone independent metastatic (D2) prostate cancer. Tumor. 84: 87-89.
51) Brattstrom D. et al., 1998, Basic fibroblast growth factor and vascular endothelial growth factor in sera from non-small cell lung cancer patients. Anticancer Res. 18: 1123-1127.
52) Bredel M. et al., 1997, Basic fibroblast growth factor expression as a predictor of prognosis in pediatric high-grade gliomas. Clin. Cancer Re.s 3: 2157-2164.
53) Mantovani A. 1994, Tumor-associated macrophages in neoplastic progression: a paradigm for the in vivo function of chemokines. Lab. Invest. 71: 5-16.
54) Graves, D.T. et al., 1991, Biochem. Pharmacol. 41: 333-337, 1991.
55) Oppenheim, J. et al., 1991, Properties of the novel proinflammatory supergene intercrine cytokine family. Annu. Rev. Immunol., 9: 617-621.
56) Polverini P. et al., 1996, How the extracellular matrix and macrophages contribute to angiogenesis-dependent diseases. Er. J. Cancer . 32A: 2430-2437.
57) O'Sullivan C. et al., 1993, Secretion of epidermal growth factor by macrophages associated with breast carcinoma. The Lancet. 342: 148-149.
58) Falcone D. J. et al.,1993, Transforming growth factor-β1 stimulates macrophage urokinase expression and release of matrix-bound basic fibroblast growth factor. J. Cell. Physiol. 155: 595-605.
59) Roger P. et al., 1994, Cathepsin D Immunostaining in Paraffin-Embedded Breast Cancer Cells and Macrophages. Human Pathol. 25: 863-871.
60) Heppner K. et al., 1996, Expression of most matrix metalloproteinase family members in breast cancer represents a tumor-induced host response. Am. J. Pathol. 149: 273-282.
61) Hildenbrand R. et al., 1995, Urokinase and macrophages in tumor angiogenesis. Br. J. Cancer 72: 818-823.
62) Toi M. et al., 1999, Significance of thymidine phosphorylase as a marker of protumor monocytes in breast cancer. Clin. Can. Res. 5:1131-1137.
63) Ishigaki S. et al., 1999, Significance of membrane type 1 matrix metalloproteinase expression in breast cancer. Jpn. J. Cancer Res. 90:516-522.
64) Toi M. et al., 1991, Interleukin-4 is a potent mitogen for capillary endothelium. Biochem. Biophys. Res. Commun .174:1287-1293.
65) Fukushi J. et al., 1998, Novel biological function of Interleukin-4: Formation of tube-like structure by vascular endothelial cells in vitro and angiogenesis in vivo. Biochem. Biophy. Res. Commu. 250:444-448.
66) Thornhill M.H. et al.,1990, IL-4 Regulates endothelial cell activation by IL-1, tumor nerosis factor, or IFN-γ1. J.Immunol. 145: 865-872.
67) Wojta J. et al., 1993, Interleukin-4 stimulates Expression of Urokinase-Type-Plasminogen activator in cultured human foreskin microvascular endothelial cells. Blood 81: 3285-3292.

68) Masinovsky B. et al., 1990, IL-4 acts synergistically with IL-1β to promote lymphocyte adhesion to microvascular endothelium by induction of vascular cell adhesion molecule-1. J. Immunol. 145: 2886-2898.
69) Yao L. et al., 1996, Interleukin 4 or oncostatin M induces a prolonged increase in P-selectin mRNA and protein in human endothelial cells. J. Exp. Med. 1,184:81-92.
70) Kuhn R.et al., 1991, Generation and analysis of interleukin-4 deficient mice. Science 254:707-710.
71) Kopf M.et al., 1993, Disruption of the murine IL-4 gene blocks Th2 cytokine responsers. Nature. 362:245-248.
72) Saleh M. et al., 1999, Effect of in situ retroviral interleukin-4 transfer on established intracranial tumors. J. Natl. Cancer Inst. 91:438-445.
73) Volpert O.V. et al., 1998, Inhibition of angiogenesis by interleukin 4. J. Exp. Med. 188:1039-1046.
74) Koch A.E. et al., 1992, Interleukin-8 as a macrophage-derived mediator of angiogenesis. Science 258:1798-1801.
75) Hu D.E. et al.,1993, Interlerkin-8 stimulates angiogenesis in rats. Inflammation 17: 135-143.
76) Singh R.K. et al., 1994, Expression of interleukin 8 correlates with the metastatic potential of human melanoma cells in nude mice. Cancer Res. 54: 3242-3247.
77) Luca M. et al., 1997, Expression of interleukin-8 by human melanoma cells up-regulates MMP-2 activity and increases tumor growth and metastasis. Am. J. Pathol. 151:1105-1113.
78) Arenberg D.A. et al., 1996, Inhibition of interleukin-8 reduces tumorigenesis of human non-small cell lung cancer in SCID mice. J. Clin. Invst. 97:2792-2802.
79) Galffy G. et al., 1999, Interleukin 8: an autocrine growth factor for malignant mesothelioma. Cancer Res. 59:367-371.
80) Kitadai Y. et al., 1998, expression of interleukin-8 correlates with vascularity in human gastric carcinomas. Am. J. Pathol .152:93-100.
81) Yoneda J. et al., 1998, Expression of angiogenesis-related genes and progression of human ovarian carcinomas in nude mice. J. Natl. Cancer Inst. 90:447-454.
82) Desbaillets I. et al., 1997, Upregulation of interleukin 8 by oxygen-deprived cells in glioblastoma suggests a role in leukocyte activation, chemotaxis, and angiogenesis. J. Exp. Med. 186:1201-1212.
83) Desbaillets I. et al., 1999, Regulation of interleukin-8 expression by reduced oxygen pressure in human glioblastoma. Oncogene 18:1447-1456.
84) Richards B.L. et al., 1997, Coexpression of interleukin-8 receptors in head and neck squamous cell carcinoma. Am. J. Surg. 174:507-512.
85) Eisma R.J. et al., 1999, Role of angiogenic factors: coexpression of interleukin-8 and vascular endothelial growth factor in patients with head and neck squamous carcinoma. Laryngoscope 109:687-693.
86) Fiorentino D. F. et al., 1985, Two types of mouse helper T cell. IV. Th2 clones secrete a factor that inhibits cytokine production by Th1 clones. J. Exp. Med. 170: 2081-2095.
87) Fiorentino D. F. et al., 1991, IL-10 acts on the antigen-presenting cell to inhibit cytokine production by Th1 cells. J. Immunol. 146: 3444-3451.
88) Moore K.W. et al., 1993, Interleukin 10. Annu.Rev. Immunol. 11: 165-190.

89) Hsu D.H. et al., 1990, Expression of IL-10 activity by Epstein-Barr virus protein BCRF1. Science 250: 830-832.

90) Hsu D.H. et al., 1992, Differencial effects of interleukin-4 and -10 on interleukin-2-induced interferon-γ synthesis and lymphokine-activated killer activity. Int. Immunol. 4: 563-569.

91) Huang S. et al., 1996, Interleukin 10 suppresses tumor growth and metastasis of human melanoma cells: potential inhibition of angiogenesis. Clin. Cancer Res. 2: 1969-1979.

92) Richter G. et al., 1993, Interleukin 10 transfected into Chinese hamster ovary cells prevents tumor growth and macrophage infiltration. Cancer Res. 53: 4134-4137.

93) Kundu N. et al., 1997, Interleukin-10 inhibits tumor metastasis, downregulates MHC class I, and enhances NK lysis. Cell Immunol. 180: 55-61.

94) Stearns M.E. et al., 1999, Interleukin 10 (IL-10) inhibition of primary human prostate cell-induced angiogenesis: IL-10 stimulation of tissue inhibitor of metalloproteinase-1 and inhibition of matrix metalloproteinase (MMP)-2/MMP-9 secretion. Clin. Cancer Res. 5: 189-196.

95) Stearns M.E. et al., 1999, Role of interleukin 10 and transforming growth factor β1 in the angiogenesis and metastasis of human prostate primary tumor lines from orthoropic implantsi in severe combined immunodeficiency mice. Clin. Cancer Res. 5: 711-720. 3.

96) Watanabe M. et al., 1997, Regulation of local host-mediated anti-tumor mechanisms by cytokines (Direct and indirect effects on leukocyte recruiment and an angiogenesis. Am. J. Pathol. 150: 1869-1881.

97) Dias S. et al., 1998, IL-12 regulates VEGF and MMPs in a murine breast cancer model. Int. J. Cancer 78: 361-365.

98) Strieter R.M. et al., 1995, Interferon γ-inducible protein 10 (IP-10), a member of the C-X-C chemokine family, is an inhibitor of angiogenesis. Biochem. Biophy. Res. Commu. 210: 51-57.

99) Arenberg D.A. et al., 1996, Interferon-γ-inducible protein 10 (IP-10) is an angiostatic factor that inhibits human non-small cell lung cancer (NSCLC) tumorigenesis and spontaneous metastases. J. Exp. Med. 184: 981-992.

100) Oppenheim J. J. et al., 1991, Properties of the novel proinflammatory supergene "intercrine" cytokine family. Annu. Rev. Immunol. 9: 617-648

101) Sarris, A.H. et al., 1993, Human interferon-inducible protein 10: expression and purification of recombinant protein demonstrate inhibition of early human hematopoietic progenitors. J. Exp. Med. 178: 1127-1132.

102) Sarris, A.H. et al., 1995, Cytokine loops involving interferon-γ and IP-10, a cytokine chenotactic for CD4+ lymphocytes: an explanation for the epidermotropism of cutaneus T-cell lymphoma? Blood 86: 651-658.

103) Valente A.J. et al.,1988, Purification of a monocyte chemotactic factor secreted by nonhuman primate vascular cells in culture. Biochemistry 27:4162-4168

104) Yosimura T. et al.,1990, Identification of high affinity receptors for human monocyte chemoattractant protein-1 on human monocytes. J. Immunol. 145: 292-297

105) Gerszten R.E. et al., 1999, MCP-1 and IL-8 trigger firm adhesion of monocytes to vascular endothelium under flow conditions. Nature 22: 398: 718-723.

106) Bian Z.M. et al., 1999, IL-4 potentiates IL-1 beta-and TNF-alpha-srimulated IL-8 and MCP-1 protein production in human retinal pigment epithelial cells. Curr. Eye. Res. 18: 349-357.

107) Koyama S. et al., 1999, Monocyte chemotactic factors released from type II pneumocyte-like cells in response to TNF-alpha and IL-1 alpha. Eur. Respir. J. 13: 820-828.

108) Biswas P. et al.,1999, Interleukin-6 induces monocyte chemotactic protein-1 in peripheral blood mononuclear cells and in the U937 cell line. Blood 91: 258-265.

109) Marumo T. et al., 1999, Vascular endothelial growth factor activates nuclear factor-kappaB and induces monocyte chemoattractant protein-1 in bovine retinal endothelial cells. Diabetes 48: 1131-1137.

110) Negus R.P. et al., 1995, The detection and localization of monocyte chemoattractant protein-1 (MCP-1) in human ovarian cancer. J. Clin. Invest. 95: 2391-2396.

111) Altenburg A. et al., 1999, CD40 ligand-CD40 interaction induces chemokines in cervical carcinoma cells in synergism with IFN-gamma. J. Immunol. 162: 4140-4147.

112) Amann B. et al., 1998, Urinary levels of monocyte chemo-attractant protein-1 correlate with tumour stage and grade in patients with bladder cancer. Br. J. Urol. 82: 118-121.

113) Gabrilovich D.I. et al., 1996, Production of vascular endothelial growth factor by human tumors inhibits the functional maturation of dendritic cells. Nat. Med. 2: 1096-1103.

114) Melder R.J. et al., 1996, During angiogenesis, vascular endothelial growth factor and basic fibroblast growth factor regulate natural killer cell adhesion to tumor endothelium. Nat. Med. 2: 992-997

MICROVESSEL DENSITY, THYMIDINE PHOSPHORYLASE EXPRESSION AND RESISTANCE OF HEAD AND NECK CANCER TO CHEMO-RADIOTHERAPY

Michael I. Koukourakis, MD
Department of Radiotherapy and Oncology, University of Thessalia, Medical School, Larisa, Greece

1. INTRODUCTION

Cancer growth, invasion and metastasis is strongly dependent on the angiogenic ability. A growing amount of clinicopathological studies confirms that intratumoural microvessel density (MVD) is one of the strongest prognostic variables in human cancer. However, vasculature is also related to the amount of haemoglobin that reaches the tumour, thus oxygenation. Moreover, during chemotherapy, the vessel density strongly defines the drug availability to the tumour. Since euoxic conditions are related to a higher response to radiotherapy (RT) and high drug availability to higher chances for the tumour to respond to chemotherapy, it would be suggested that high microvessel density is a marker of a better response to cytotoxic therapy.

2. MATERIALS AND METHODS

We examined immunohistochemically 2μm tissue sections of paraffin embedded surgically removed material from 98 primary squamous cell cancer lesions of the head and neck area. All patients had locally advanced disease and were treated with induction platinum/5-fluorouracil based

Angiogenesis: From the Molecular to Integrative Pharmacology
Edited by Maragoudakis, Kluwer Academic / Plenum Publishers, New York, 2000

chemotherapy followed by radical radiotherapy or concurrent platinum chemo-radiotherapy (70Gy, 2Gy/fraction, in 7 weeks).

The JC70 monoclonal antibody (Dako) recognising the CD31 (PECAM-1) endothelial cell membrane antigen was used for microvessel staining using the alkaline phosphatase/anti-alkaline phosphatase (APAAP) procedure. Thymidine phosphorylase (TP) expression was assessed with the PGF-44c MAb (Oxford, UK) (Koukourakis et al 1997).

The changes of MVD and of TP expression induced by radiotherapy were assessed in a series of 14 patients treated with RT alone. Biopsies were performed before the beginning and following the delivery of 20Gy of RT.

3. RESULTS AND DISCUSSION

3.1 MVD and treatment outcome.

Patients were divided in 4 groups according to the MVD: Low MVD, Medium MVD, High MVD and very high MVD. The complete response (CR) rate to induction chemotherapy and to radical radiotherapy was very low in cases with low MVD (0% and 45% respectively). This is in accordance with the hypothesis that poor vascularisation correlates with poor oxygen and drug availability and therefore poor response to RT and chemotherapy. Cases with medium MVD had a very high CR rate to induction chemotherapy (45%) and to radical RT (80%). However, further increase of the MVD was associated with a rapid decrease of the CR rate reaching 8% and 40% following induction chemotherapy and RT respectively (Figure 1). Survival analysis showed that 5 year local disease free survival was significantly better in medium MVD cases (70%). Patients with low MVD and high/very high MVD had a very low 5 year survival rate (10-30%).

It seems therefore that a medium degree of MVD allows oxygen and drugs to reach adequate concentrations for a high rate of complete tumour regression and local curability. Although oxygenation and drug availability is much higher in cases with high MVD, this later group of patients had a very poor survival. Analysis of partial response (PR; shrinkage by 50-90%) rate showed that although tumours with high MVD had a very low CR rate, they had the highest PR rate. This shows that high MVD cases respond indeed to chemotherapy and radiotherapy but an additional parameter

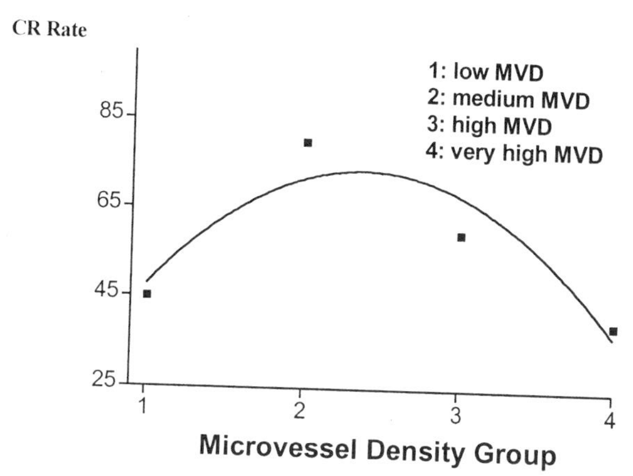

Figure 1. Complete response rate following radical radiotherapy of locally advanced squamous cell carcinoma of the head and neck according to the microvessel density.

present in these cases prevents the tumours from entering a CR, which leads to rapid tumour recurrence and short survival. Therefore, intense angiogenic pathway activation allows the tumour to survive and relapse despite the good oxygen and drug availability.

3.2 Thymidine phosphorylase expression

High nuclear TP expression was associated with high MVD. A significant association of TP expression with cases of very high MVD was observed (Figure 1). Tumours that reached CR following induction chemotherapy of RT had a significantly lower TP nuclear reactivity. Strong nuclear TP expression was associated with poorer local relapse free survival. Since TP has been shown not to relate to response to platinum (Fujieda 1998) but also to relate to a better response to 5-fluorouracil, it may be suggested that the association of TP expression with resistance to cytotoxic therapy is unlikely to be due to a DNA repair role. The strong association of TP with high MVD suggests that angiogenesis related pathways may be involved in the low radiocurability of squamous carcinomas overexpressing TP.

3.3 MVD, TP before and following 20Gy of RT

Small islets of viable tumour cells sparse among large areas of degenerating cancer tissue were observed in biopsies following 20Gy of RT.

In cases that did not reach CR, these islets showed a strong nuclear TP expression and a high microvessel density, while TP expression and vessels were absent in the areas of degenerating carcinoma.

4. IMPLICATIONS

MVD could therefore be used to distinguish three groups of patients with squamous cell head and neck cancer that should be treated with different radiotherapy regimens: A. Patients with low MVD should be treated with cytoprotective agents such as amifostine (Koukourakis et al, 1998) together with larger fractions of radiotherapy to overcome hypoxia related resistance without increasing late sequel. Bioreductive drugs (i.e tirapazamine) that are activated under hypoxic conditions may prove of help. B. Patients with medium MVD are radiocurable with conventionally or slightly accelerated concurrent chemoradiotherapy. C. Patients with high MVD are not curable with currently used regimens. Acceleration of radiotherapy to deliver the whole radiation dose within 3 weeks would prevent angiogenic tumour regeneration during the course of radiotherapy. Concurrent use of drugs that block cells to the G2/M radiosensitive phase (i.e taxanes) may be of value. Antiangiogenic therapy or drugs activated by TP (in TP overexpressing cases) may substantially improve the efficacy of RT in this particular subset of patients (Mauceri et al 1998).

5. CONCLUSION

Several studies in human cancers suggest a linear correlation between the degree of tumour angiogenesis and survival after surgery (Giatromanolaki et al., 1997). The present results show that response and microvessel density follow a bell-shaped relation, the response being better in cases with an intermediate vascular grade. This may be a consequence of 2 vasculature-dependent factors, i.e., the drug/oxygen availability and the ability of cancer cells to undergo rapid repopulation in optimally oxygenated conditions. Thymidine phosphorylase, a potent angiogenic factor in head and neck cancer, is involved in this angiogenesis related resistance to cytotoxic therapy. Combination of anti-angiogenic agents, thymidine phosphorylase inhibitors or drugs activated by TP (Capecitabin) with radiotherapy may be of importance in the eradication of highly angiogenic head and neck carcinomas.

ACKNOWLEDGMENTS

The work is supported by the Tumour and Angiogenesis Research Group and the Imperial Cancer Research Fund.

REFERENCES

Fujieda S, Sunaga H, Tsuzuki H, et al (1998). Expression of platelet-derived endothelial cell growth factor in oral and oropharyngeal carcinoma. *Clin Cancer Res.,* 4:1583-1590

Giatromanolaki A, Koukourakis M, Theodossiou D, et al (1997). Comparative evaluation of angiogenesis assessment with anti-Factor VIII and anti-CD31 immunostaining in non small cell lung cancer.*Clin Cancer Res* 3:2493-2500

Koukourakis MI, Giatromanolaki A, O'Byrne K, et al (1997). Platelet-Derived Endothelial Cell Growth Factor expression correlates with tumour angiogenesis and prognosis in non-small cell lung cancer. *Br J Cancer* 4:477-481

Koukourakis MI. The use of Ethyol in chemotherapy and radiation therapy. In «Cytoprotection in Cancer Therapy: Improving the Therapeutic Index of Cancer Treatment.», Ethyol Investigators Meeting; ASCO May 1998, Los Angeles, California, USA.

Mauceri HJ, Hanna NN, Beckett MA, et al (1998). Combined effects of angiostatin and ionizing radiation in antitumor therapy. Nature, 394: 287-291

DIFFERENTIAL EXPRESSION OF ANGIOGENIC AND OF VASCULAR SURVIVAL FACTORS IN NON-SMALL CELL LUNG CANCER (NSCLC)

Alexandra Giatromanolaki, MD
Department of Pathology, Democritus University of Thrace,Alexandroupolis, Greece

1. INTRODUCTION

It is well established that tumour growth depends on the presence of a variety of angiogenic factors such as Vascular Endothelial Growth Factor (VEGF), Platelet-Derived Endothelial Cell Growth Factor (PD-ECGF) and basic Fibroblast Growth Factor (bFGF). These factors are involved in the endothelial cell proliferation, migration and tube formation. The present study provides strong evidence that growth factors responsible for the maintenance of the integrity of the newly formed vasculature is also an important process during tumour growth and progression.

1.1 Initial Comments

The assumption that the formation of new vessels is the only important vasculature related parameter is a static view of the live process of tumour growth. Endothelial cell migration and vessel formation is indeed one step. Immediately following their formation, the endothelial cells composing the vessels undergo a dynamic process of proliferation and apoptosis. This process is the so-called remodelling of the established vasculature. A balance between proliferation and apoptosis will allow the vessels to survive. Prevalence of the apoptotic pathway will lead to vascular degradation.

Angiogenesis: From the Molecular to Integrative Pharmacology
Edited by Maragoudakis, Kluwer Academic / Plenum Publishers, New York, 2000

The tumour grows forming sheaths of cancer tissue. The invading front becomes an inner layer within a while as the tumour continues to invade and grow. Endothelial cells are attracted from the normal surrounding and new vessels are formed. This layer with the newly formed vessels consists the invading front of the tumour. Later on, this invading front will be replaced by cancer cells that will grow around the vessels and a new invading front will appear.

In the invading front the vessel density depends on growth factors related to endothelial cell migration, proliferation and vessel formation. As long as the tumour maintains its specific angiogenic phenotype, there is no reason for the vessel density to change from the one invading front to the subsequent one. However, as soon as one invading layer becomes an inner tumour area, the vessel density depends on multiple processes such as the ability of endothelial cells to migrate to inner tumour areas or the expression of growth factors related to the proliferation and apoptosis of the endothelial cells composing the already formed vessels. In a study by Fox et al [1993], single endothelial cell proliferation was found to occur in the periphery of breast carcinomas while inner vascularisation was shown to undergo a continuous remodelling.

2. MATERIALS AND METHODS

We examined immunohistochemically 5μm tissue sections of paraffin embedded surgically removed material from 178 primary lung lesions, histologically shown to be non-small cell lung carcinoma. 116 were squamous cell carcinomas and 62 were adenocarcinomas. The JC70 monoclonal antibody (Dako), recognising the CD31 (PECAM-1) endothelial cell membrane antigen, was used for microvessel staining on 5μm paraffin embedded sections using the alkaline phosphatase/anti-alkaline phosphatase (APAAP) procedure. VEGF and PD-ECGF expression was assessed with the VG1 monoclonal antibody recognising the 121, 165 and 189 isoforms of VEGF and the PGF-44c MAbs (Oxford, UK) respectively (Giatromanolaki et al 1998, Koukourakis et al 1997). We also used the LH39 MAb, which recognises an epitope in the lamina lucida of vessels (ICRF, UK).

The invading front was identified and microvessel counting was performed in x200 optical field in the normal tissue adjacent to the tumour and in the tumoural invading layer. The field was thereafter changed by moving perpendicularly to the invading front towards the tumour centre and microvessel counting was performed in two consecutive inner areas. Each field covered a 2mm tissue thickness. A percentage of 50% of cancer cells with reactivity for VG1 and PGF-44c was required to group a case as positive for VEGF and PD-ECGF respectively. Double staining with LH39 and anti-CD31 was used to assess the vascular maturation index (VMI=double stained LH39-CD31 vessels / single CD31 stained).

3. RESULTS

3.1 The EDVIN score

According to the microvessel score (MS) in the invading front and in inner areas we grouped our cases in three EDVIN (edge vs. inner) types of vascularisation. The EDVIN 1 comprises cases with low MS throughout the tumour (low angiogenic ability), the EDVIN 2 comprises cases with high MS in the invading front but low in inner tumour areas (high angiogenic but low vascular survival ability) and the EDVIN 3 comprises cases with high MS in both the invading front and inner areas (high angiogenic and high vascular survival ability).

Both EDVIN 2 and 3 cases were associated with high incidence of node involvement, which shows that high angiogenic ability is required for the lymphatic spread of the NSCLC. Necrosis was significantly less frequent in EDVIN 3 cases showing that impaired vascular survival ability accounts for the degree of tumour necrosis. No association was found with T-stage, histological grade and histological type. EDVIN 1 patients had a significantly better outcome, while EDVIN 3 cases had the worse prognosis. The EDVIN score was the only parameter approaching prognostic significance at multivariate analysis.

3.2 VEGF and PD-ECGF expression

Expression of VEGF was significantly associated with EDVIN 3 cases, which shows that VEGF is an angiogenic and a vascular survival factor. On the contrary, PD-ECGF was more frequent in EDVIN 2 cases showing that although PD-ECGF is angiogenic, it does not sustain the maintenance of the newly formed vasculature. It is therefore suggested that angiogenic factors differ as far as their vascular survival properties are concerned.

3.3 VEGF and PD-ECGF expression

Expression of PD-ECGF, shown to be associated with EDVIN 2 score, was significantly associated with low vascular maturation index, thus vessels with immature lamina lucida. This was not observed for VEGF expressing cases. This observation suggests that vascular maturation is an important step in the survival of the newly formed vessels.

3.4 Single endothelial cell score

The density of single endothelial cells was assessed in the invading front and in inner areas of the tumours. The density was rapidly decreasing in both EDVIN 2 and 3 cases showing that endothelial cell migration in inner areas is unlikely to occur and does not contribute to the survival of the new vessels.

4. CONCLUSIONS

Vascular survival is a process that follows angiogenesis. Tumours largely differ as for their angiogenic and vascular survival ability is concerned. Although PD-ECGF is an angiogenic factor, it had low vascular survival abilities, which is in accordance with the low maturation of vessels produced under PD-ECGF stimulation. On the contrary VEGF is both an angiogenic and vascular survival factor. Further studies are required to assess the clinical role of these observations.

ACKNOWLEDGMENTS

The work is supported by the Tumour and Angiogenesis Research Group and the Imperial Cancer Research Fund.

REFERENCES

Fox SB, Gatter KC, Bicknell R, Going JJ, Stanton P, Cooke TG, Harris AL 1993. Relationship of endothelial cell proliferation to tumour vascularity in human breast cancer. *Cancer Res* 53:4161-4163

Giatromanolaki A, Koukourakis MI., Kakolyris S, Turley H, O'Byrne K, Scott P.A.E., Pezzella F, Georgoulias V, Harris AL,Gatter KC 1998. Vascular endothelial growth factor, wild-type p53 and angiogenesis in early operable non-small cell lung cancer. *Clin Cancer Res* 4:3017-3024

Koukourakis MI, Giatromanolaki A, O'Byrne K, Comley M, Whitehouse R, Talbot DC,Gatter KC, and Harris AL 1997. Platelet-Derived Endothelial Cell Growth Factor expression correlates with tumour angiogenesis and prognosis in non-small cell lung cancer. *Br J Cancer* 4:477-481

THYMIDINE PHOSPHORYLASE ACTIVITY IN NORMAL, HYPERPLASTIC AND NEOPLASTIC ENDOMETRIUM – CORRELATION WITH INTRATUMORAL ANGIOGENESIS

EFTHIMIOS SIVRIDIS
Department of Pathology, Democritus University of Thrace, Alexandroupolis, Greece

1. INTRODUCTION

Thymidine phosphorylase (TP), also known as platelet-derived endothelial cell growth factor (PD-ECGF), is specifically involved in the reversible dephosphorylation of thymidine to thymine and 2-deoxy-D-ribose-1-phosphate. It is a potent angiogenic factor shown to induce endothelial cell migration and proliferation (Moghaddam et al 1995).

The enzyme is known to be expressed in the normal endometrium (Fox et al 1995, Zhang et al 1997), but a similar activity was disputed in endometrial adenocarcinomas (Zhang et al 1997, Fujiwaki et al 1998). Furthermore, no information is at present available regarding the production of the enzyme in the various forms of endometrial hyperplasia, while information on the role of TP in relation to angiogenesis in these tissues is incomplete.

2. THYMIDINE PHOSPHORYLASE IN NORMAL AND HYPERPLASTIC ENTOMETRIUM

We explored the expression of thymidine phosphorylase in the endometrium by an immunohistochemical technique, using a specific monoclonal antibody (P-GF.44C) to thymidine phosphorylase (Fox *et al.* 1995). This antibody detected platelet-derived endothelial cell growth factor and thymidine phosphorylase (PD-ECGF/TP) activity. Our investigation revealed that thymidine phosphorylase activity is invariably expressed in the normal, hyperplastic and neoplastic endometrium (Sivridis et al 1999a).

In the normal endometrium, thymidine phosphorylase expression has well defined cellular and tissue patterns which are dependent on the phase of the menstrual cycle. Thus, in the early and mid proliferative phase, TP was expressed exclusively in the basal endometrium, a site where neoangiogenesis is presumed to occur. The expression was cytoplasmic in the glandular epithelial cells and nuclear in the stromal cells. It was invariably patchy. In the late proliferative phase endometrium, the reactivity of the epithelial cells extended towards the lower third of the functionalis. This immunohistochemical picture remained, by and large, unchanged during the early and mid secretory phase endometrium, however, in the late secretory phase of the cycle, a mixed cytoplasmic/nuclear thymidine phosphorylase activity was uniformly expressed by every single epithelial cell in every single endometrial gland throughout the entire endometrium. The reaction was strong, in contrast to the remaining phases of the menstrual cycle, where the intensity of the staining was, in general, weak and varied within individual cases. Stromal cell reactivity was not a feature.

In simple endometrial hyperplasia TP expression showed subcellular staining patterns identical to normal proliferative endometrium, with a distribution which was usually confined to a few, rather weakly proliferating, glands and to the adjacent periglandular stroma of the deep endometrium (Sivridis *et al.* 1999a). With regard to complex and atypical hyperplasias, most, of the few, cases studied showed a more extensive distribution of the enzyme and a mixed cytoplasmic/nuclear immunostaining, but stromal TP expression was less apparent.

In all cases of normal and hyperplastic endometrium, scattered macrophages were stained positively for TP, without following a particular cyclical pattern, while in approximately 10% of the cases a number of endothelial and myometrial cells expressed a similar enzymatic activity.

2.1 Conclusions

Thymidine phosphorylase is consistently expressed in the normal and hyperplastic endometrium, suggesting a role in physiological and pathological angiogenesis. In the normal endometrium menstrual cycle related changes occur. In the endometrial hyperplasias TP distribution is at random. Angiogenesis seems to be the result of a co-ordinated nuclear/ cytoplasmic TP activity. It is induced through a paracrine pathway.

3. THYMIDINE PHOSPHORYLASE IN NEOPLASTIC ENTOMETRIUM

In endometrial carcinomas, TP was expressed in the tumour itself (cancer cells) and in the host tissues (stromal fibroblasts, myometrial cells) (Sivridis *et al.* 1999b). It was also expressed in the tumour - associated macrophages (TAMs). The pattern of staining was mainly nuclear or mixed nuclear/ cytoplasmic, and only exceptionally was purely cytoplasmic. An exclusively cytoplasmic pattern of staining was established for the foamy macrophages of the stroma.

Thymidine phosphorylase positive cancer cells were seen scattered among the neoplastic epithelium. A "high" TP expression in these cells was a rare event, with only 3.2% of endometrial carcinomas reacting with anti-TP in more than 50% of the neoplastic cell population, and 12% of the cases expressing the enzyme in more than 10% of the total cancer cells (Sivridis *et al.* 1999b). By contrast, TP fibroblastic reactivity in more than 50% of the fibroblasts was observed in over 1/3 of the endometrial carcinomas, and, to an equal extent, reactivity was noted at the cells confronting the invading tumour front (stromal fibroblasts and, mainly, myometrial cells) Foamy macrophages were present in approximately 1/4 of the carcinomas studied and, in all cases, they were reactive to P-GF.44C. A prominent lymphocytic response was recognised in approximately 1/5 of the cases, but lymphocytes remained unreactive to monoclonal anti-TP.

A strong TP expression in >10% of the cancer cells was not significantly associated with any of the several histopathological parameters studied, including tumour type and grade, depth of myometrial invasion, lymph-vascular space penetration and FIGO stage of disease. Similarly, a "high" TP reactivity in cancer cells was not statistically associated with either angiogenesis or bcl-2 and p53 oncoproteins (Sivridis *et al.* 1999b).

A high TP reactivity in >50% of the stromal fibroblasts was significantly associated with the presence of TP-laden foamy macrophages and an intense lymphocytic response, but not with angiogenesis.

A high TP reactivity in >50% of the host cells (stromal fibroblasts/ myometrial cells) confronting the invading tumour front was associated with a marked lymphocytic response, and the adverse prognostic factors of high histological grade, deep myometrial invasion, non-endometrioid type of carcinomas and with advanced stage of disease (Sivridis *et al.* 1999b). These associations were of statistical significance. Equally, within stage I endometrioid carcinomas, a high stromal/myometrial cell reactivity at the invading front was significantly associated with high tumour grade and deep myometrial invasion. There was no statistical association with angiogenesis.

3.1 Conclusions

Thymidine phosphorylase is primarily produced by fibroblasts, and only to a limited extent by cancer cells. The expression is probably regulated by cytokines, released by lymphocytes and macrophages, being present in large numbers in the stroma and the invading tumour front. Thymidine phosphorylase in cancer cells is deprived of any practical significance, since no significant association with known histological parameters, the oncoproteins p53, bcl-2 or angiogenesis was noted. Thymidine phosphorylase is neither a bcl-2 nor p53 dependent function in endometrial carcinomas. High TP activity at the invading tumour front may promote tumour invasion and progression for all types of endometrial cancer, and for stage 1 endometrioid adenocarcinoma; it signifies a more aggressive tumour. There is no significant association between high TP expression in cancer or host cells and angiogenesis.

4. ANGIOGENESIS IN THE ENDOMETRIUM

Angiogenesis is the phenomenon of new blood vessels formation from a preexisting vascular network (Folkman 1985). It is a physiologically and pathologically occurring process. Tumour growth, invasion and metastases and angiogenesis dependent.

In the uterine mucosa, it was shown that the stroma of a normal mid-secretory and hyperplastic endometrium (simple, complex and atypical hyperplasia) was more vascular than that of proliferative endometrium. Furthermore, the stroma of an adenocarcinoma, though reduced in proportion to the epithelium, was more vascular than that of normal or hyperplastic endometrium (Morgan *et al.* 1996a).

In endometrial carcinomas most of the few studies on the subject reported a negative correlation of microvessel density (MVD) with tumour type and grade, depth of myometrial invasion, lymphatic vascular space invasion, state of the endometrium adjacent to an adenocarcinoma (whether atrophic or hyperplastic) and FIGO stage of disease (Morgan *et al.* 1996a and 1996b., Giatromanolaki *et al.* 1999). One study showed a positive association with histological grade, depth of myoinvasion and lymphatic vascular space invasion (Kaku *et al.* 1996), and another study demonstrated an association with the state of the adjacent non-neoplastic endometrium (whether atrophic or hyperplastic) (Morgan *et al.* 1996b). There was no association between MVD and nuclear p53 overexpression (Giatromanolaki *et al.* 1999). By contrast, a strong cytoplasmic and/or perinuclear expression of bcl-2 in more than 10% of neoplastic cells was more frequent in endometrial carcinomas of low MVD (Giatromanolaki *et al.* 1999). With regard to intratumoral angiogenesis and prognosis in endometrial carcinomas there is a general agreement that low MVD is connected with a favourable survival (Kirschner *et al.* 1996, Salvesen *et al.* 1998). Using the JC70 monoclonal antibody recognising CD31 and the APAAP procedure, we have also shown that low MVD (<15), together with early stage of disease and the absence of lymphatic vascular space invasion, is associated with an improved survival (univariate analysis) (Giatromanolaki *et al.* 1999). In multivariate analysis, however, the only independent variable noted was FIGO stage of disease. Within stage I endometrioid adenocarcinomas, only intratumoral angiogenesis was associated with prognosis (univariate analysis): high MVD (>30) cases had a significantly worse prognosis compared to medium MVD (15-30). Low MVD adenocarcinomas, on the other hand, were associated with an intermediate prognosis, indicating that other factors, such as hypoxia and related mechanisms, may also be important (Giatromanolaki *et al.* 1999).

4.1 Conclusions

Thymidine phosphorylase is not a major angiogenic factor in endometrial cancer, as no association with cancer cells, fibroblasts or myocytes at the invading tumour front was noted. Other angiogenic factors, such as VEGF, bFGF, may be important. The loss of bcl-2 function may be associated with the up-regulation of angiogenesis in this tumour, just as in other human malignancies. High MVD seems to be a feature of a more aggressive subtype of endometrial cancer, i.e., carcinomas arising in an atrophic endometrium, but does not correlate with other prognostic variables. Intratumoral angiogenesis is a putative prognostic indicator for all types and stages of endometrial cancer. Within stage I endometrial carcinomas, MVD is the only parameter associated with prognosis.

ACKNOWLEDGMENT

This study was supported financially by the Tumour Angiogenesis Research Group and the Imperial Cancer Research Fund.

REFERENCES

Folkman, J., 1985, Tumor angiogenesis. *Adv. Cancer Res.* 43: 175-203.

Fox, S.B., Moghaddam, A., Westwood, M., Turley, H., Bicknell, R., Gatter, K.C., Harris, A.L., 1995, Platelet-derived endothelial cell growth factor / thymidine phosphorylase expression in normal tissues: an immunohistochemical study. *J. Pathol.* 176: 183-190.

Fujiwaki, R., Hata, K., Iida, K., Koike, M., Miyazaki, K., 1998, Immunohistochemical expression of thymidine phosphorylase in human endometrial cancer. *Gynecol. Oncol.* 68: 247-252.

Giatromanolaki, A., Sivridis, E., Koukourakis, M.I., Gatter, K.C., Harris, A.L., 1999, Intratumoral angiogenesis: a new prognostic indicator for stage I endometrial adenocarcinomas? *Oncol. Res.* (in press)

Kaku, T., Kamura, T., Kinukawa, N., Kobayashi, H., Sakai, K., Tsuruchi, N., Saito, T., Kawauchi, S., Tsuneyoshi, M., Nakano, H., 1997, Angiogenesis in endometrial carcinoma. *Cancer* 80: 741-747.

Kirschner, C.V., Alanis-Amezcua, J.M., Martin, V.G., Luna, N., Morgan, E., Yang, I.J., Jordan, E.L., 1996, Angiogenesis factor in endometrial carcinoma: a new prognostic indicator? *Am. J. Obstet. Gynecol.* 174, 1882-1884.

Moghaddam, A., Zhang, H.-T., Fan, T.-P.D., Hu, D.-E., Lees, V.C., Turley, H., Fox, S.B., Gatter, K.C., Harris, A.L., Bicknell, R., 1995, Thymidine phosphorylase is angiogenic and promotes tumor growth. *Proc. Nat.l Acad. Sci.* 92: 998-1002.

Morgan, K.G., Wilkinson, N., Buckley, C.H., 1996a, Angiogenesis in normal, hyperplastic, and neoplastic endometrium. *J. Pathol.* 179, 317-320.

Morgan, K.G., Wilkinson, N., Buckley, C.H., 1996b, Angiogenesis in endometrial carcinoma. *Int. J. Gynecol. Cancer* 6, 385-388.

Salvesen, H.B., Iversen, O.E., Akslen, L.A., 1998, Independent prognostic importance of microvessel density in endometrial carcinoma. *Br. J. Cancer* 77, 1140-1144.

Sivridis, E., Giatromanolaki, A., Koukourakis, M., Bicknell, R., Harris, A.L., Gatter, K.C., 1999a, Thymidine phosphorylase expression in normal and hyperplastic endometrium. Submitted for publication.

Sivridis, E., Giatromanolaki, A., Koukourakis, M., Bicknell, R., Harris, A.L., Gatter, KC., 1999b, Thymidine phosphorylase expression in endometrial carcinomas. Submitted for publication.

Zhang, L., Mackenzie, I.Z., Rees, M.C.P., Bicknell, R., 1997, Regulation of the expression of the angiogenic enzyme platelet-derived endothelial cell growth factor / thymidine phosphorylase in endometrial isolates by ovarian steroids and cytokines. Endocrinology 138: 4921-4930.

Preclinical Developments of Angiogenesis Inhibitors

INHIBITION OF VEGF SIGNAL TRANSDUCTION

Identification of ZD4190

Stephen R. Wedge and Donald J. Ogilvie
Cancer and Infection Research Department, AstraZeneca Pharmaceuticals, Alderley Park, Macclesfield, Cheshire, SK10 4TG, U.K.

1. INTRODUCTION

There is a significant unmet need for cytostatic (i.e. non-cytotoxic) therapy to stabilise or slow the progression of solid tumour disease, particularly in non-hormone dependent cancers. Anti-angiogenics are one approach which might satisfy this need. Experience from the anti-hormonal cytostatics suggests that *chronic* dosing will be required and hence oral delivery of drug is preferable. Side effects will also need to be compatible with the clinical benefit.

1.1 VEGF receptor tyrosine kinase (VTK) activity as a target

There is evidence that VEGF contributes to tumour growth through the promotion of both angiogenesis and vascular permeability (Folkman 1995, Senger *et al.*, 1993), and sequestration of VEGF with antibody has been shown to reduce tumour growth in animal models (Kim *et al.*, 1993). There are several technical options for modulation of VEGF activity. We have sought to abate VEGF signalling by inhibiting VEGF receptor-associated tyrosine kinase (VTK) activity. The *in vitro* properties of such a compound would include potent inhibition of VTK enzyme(s) in isolation and in endothelial cells at concentrations which do *not* show direct effects on the

Angiogenesis: From the Molecular to Integrative Pharmacology
Edited by Maragoudakis, Kluwer Academic / Plenum Publishers, New York, 2000

normal growth of endothelial or tumour cells. This ensures that anti-tumour effects *in vivo* can be attributed to inhibition of VEGF signalling rather than a direct effect on tumour cell proliferation.

2. SCREENING FOR VTK INHIBITORS

In order to screen for VTK inhibitors, the cytoplasmic domain of the VEGF receptor Flt (cFlt) was cloned and expressed. The extracted enzyme was intrinsically active. Screening of a panel of kinase inhibitors revealed a distinct selectivity profile for cFlt compared to another tyrosine kinase, EGFR. Hints of a structure-activity relationship (SAR) in an anilinoquinazoline series were confirmed by robotic synthesis. Further medicinal chemistry led to the identification of highly potent cFlt inhibitors which were also active against another VEGF receptor, KDR.

In order to satisfy the perceived clinical requirements of an anti-angiogenic (see above), several *in vivo* attributes were considered to be necessary. A successful VTK compound would need to have PK compatible with chronic oral dosing, anti-tumour activity at well-tolerated doses and evidence to support the proposed mode of action *in vivo*. Unfortunately, neither the degree nor duration of VTK inhibition required to achieve anti-tumour activity in animal models (or man) was known. Furthermore, the best early VTK inhibitors identified had suboptimal bioavailability in mice.

Further medicinal chemistry led to a series of compounds with much improved bioavailability though this was accompanied by some loss of potency. Encouragingly, consistent with the proposed anti-angiogenic mechanism, a prototype compound was found to be active in *all* human tumour xenograft models tested (n=9). Comparison of a series of compounds revealed that the best anti-tumour activity correlated with sustained blood levels of drug. This indicates that *sustained* inhibition of VEGF signalling is required - at least in a rapidly growing xenograft model. ZD4190 (*N*-(4-bromo-2-fluorophenyl)-6-methoxy-7-[2-(1*H*-1,2,3-triazol-1-yl)ethoxy]quinazolin-4-amine) was selected from this group of compounds for clinical development.

2.1 ZD4190 profile

ZD4190 is a potent (IC50 ~0.05uM) inhibitor of VTK enzyme and VEGF signalling in HUVECs but is only active against tumour cell proliferation at higher concentrations (IC50 >10uM). In xenograft models, ZD4190 was significantly active (50 mg/kg/day) against all tumours tested and was well tolerated during chronic dosing (28 days). In all cases, tumours were established (~0.5cm^3) before dosing was commenced. Sustained inhibition of growth has been demonstrated in one particularly aggressive prostate tumour xenograft model (PC-3) with daily dosing of ZD4190 for 10 weeks. In line with expectations, when therapy was withdrawn tumour growth resumed at the normal rate after a short delay. Acute effects of ZD4190 on tumour vascular permeability have also been demonstrated using dynamic contrast medium-enhanced magnetic resonance imaging (Wedge *et al.*, 1999).

Further rat pharmacology provides support for the mode of action of ZD4190. We have previously described the use of the anaesthetised rat to show the acute hypotensive effects of large doses of VEGF or bFGF (Curwen and Ogilvie, 1997). Orally dosed ZD4190 (25 mg/kg) inhibits the effects of VEGF, but not bFGF, in this model indicating that this compound can inhibit VEGF activity *in vivo*. During chronic dosing in the immature (growing) rat, ZD4190 produces epiphyseal hypertrophy at the femur growth plate. This is consistent with the inhibition of angiogenesis which is a critical step in the conversion of cartilage to bone during growth. Similar effects have also recently been described with a humanised anti-VEGF monoclonal antibody in young primates (Ryan *et al.*, 1999).

3. CONCLUSION

In summary, we have identified ZD4190, a potent inhibitor of VTK activity *in vitro* with pharmacokinetic properties compatible with chronic oral administration. This compound significantly inhibits the growth of a range of histologically distinct tumour xenografts *in vivo*, and at doses which are unlikely to have any direct effect on tumour cells. This and other *in vivo* evidence supports the proposed mode of action of ZD4190.

ZD4190 is one of a series of compounds in clinical development for the treatment of cancer and potentially other disorders.

ACKNOWLEDGMENTS

This work was accomplished with considerable Bioscience and Chemical Team effort. In particular, contributions made by M. Dukes, J. Kendrew, J.O.Curwen, L.F.Hennequin, E.S.E. Stokes, P.Ple, A.P. Thomas and C. Johnstone are to be acknowledged.

REFERENCES

Curwen, J.O. and Ogilvie, D.J. , 1997, Production of the angiogenic factors VEGF and bFGF at tumour sites may also confer an acute haemodynamic advantage to the tumour. *Br. J. Cancer* **75**: P89.

Folkman J., 1995, Angiogenesis in cancer, vascular, rheumatoid and other disease. *Nature Medicine*, **1**: 27-31.

Kim, K.L., Li, B., Winer, J., Armanini, M., Gillett, N., Phillips, H.S., and Ferrara, N., 1993, Inhibition of vascular endothelial growth factor-induced angiogenesis suppresses tumor growth in vivo. *Nature* **362**: 841-844.

Ryan, A.M., Eppler, D.B., Hagler, K.E., Bruner, R.H., Thomford, P.J., Hall, R.L., Shopp, G.M., and O'Niell, C.A., 1999, Preclinical safety evaluation of rhuMAbVEGF, an antiangiogenic humanised monoclonal antibody. Tox. Path., **27**: 78-86.

Senger, D.R., Van De Water, L., Brown, L.F., Nagy, J.A., Yeo, K.T., Yeo, T.K., Berse, B., Jackman, R.W., Dvorak, A.M., and Dvorak, H.F., 1993, Vascular permeability factor (VPF,VEGF) in tumor biology. *Cancer and Metastasis Reviews* **12**: 303-324.

Wedge, S.R., Waterton, J.C., Tessier, J.J., Checkley, D., Dukes, M., Kendrew, J. and Curry, B., 1999, Effect of the VEGF receptor tyrosine kinase inhibitor ZD4190 on vascular endothelial permeability. Proc. Amer. Assoc. Cancer Res. **40**: 2741.

COMBRETASTATINS NOVEL VASCULAR TARGETING DRUGS FOR IMPROVING ANTI-CANCER THERAPY

COMBRETASTATINS AND CONVENTIONAL THERAPY

M. R. Horsman, R. Murata, T. Breidahl, F. U. Nielsen, R. J. Maxwell, H. Stødkiled-Jørgensen and J. Overgaard.
Danish Cancer Society, Department of Experimental Clinical Oncology, Aarhus University Hospital and MR-Centre, Skejby Hospital, Aarhus, Denmark

Abstract Combretastatins are a new class of compounds that appear to have anti-tumour activity as a result of specifically targeting the vasculature of tumours. The aim of this study was to investigate the potential of combretastatin A-4 disodium phosphate (CA4DP) to induce vascular effects in a C3H mouse mammary carcinoma, and to see if the anti-tumour response could be improved by combining the drug with conventional anti-cancer therapies. It was found that CA4DP (250 mg/kg) significantly decreased tumour perfusion within 30 minutes after injection and maintained this decrease for several hours, although there was a return to normal by 24 hours. Similar changes were seen in the tumours bioenergetic and oxygenation status. The drug also significantly increased tumour necrosis and had a small inhibitory effect on tumour growth. It was also able to enhance the tumour response to radiation and hyperthermia, when given at the same time or 30 minutes after the radiation and hyperthermia, respectively. Giving the drug 1 hour after cisplatin injection only resulted in a tumour response that was no greater than additive. These results confirm the anti-vascular effects of CA4DP and demonstrate its potential to enhance the anti-tumour activity of conventional therapy.

Introduction

A critical aspect in the development and continued growth of most solid tumours is the presence of a functional vascular supply. Indeed, it is known that most solid tumours are unable to grow beyond 1-2 mm in diameter without evolking a blood supply (Folkman, 1976; 1990). This vascular supply can be aquired, in part, by incorporation of existing blood vessels, however, it is now well established that the majority of tumour blood vessels are newly formed as a result of angiogenesis, which is a highly complex process involving a number of steps that are triggered by the release of specific growth factors (Paweletz and Knierim, 1989; Blood and Zetter, 1990).

The importance of the tumour vasculature makes it an attractive target for therapy. Most of the research in this area has focused on preventing the growth of the neo-vasculature, so called anti-angiogenesis therapy (Mahadevan and Hart, 1990; Moses and Langer, 1991). The list of compounds that have been reported to possess anti-angiogenic activity is extensive and many of these agents have even progressed to preliminary clinical testing (NCI report, 1998). An alternative approach for targeting the tumour vasculature involves the use of agents that can actually damage the already established tumour vessels. This is not a new concept since it was first demonstrated with the tubulin binding agent colchicine back in the 1940's (Ludford, 1945). Since then a number of agents have been shown to be capable of damaging tumour vessels. These include hyperthermia, photodynamic therapy, interleukins, interferon, tumour necrosis factor, flavone acetic acid (FAA), dimethylxanthenone acetic acid (DMXAA), the dolestatins and combretastatins (Baugeley et al., 1991; Denekamp and Hill, 1991; Horsman, 1994; Chaplin et al. 1996).

Combretastatins are a new group of drugs, originally derived from the African Bush Willow *Combretum caffrum*, which preliminary data suggests possess anti-tumour activity that involves a vascular component (Chaplin et al., 1996). The lead compound in this series, combretastatin A-4 disodium phosphate (CA4DP), has been shown to induce vascular damage that leads to the development of substantial necrosis in a range of animal tumours (Dark et al., 1997; Beauregard et al., 1998; Horsman et al., 1998; Li et al., 1998; Zhao et al., 1999). Furthermore, it produces these effects at drug doses that are well below the maximum tolerated dose in such animals and as a result the drug is now undergoing phase I clinical testing.

Despite inducing such profound damage in tumours the drug is unable to kill all the tumour cells, and the tumour is able to regrow from the surviving population. This suggests that the likely clinical application of CA4DP will be when combined with a treatment that can effectively kill these surviving cells. The aim of this study was to investigate the potential of CA4DP to induce vascular damage in a C3H mouse mammary carcinoma, and to see whether it could be combined with more conventional therapies, such as radiation chemotherapy and hyperthermia, to improve the anti-tumour response.

Materials and Methods.

Animal and tumour model.

A C3H mammary carcinoma grown in the right rear foot of 10-14 week-old female CDF1 mice was used in all experiments. Details of its derivation and maintenance have been previously described (Overgaard, 1980). Experimental tumours were produced from dissected large flank tumours, by mincing the tumour material with scissors and injecting 5-10 μl subcutaneously into the dorsal surface of the foot of the experimental animals. All experiments were performed on restrained, but non-anaesthetised animals when tumours had reached a volume of about 200 mm^3, which generally occurred 2-3 weeks after challenge, and was determined by the formula: $D1 \times D2 \times D3 \times \pi/6$ (where the D values represent three orthogonal diameters). All experiments were performed under Nationally approved guidelines for animal welfare.

Drug preparation.

CA4DP was supplied by Oxigene Europe AB (Lund, Sweden) and prepared fresh before each experiment by dissolving in saline. Cisplatin was supplied in solution by Teva Pharmaceutical Industries and diluted to the final concentrations in saline. All drug injections were given intraperitoneally (i.p.) at a constant injection volume of 0.02 ml/g mouse body weight.

Tumour perfusion estimates.

Tumour blood perfusion was assessed using the $^{86}RbCl$ extraction technique first described by Sapirstein (1958). Basically, mice were intravenously injected with 0.1 ml RbCl and 90 seconds later the mice were sacrificied by cervical dislocation and the tumour excised. Care was taken to separate the tumour from other tissues and to wipe off any excess blood. The tumour material was then transferred to pre-weighed counting tubes, reweighed and radioactivity in the samples counted on a Packard Gamma counter. Tails of injected mice were also excised and counted to check on residual activity at the site of injection. Samples were rejected if the tail counts were >10% of the injected solution. Radioactive counts per minute were expressed as percent injected/g of tissue.

Bioenergetic status.

Tumour bearing mice were restrained in specially constructed perspex jigs and the tumour bearing leg exposed and attached to the jig with tape without interferring with blood flow to the foot. An i.p. line was then inserted into the mice and a 9 mm two-turn surface coil placed over the tumour. Mice were then placed inside a 7-Tesla horizontal bore magnet and phosphorous spectra collected (pulse aquire, TR = 6 s; 160 averages). Data analysis involved time domain fitting (Varpro) using the MRUI analysis package running under Matlab. Results were expressed as the ratio of the peaks of inorganic phosphate (Pi) to β-nucleoside triphosphate (β-NTP).

Oxygenation status.

Measurements of tumour oxygen partial pressure (pO_2) distributions were made using a computerised fine-needle polarographic oxygen electrode probe (Eppendorf, Hamburg, Germany), the details of which have been described before (Kallinowski et al., 1990). Mice were restrained and the tumour bearing leg exposed and immobilised as described above. The electrode was inserted up to a depth of about 1 mm into the tumour and then moved automatically through the tissue in 0.7 mm increments, followed each time by a 0.3 mm backward step prior to measurement. An average of 5 repeated parallel insertions were made in each tumour, giving an average of 56 pO_2 values. Results were expressed as the percentage of pO_2 values ≤ 2.5 mmHg.

Necrotic fraction.

Tumour bearing mice were killed and the tumours excised. They were then fixed in formalin and haematoxylin-stained histological sections made. A randomly selected section was cut and then two additional, equally spaced sections were produced and examined under a projecting microscope. Each section was scanned and each field of vision was evaluated with a test system containing equidistant-spaced points. For each field of vision, the total number of points hitting the tumour (nT) and those hitting only necrosis (nN) were recorded, and the necrotic fraction was expressed by $\Sigma nN / \Sigma nT$.

Radiation treatment.

Irradiations were given with a conventional therapetic X-ray machine (250 kV; dose rate 2.3 Gy/min), dosimetry being accomplished by use of an integrating chamber. Tumours only were irradiated. This was achieved by restraining the animals in jigs, exposing the leg as described above, and then shielding the animals body with 1 cm of lead. To secure homogeneity of the radiation dose the tumours were immersed in a water bath with about 5 cm of water between the X-ray source and the tumour. Tumour response was assessed using a local tumour control assay. This involved observing animals on a weekly basis upto 90 days post-treatment and recording the percentage of animals in each treatment group showing local tumour control.

Cisplatin treatment.

Mice were injected with Cisplatin and tumour response determined using a growth delay assay. This involved measuring tumour volume fives times a week and then calculating the time taken for tumours to grow to three times the initial treatment volume.

Hyperthermia treatment.

Tumour bearing mice were restrained in jigs and their tumour bearing legs exposed and immobilised as previously described. Hyperthermia was delivered by immersing the leg into a circulating water bath (type TE 623, Heto, Birkerød, Denmark) stabilised to $\pm$ 0.2°C of the adjusted temperature. The water bath was covered with a lucite plate with holes allowing immersion of the foot approximately 1 cm below the water surface. Previous measurements of intratumoral temperature have shown stabilisation within a few minutes to approximately 0.2°C below the water bath temperature (Horsman et al., 1989). The temperature of the water bath was thus adjusted to 0.2°C above the required tumour teperature of 42.5°C. All temperature measurments were calibrated against a certified precision mercury thermometer. Tumour response to heat was assessed using the tumour growth delay assay as previously described.

Results.

The pathophysiological changes induced by CA4DP in this C3H mammary carcinoma are summarised in table 1. Within 30 minutes after drug injection blood perfusion was significant reduced by greater than 50%. This decrease was maintained for at least 3 hours, although by 24 hours after injection it had recovered to normal levels. Similar changes were observed with bioenergetic status, although the decrease did not become significant until after 1 hour and was not maximal until 3 hours.

With tumour oxygenation there was a substantial decrease within 1 hour although it was not significant until after 3 hours. There was also recovery by 24 hours after treatment, but there was a suggestion that the tumours were still slightly more hypoxic than that seen in the control tumours. This probably reflects the fact that although perfusion had returned to normal there was a significant increase in the degree of necrosis at 24 hours than compared to that seen in control tumours or 3 hours after drug injection.

The radiation response of this tumour to single doses of radiation is illustrated in figure 1. The TCD50 dose (the radiation dose required to control 50% of treated tumours), with 95% confidence intervals, was calculated to be

Table1. Pathophysiological changes induced in a C3H mammary carcinoma following CA4DP (250 mg/kg; i.p.) treatment.

Parameter	Time after treatment (hours)[1]				
	Control	0.5	1	3	24
Blood perfusion[2]	4.0±0.4	1.9±0.2*	1.4±0.2*	1.7±0.4*	3.8±0.5
Bioenergetic status[3]	1.2±0.1	0.9±0.1	0.8±0.1*	0.6±0.1*	1.4±0.1
Oxygen status[4]	43±6	n.d.	68±10	72±7*	55±6
Necrosis[5]	10±4	n.d.	n.d.	11±3	38±3*

Note: [1]All values are means ± 1 S.E (n = 6-10 mice/group).
[2]RbCl uptake as % injected/g tumour.
[3]β-NTP /Pi ratio.
[4]Percentage of pO_2 values ≤ 2.5 mmHg.
[5]Percentage of necrosis.
*Significantly different from controls (Student's t-test; $p<0.05$).
n.d.: Not done.

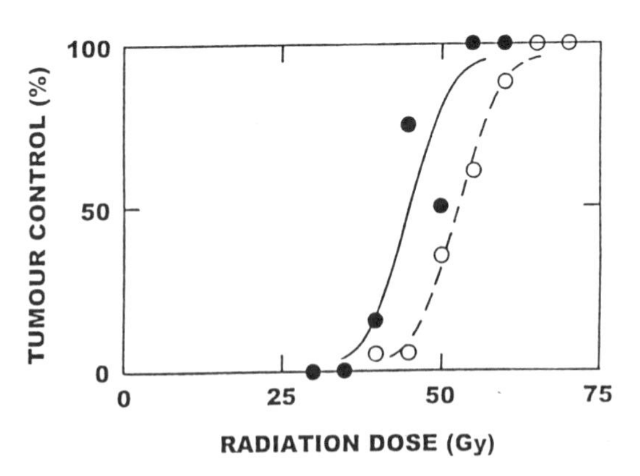

Figure 1: The effect of CA4DP on the radiation response of a C3H mouse mammary carcinoma. Tumours were irradiated with either radiation alone (O) or at the same time as i.p. injecting mice with 250 mg/kg CA4DP (●). Results are for an average of 14 mice/group. Lines were drawn following logit analysis.

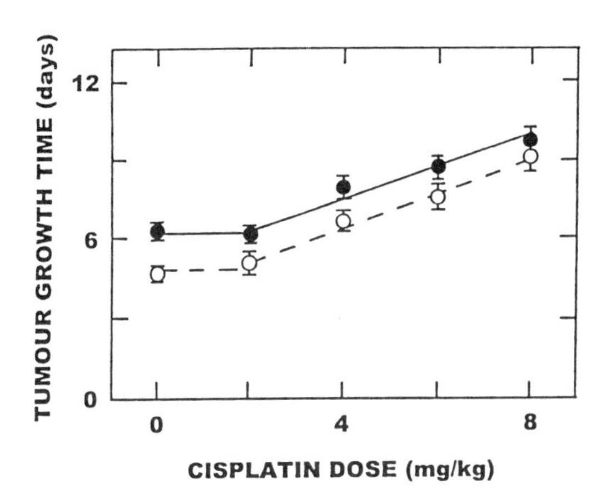

Figure 2: The effect of CA4DP on the response of a C3H mouse mammary carcinoma to cisplatin. Mice were i.p. injected with either cisplatin alone (O) or cisplatin followed 1-hour later with 250 mg/kg CA4DP (●). Results show means ± 1 S.E. from an average of 20 mice/group.

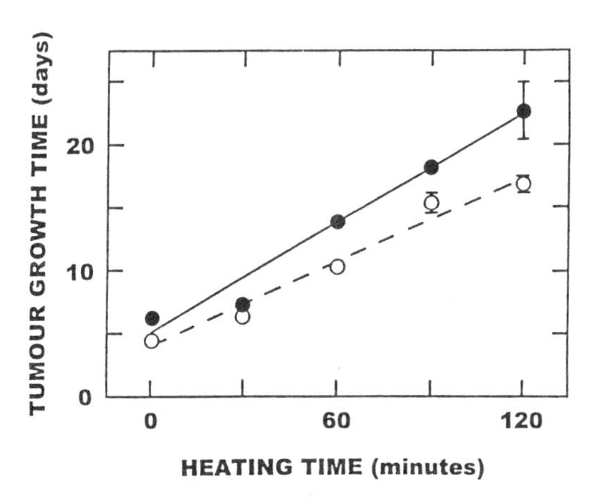

Figure 3: The effect of CA4DP on the response of a C3H mouse mammary carcinoma to hyperthermia (42.5°C). Treatments were either heat alone (O) or heat 30 minutes after an i.p. injection with 250 mg/kg CA4DP (●). Results show means ± 1 S.E. from an average of 27 mice/group.

52 Gy (50-55) for radiation alone. Irradiating tumours and giving a single i.p. injection of CA4DP at the same time significantly (Chi-squared test; $p<0.05$) reduced the TCD50 dose to 45 Gy (42-48).

Figure 2 shows the effect of CA4DP on the tumour response to cisplatin. For cisplatin alone there is no apparent effect on tumour growth until doses above 2 mg/kg are used. Above this dose there is a linear relationship between the drug dose and tumour growth time. CA4DP on its own has a small inhibitory effect on tumour growth time. Treating mice with CA4DP 1 hour after giving cisplatin results in a parallel shift in the cisplatin dose response curve.

The ability of CA4DP to enhance the response of this C3H mammary carcinoma to hyperthermia is shown in figure 3. A linear relationship was seen between the heating time at 42.5^{o}C and tumour growth time. Injecting CA4DP 30 minutes prior to heating results in a 1.3-fold increase in the slope of the heat dose response curve.

Discussion.

The pathophysiological changes induced by CA4DP in this C3H mammary carcinoma are indicative of some sort of vascular effect. Tumour perfusion decreased very soon after drug treatment and there were corresponding changes in bioenergetic status and oxygenation. However, the effects were somewhat transient in nature, since within 24 hours after injecting CA4DP most of the pathophysiological changes had dissappeared. This suggests that in this tumour the drug is having an effect that does not involve vascular damage *per se* or that any vascular damage that is induced is rapidly repaired. Despite the transient nature of the vascular changes we observed, they were still sufficent to significantly increase the level of necrosis.

Other studies have reported the ability of CA4DP to induce vascular effects in murine tumours (Chaplin et al., 1996; Dark et al., 1997; Li et al., 1998; Beauregard et al., 1998; Tozer et al., 1999; Zhao et al., 1999). But, all these studies used drug doses around 100 mg/kg and the effects seen were more profound than we obtained at the higher 250 mg/kg dose. This lower dose of 100 mg/kg can decrease bioenergetic and oxygenation status in our C3H mammary carcinoma, as well as significantly increase tumour necrosis (Maxwell et al., 1998; Horsman et al., 1998), but the effects were less than we found with 250 mg/kg. Why our C3H mammary carcinoma should be less sensitive to CA4DP than those other tumour models is not entirely clear. One possibility could be related to the levels of nitric oxide (NO), which has been shown to be important in influencing the activity of CA4DP (Tozer et al.,

1999), but whether NO levels in our tumour are higher than those found in the other tumour types is not known.

Although the pathophysiological changes we observed after giving 250 mg/kg CA4DP were not that large and in some instances reversible, they were clearly sufficient to have a small inhibitory effect on tumour growth (figures 2 and 3) and this dose is still well below the estimated maximum tolerated dose of approximately 1000 mg/kg in mice (Dark et al., 1997). It was also capable of enhancing the activity of more conventional therapies. For radiation we administered CA4DP at the same time as irradiating. This was to ensure that the radiation was complete before the drug induced reduction in tumour oxygenation occurred, an effect that could have decreased the effectiveness of the radiation treatment. Using this schedule we found a significant 1.2-fold enhancement of the radiation-induced local tumour control. A number of other studies have reported the ability of vascular damaging agents to enhance radiation response. This has included colchicine (Ludford, 1945), DMXAA (Wilson et al., 1998) and CA4DP (Li et al., 1998; Chaplin et al., 1999). Those previous studies used either clonogenic survival or tumour regrowth as the endpoints. Our study now demonstrates that similar effects can be seen using the more clinically relevant endpoint of local tumour control.

We also combined CA4DP with cisplatin. In that study we gave the CA4DP 1 hour after injecting the cisplatin, simply to allow sufficient time for the cisplatin to be taken up by the tumour before blood flow was shut down. However, there was no enhancement of the cisplatin effect (figure 2). Instead, the result of the combination appeared to be due to a simple additive effect of the CA4DP and cisplatin alone. One other previous study has combined CA4DP and ciplatin (Chaplin et al., 1999) and in that study there was a suggestion that a much shorter time interval of only 15 minutes may have been slightly better. Even though the effect of CA4DP and cisplatin was only additive, it meant that we could see the same anti-tumour effect with the combination as with cisplatin alone, but with a 25% reduction in the cisplatin dose (ie, CA4DP + 6mg/kg cisplatin was equivalent to 8 mg/kg cisplatin alone). Maintaining the anti-tumour effect, but reducing the cisplatin dose, could result in less systemic toxicity from the cisplatin.

The last conventional therapy we combined CA4DP with was hyperthermia. Numerous *in vivo* studies have clearly shown that the response of solid tumours to heat can be significantly increased if blood flow to those tumours was compromised either by clamping (Suit, 1975; Hill and Denekamp, 1978; Wallen et al., 1986) or physiological drug manipulation (Crile, 1963; Horsman et al., 1989; Kalmus et al., 1990). There is also evidence that such an enhancement can occur following vascular damage with FAA (Horsman, et al., 1991; 1996; Sakaguchi et al., 1992). Our own results with CA4DP (figure 3) clearly show that heating tumours 30 minutes after drug injection, a time when the vascular effects were at or near to maximal, can also substantially enhance the tumour response to heat. This enhancement of heat by vascular damaging agents has been shown to be the result of a better tumour heating and an increase in the adverse environmental conditions within the tumour leading to an increase in tumour damage by the heat (Horsman et al., 1996).

In conclusion, this study has shown the abilty of CA4DP to induce significant anti-vascular effects in this C3H mammary carcinoma, which although were less substantial than those effects previously reported for other tumours, they were still sufficient to increase tumour necrosis and inhibit tumour growth. More importantly, we were able to demonstrate that CA4DP could be used to significantly enhance the anti-activity of more conventional therapies. Since it is most likely that the clinical application of CA4DP, and other vascular damaging agents for that matter, will be when combined with such conventional therapies, our data clearly support the continued clinical testing of CA4DP.

Acknowledgments.

The authors would like to thank Ms. M. Simonsen, Ms. I.M. Johansen, Ms. D. Grand, Ms. M. Andersen, Ms. P. Schjerbeck and Mr. M. Johanssen for expert technical assistance. This study was supported by Oxigen Europe AB, the Danish Cancer Society and the Danish Medical Research Council.

References.

Baugeley, B.C., Holdaway, K.M., Thomsen, L.L., Zhuang, L. And Zwi. L.J.: Inhibition of growth of colon 38 adenocarcinoma by vinblastine and colchicine: evidence for a vascular mechanism. Eur. J. Cancer 27: 482-487, 1991.

Beauregard, D.A., Thelwall, P.E., Chaplin, D.J., Hill, S.A., Adams, G.E. and Brindle, K.M.: Magnetic resonance imaging and spectroscopy of combretastatin A4 prodrug-induced disruption of tumour perfusion and energetic status. Br. J. Cancer 77: 1761-1767, 1998.

Blood, C.H. and Zetter, B.R.: Tumor interactions with the vasculature: angiogenesis and tumor metastasis. Biochim. Biophys. Acta 1032: 89-118, 1990.

Chaplin, D.J., Pettit, G.R. and Hill, S.A.: Anti-vascular approaches to solid tumour therapy: evaluation of combretastatin A4 phosphate. Anticancer Res. 19: 189-196, 1999.

Chaplin, D.J., Pettit, G.R., Parkins, C.S. and Hill, S.A.: Anti-vascular approaches to solid tumour therapy: evaluation of tubulin binding agents. Br. J. Cancer 74 (Suppl. XXVII): S86-S88, 1996.

Crile, G.: The effects of heat and radiation on cancers implanted on the feet of mice. Cancer Res. 23: 372-380, 1963.

Dark, G.G., Hill, S.A., Prise, V.E., Tozer, G.M., Pettit, G.R. and Chaplin, D.J.: Combretastatin A-4, an agent that displays potent and selective toxicity towrad tumour vasculature. Cancer Res. 57: 1829-1834, 1997.

Denekamp, J. and Hill, S.: Angiogenic attack as a therapeutic strategy for cancer. Radiother. Oncol. 20 (Suppl.): 103-112, 1991.

Folkman, J.: The vascularization of tumors. Sci. Amer. 234: 58-73, 1976.

Folkman, J.: What is the evidence that tumors are angiogenesis dependent? J. Natl. Cancer Inst. 82: 4-6, 1990.

Hill, S.A. and Denekamp, J.: The effect of vascular occlusion on the thermal sensitization of a mouse tumour. Br. J. Radiol. 51: 997-1002, 1989.

Horsman, M.R.: Modifiers of tumour blood flow. IN: Hyperthermia and Oncology, Vol. 4. Urano, M. And Douple, E.B. (Eds.). VSP, Utrecht, pp. 259-283, 1994.

Horsman, M.R., Chaplin, D.J. and Overgaard, J.: The effect of combining flavone acetic acid and hyperthermia on the growth of a C3H mammary carcinoma *in vivo*. Int. J. Radiat. Biol. 60: 385-388, 1991.

Horsman, M.R., Christensen, K.L. and Overgaard, J.: Hydralazine-induced enhancement of hyperthermic damage in a C3H mammary carcinoma *in vivo*. Int. J. Hyperthermia 5: 123-136, 1989.

Horsman, M.R., Ehrnrooth, E., Ladekarl, M. and Overgaard, J.: The effect of combretastatin A-4 disodium phosphate in a C3H mouse mammary carcinoma and a variety of murine spontaneous tumors. Int. J. Radiat. Oncol. Biol. Phys. 42: 895-898, 1998.

Horsman, M.R., Sampson, L.E., Chaplin, D.J. and Overgaard, J.: The *in vivo* interaction between flavone acetic acid and hyperthermia. Int. J. Hyperthermia 12: 779-789, 1996.

Kallinowski, F., Zander, R., Hoeckel, M. and Vaupel, P.: Tumor tissue oxygenation as evaluated by computerised-pO_2-histography. Int. J. Radiat. Oncol. Biol. Phys. 19: 953-961, 1990.

Kalmus, J., Okunieff, P. and Vaupel, P.: Dose-dependent effects of hydralazine on microcirculatory function and hyperthermic response of murine FSaII tumors. Cancer Res. 50: 15-19, 1990.

Li, L., Rojiani, A. and Siemann, D.W.: Targeting the tumor vasculature with combretastatin A-4 disodium phosphate: effects on radiation therapy. Int. J. Radiat. Oncol. Biol. Phys. 42: 899-903, 1998.

Ludford, R.J.: Colchicine in the experimental chemotherapy of cancer. J. Natl. Cancer Inst. 6: 89-101, 1945.

Mahadevan, V. and Hart, I.R.: Metastasis and angiogenesis. Rev. Oncol. 3: 97-103, 1990.

Maxwell, R.J., Nielsen , F.U., Breidahl, T., Stødkilde-Jørgensen, H. and Horsman, M.R.: Effects of combretastatin on murine tumors monitored by ^{31}P MRS, ^{1}H MRS and ^{1}H MRI. Int. J. Radiat. Oncol. Biol. Phys. 42: 891-894, 1998.

Moses, M.A. and Langer, R.: Inhibitors of angiogenesis. Biotechnology 9: 630-634, 1991.

NCI report: Inhibitors of angiogenesis enter Phase III testing. J. Natl. Cancer Inst. 90: 960-962, 1998.

Overgaard, J.: Simultaneous and sequential hyperthermia and radiation treatment of an experimental tumor and its surrounding normal tissue *in vivo*. Int. J. Radiat. Oncol. Biol. Phys. 6: 1507-1517, 1980.

Paweletz, N. and Knierim, M.: Tumor-related angiogenesis. Crit. Rev. Oncol. Hematol. 9: 197-242, 1989.

Sakaguchi, Y., Maehara, Y., Baba., H., Kusumoto, T., Sugimachi, K. and Newman, R.A.: Flavone acetic acid increases the antitumor effect of hyperthermia in mice. Cancer Res. 52: 3306-3309, 1992.

Sapirstein, L.A.: Regional blood flow by fractional distribution of indicators. Amer. J. Physiol. 193: 161-168, 1958.

Suit, H.D.: Hyperthermia in the treatment of tumours. Proceedings of the International Symposium on Cancer Therapy by Hyperthermia and Radiation, Washington, 28-30 April (American College of Radiology), pp. 107-114, 1975.

Tozer, G.M., Prise, V.E., Wilson, J., Locke, R.J., Vojnovic, B., Stratford, M.R.L., Dennis, M.F. and Chaplin, D.J.: Combretastatin A-4 phosphate as a tumor vascular-targeting agent: early effects in tumors and normal tissues. Cancer Res. 59: 1626-1634, 1999.

Wallen, C.A., Colby, T.V. and Stewart, J.R.: Cell kill and tumor control after heat treatment with and without vascular occlusion in RIF-1 tumors. Radiat. Res. 106: 215-223, 1986.

Wilson, W.R., Li, A.E., Cowan, D.S.M. and Siim, B.G.: Enhancement of tumor radiation response by the antivascular agent 5,6-dimethylxanthenone-4-acetic acid. Int. J. Radiat. Oncol. Biol. Phys. 42: 905-908, 1998.

Zhao, S., Moore, J.V., Waller, M.L., McGown, A.T., Hadfield, J.A., Pettit, G.R. and Hastings, D.L.: Positron emission tomography of murine liver metastases and the effects of treatment by combretastatin A-4. Eur. J. Nucl. Med. 26: 231-238, 1999.

Clinical Applications

THERAPEUTIC ANGIOGENESIS FOR ISCHEMIC HEART DISEASE

[1]Armin Helisch and [2]J. Anthony Ware
[1]Department of Experimental Cardiology, Max-Planck-Institute for Physiological and Clinical Research, Bad Nauheim, Germany: [2]Division of Cardiovascular Medicine, Departments of Medicine and Molecular Pharmacology, Albert Einstein College of Medicine, Bronx, NY

Key words: Angiogenesis, arteriogenesis, ischemic heart disease, therapy, growth factors

Abstract: The de-novo formation of vessels (angiogenesis) and the remodelling of preexisting collateral vessels (arteriogenesis) are processes that occur naturally in ischemic heart disease. Promoting these processes by administration of various substances or other physical stimuli (therapeutic angiogenesis) may provide a future strategy for the treatment of ischemic vascular diseases. Mechanisms of angiogenesis and arteriogenesis, as well as trials of therapeutic angiogenesis in animal models and humans are reviewed.

1. INTRODUCTION

Ischemic heart disease is a major cause of morbidity and the leading cause of death in the Western world. Many affected patients will at some point require a revascularizing procedure to ameliorate symptoms not sufficiently controlled by medications or also to improve prognosis, e.g. in patients with significant left main coronary artery stenosis. Unfortunately, especially in patients with severe diffuse coronary heart disease revascularization by balloon angioplasty or bypass surgery can be difficult or even impossible. Incomplete revascularization, on the other hand, is associated with a worsened post-operative outcome.

There are naturally occurring compensatory mechanisms in chronic ischemic vascular diseases as suggested by the development of angiographically visible collateral vessels in many patients. These appear to be mainly due to the remodelling of preexisting small collateral vessels, a process for which Wolfgang Schaper has proposed the term "neoarteriogenesis" [1]. However, true "angiogenesis" has also been observed, defined as the growth of new vessels from preexisting ones. This in turn has been differentiated from the development of new vessels directly from mesodermal endothelial cell precursors, i.e. "vasculogenesis" [2]. While vasculogenesis is an essential part of the embryological development of blood vessels, more recent evidence suggests that it also contributes to postnatal neovascularization [3,4].

The sprouting of capillaries from preexisting vessels (angiogenesis) does not only occur in ischemic vascular diseases. It also is a normal and necessary process supplying the tissues of growing organisms with nutrients and oxygen. Wound healing and the cyclic variations of the endometrium are other examples in which angiogenesis plays a role. On the other hand, angiogenesis is also part of the pathogenic mechanism of tumors, hemangiomas, proliferative retinopathies, and inflammatory diseases like rheumatoid arthritis and psoriasis [5]. It also appears be involved in the growth of atherosclerotic plaques [6] and possibly also their instability [7].

In recent years there has been a growing interest in learning how to augment these naturally occurring revascularization processes to treat peripheral vascular disease and ischemic heart disease and whether it is possible to do so without major adverse effects. The underlying concepts and the results of animal and human studies are reviewed in this chapter.

2. ANGIOGENESIS - A SEQUENCE OF EVENTS

The vasculature of a 70 kg adult is lined by approximately 1000 m^2 of usually quiescent endothelial cells with a very low turnover rate that can exceed 1000 days [5]. Upon activation, e.g. by growth factors released during ischemia, a local "inflammatory" reaction can often be observed with an increased vascular permeability, vasodilatation and accumulation of monocytes and macrophages. These latter cells release more cytokines and growth factors and thereby cause accumulation of additional inflammatory cells. They also release enzymes that promote the proteolytic degradation of the underlying extracellular matrix and basal membrane. Such degradation causes the endothelial cells to detach from their neighbouring cells and the underlying matrix, followed by chemotactic migration and proliferation. Subsequently, formation of a lumen occurs and eventually maturation and growth of the newly formed vessel [2, 8, 9]. In vessels larger than capillaries, vascular smooth muscle cells may have to proliferate and migrate as well. It is unclear at this point, whether small vessel angiogenesis in the adult organism can be followed by an arteriogenic remodelling process resulting in larger "de-novo" vessels or whether arteriogenesis is limited to preexisting rudimentary collateral vessels. Recent observations suggest that circulating endothelial cell precursors released from the bone marrow participate in the formation of new vessels in the setting of tissue ischemia [3, 4], wound healing, tumor vascularization and even the cyclic changes of the ovaries and the endometrium [3]. It may therefore be that angiogenesis, arteriogenesis and vasculogenesis are all components of a very complex and integrated process.

3. THE YIN AND YANG OF ANGIOGENESIS

Angiogenesis appears to be a tightly regulated process. Both, mechanical factors and multiple endogenous substances have been identified that inhibit inappropriate endothelial cell proliferation following mitogenic stimulation [5]. Most of the known angiogenesis inhibitors circulate in the blood; some have also been detected in the matrix around endothelial cells. They include platelet factor 4, thrombospondin-1 and -2, tissue inhibitors of metalloproteinases and interferon α [5, 10]. It is unclear, whether the endothelium- specific inhibitors angiostatin, a 38 kD fragment of plasminogen [5], and endostatin, a 20 kD fragment of collagen XVIII [11], are present in physiologically active amounts in organisms without malignant tumors. From bovine hearts an 11 kD inhibitor of endothelial and smooth muscle cell proliferation and angiogenesis with homology to the B-cell translocator gene btg-1 has been isolated [1, 12]. Neighbouring endothelial cells and the surrounding basal lamina limit excessive endothelial cell growth. Pericytes may regulate or restrain endothelial cell proliferation. Angiogenesis can also be inhibited by sequestration of angiogenic factors in the extracellular matrix; furthermore, changes of endothelial cell shape may decrease their sensitivity to growth factors [5].

On the other hand, many endogenous factors that promote angiogenesis have been identified. These include various growth factors that interact with receptor tyrosine kinases (see below), angiogenin, granulocyte colony stimulating factor, interleukin-8 and proliferin [5, 13, 14]. The balance of these inhibitory and mitogenic influences determines whether angiogenesis occurs.

4. EFFECTS OF ISCHEMIA

Angiogenesis has been observed in many animal models of acute and chronic tissue ischemia and occurs naturally following ischemia in patients. The decrease or lack of blood flow leads to a local deficiency of oxygen and glucose. In various cell types, hypoxia has been shown to induce increased expression of multiple growth factors, including PDGF (types A & B), placental growth factor (PlGF), fibroblast growth factors (FGF) I and II, transforming growth factor (TGF)-β1, vascular endothelial growth factor (VEGF) [15, 16], as well as the VEGF receptors KDR/Flk-1 (after an initial downregulation) [17] and flt-1 [14]. Upregulation of VEGF occurs within 6 hours in vivo and is due to both increased transcription and increased mRNA stability. Hypoglycemia exerts similar effects [18, 19]. These transcriptional changes appear to be mediated through activation of hypoxia-inducible factor 1 (HIF-1), a heterodimeric helix-loop-helix transcription factor with an α- (HIF-1 α) and a β- arylhydrocarbon-receptor nuclear translocator (ARNT) component that was first found to regulate the expression of the erythropoietin gene [20]. $ARNT^{-/-}$ mice embryos are not viable beyond embryonic day 10.5, because of defective angiogenesis of the yolk sac and branchial arteries [18]. Even more important than the transcriptional changes may be a hypoxia/hypoglycemia-induced increase of VEGF mRNA stability [19] which appears to be mediated by an untranslated 3' region of the VEGF gene [21].

Within one hour of an acute infarction the mRNA levels of VEGF (275%) and its receptors flk-1 (375%) and flt-1 (400%) rise throughout the entire rat heart [22]. Thereafter, this increase is limited mainly to the peri-infarct area in which angiogenesis is taking place. In-situ hybridization reveals markedly elevated amounts of VEGF expression in the peri-infarct zone myocytes (at 6 hours), upregulation of flk-1 and flt -1 in the endothelial cells that form microvessels at the infarct edge (6 and 24 hours), and also as late as six weeks in the new vessels growing into the infarct area [22]. The transient increase of VEGF expression in the non-affected myocardium may at least in part be related to stretch induced TGF-β expression [23] which illustrates that there may be other inducers of gene expression and angiogenesis than local tissue hypoxia and hypoglycemia during myocardial ischemia in vivo.

As already pointed out, revascularization in the heart does not only occur by angiogenesis but also from the recruitment and subsequent enlargement of preexisting arterioles with a well developed tunica media, i.e. arteriogenesis. In dogs this may result in an about 20-fold increase in collateral vessel diameter and a 50-fold increase in collateral vessel tissue mass [1]. Even though arteriogenesis is almost exclusively seen in ischemic settings, the local effects of ischemia do not easily explain it. The stem and midzone segments of these collaterals, which become enlarged, are usually not adjacent to ischemic tissue. In addition, there is some experimental evidence in dogs that, after repetitive brief occlusions of the left circumflex artery, more collateral vessel development was reached in those dogs with more preexisting collaterals and therefore less ischemia [1]. The regulatory mechanisms for arteriogenesis are quite unclear. However, mechanical/hemodynamic factors, like increased hydrostatic pressure [24, 25] and shear stress [26] resulting from increased blood flow through collaterals, may play a role. Increased shear stress may also lead to increased adhesion of circulating blood cells, such as monocytes, to endothelial cells. The former appears to invade the subintimal space and release various growth factors, like TNF-α, FGF-1 and FGF-2 [27]. These growth factors promote endothelial and smooth muscle cell migration and proliferation as well as other aspects of the remodelling process, including destruction of the internal elastic membrane and smooth muscle cell apoptosis. Thus, the stimuli for angiogenesis and neoarteriogenesis may be quite different, even though both occur in the ischemic heart and both processes are promoted by many of the same growth factors.

As already mentioned, bone-marrow-derived endothelial cell precursors also seem to participate in the revascularization of ischemic tissue [3, 4]. The regulatory pathways for this, however, are unknown at this point.

5. ANGIOGENIC GROWTH FACTORS

Two growth factor families whose members have come into widespread use for therapeutic angiogenesis trials are the fibroblast growth factors (FGF) and the vascular endothelial growth factors (VEGF).

5.1 Fibroblast Growth Factor Family

Fibroblast growth factors comprise a growing family of structurally related polypeptides (FGF-1 to FGF-18) [13]. They exhibit a high degree of cross-species homology [28]. They all require cell surface heparan sulfate proteoglycans to bind to their tyrosine kinase receptors. The mitogenic effect of FGF-7 to FGF-10 appears to be limited to epithelial cells. The prototype members of the group, FGF-1 (acidic), and FGF-2 (basic) are the best characterized and, while having mitogenic and differentiating effects on a variety of cell types, they have been shown to have direct angiogenic effects on endothelial cells and smooth muscle cells. Secreted members of the FGF family appear to remain bound to cells and the extracellular matrix. Therefore, similar to VEGF, they have a very short half-life of about 3 minutes after intravenous injection [29]. The intravascular administration of various growth factors is limited by hypotension, due in part to the release of nitric oxide [30, 31]. It is of interest, however, that the angiogenic potential of FGF-2 may be independent of its effect on nitric oxide (NO) release. This is in contrast with the inhibition of VEGF-induced angiogenesis by NO synthase inhibition in a rabbit model in which cells from a VEGF transfected breast cancer cell line were implanted in the cornea [32].

5.2 Vascular Endothelial Growth Factor Family and Related Growth Factors

Vascular Endothelial Growth Factor (VEGF) has attracted much attention for therapeutic use, because, unlike the case with FGF-1 and FGF-2, its mitogenic effect is limited to endothelial cells, at least theoretically, because its receptors have a much more limited range of tissue expression. VEGF occurs in several isoforms with different numbers of amino acid residues (in humans: $VEGF_{121}$, $VEFG_{165}$, $VEGF_{189}$, $VEGF_{206}$) resulting from alternative splicing of its gene, which has been located on the human chromosome 6p21.3 [33]. $VEGF_{165}$, a basic, heparin binding 45,000 Dalton glycoprotein, is the predominant subtype produced by a variety of cells. $VEGF_{206}$ so far has only been found in a human fetal liver cDNA library [14]. $VEGF_{121}$ is the only one of these isoforms that is mildly acidic; it does not bind to heparin at all, which correlates with a decrease of mitogenic activity in an endothelial cell growth assay [34]. The isoforms also differ significantly in their bioavailabilities. While $VEGF_{121}$ is freely diffusible, $VEGF_{165}$ is secreted, but a significant fraction of it remains bound to the endothelial cell surface and the extracellular matrix. $VEGF_{189}$ and $VEGF_{206}$ are almost completely sequestered in the extracellular matrix [14]. In addition to the mitogenic effect on endothelial cells VEGF has stimulating effects on the expression of urokinase-type and tissue-type plasminogen activators and the metalloproteinase collagenase in endothelial cells. These proteases promote matrix dissolution, another important step in the angiogenic process. It also increases vascular permeability by inducing endothelial cell fenestrations, causes vasodilation (probably mainly due to NO release) and may attract monocytes [14]. The effects of VEGF on endothelial cells are mediated by two tyrosine kinase receptors, flt-1 (fms-like tyrosine kinase) or VEGFR-1, a high affinity receptor, and KDR (kinase-domain region)/ flk-1 (fetal liver kinase-1 in mice) or VEGFR-2, a lower affinity receptor. An additional mechanism by which VEGF induces neovascularization may be the mobilization of endothelial progenitor cells from the bone marrow [35, 36].

VEGF related factors include placental growth factor-1 ($PlGF_{131}$) and -2 ($PlGF_{152}$), VEGF-B, VEGF-C/ VEGF-related peptide (VRP) [14], VEGF-D [37] and VEGF-E [38]. All of these have various degrees of mitogenic effects on endothelial cells. VEGF-C has recently been reported to be clearly angiogenic in ischemic tissue in vivo [39] VEGF-E was shown to be angiogenic in a rabbit corneal assay [38]. PlGF, like VEGF a dimeric protein, may only be angiogenic in vivo when heterodimerized with VEGF [40, 41], even though in one study PlGF-1 alone was found to be angiogenic in in-vivo assays as well [42].

6. ANIMAL STUDIES OF THERAPEUTIC ANGIOGENESIS

As noted above, new vessel formation occurs naturally following ischemia or infarction. Unfortunately, it is rare for blood flow to be completely restored by these mechanisms of revascularization in coronary heart disease and peripheral vascular disease in humans. Therefore, many investigators have sought to increase this response by the administration of exogenous growth factors, and first tested this possibility in animal models.

6.1 Choice of Animal

The choice of the animal for these experiments is important for the interpretation of the results, and to determine how applicable they are for patients. Some animal models have a better supply of preexisting arteriolar collaterals than others. Rats, rabbits and pigs have end-arteries without arteriolar connections; however, dogs, cats, and especially guinea pigs, have an abundant arteriolar network. For that reason, acute coronary artery occlusion leads to rapid transmural infarcts in rats, rabbits and pigs, to non-transmural infarctions in dogs and cats, and to no infarcts at all in guinea pigs [1]. On the other hand, only very limited small vessel angiogenesis has been detected in the canine model of chronic progressive ischemia, in contrast to the significant degree seen in a similar pig model. Therefore, some animal models may be more suitable for studying small vessel angiogenesis, while others are preferable, if the goal is to mimic the neoarteriogenesis seen in patients.

6.2 Methods of Creating Ischemia

Various methods of creating myocardial ischemia in animals have been employed. Ameroid occluders are used to create a slowly progressive stenosis that becomes occlusive after 10-17 days. An ameroid occluder consists of modified casein that swells with hydration. A surrounding metal ring prevents outward expansion and directs the volume gain to the central lumen of the occluder, thereby gradually constricting the vessel [43]. Pneumatic occluder cuffs, usually with Doppler flow probes, have been used to create repetitive brief periods of ischemia and to monitor blood flow at the same time. The injection of plastic microspheres via a coronary catheter allows a less invasive and, depending on the size of the microspheres, a more peripheral mode of coronary occlusion or partial occlusion. In addition, acute infarction models involving ligation of a coronary artery have been used in a variety of animals. As this procedure is also feasible in mice [44], it should provide a valuable means of studying peri-infarct angiogenesis and the effect of growth factor therapies in genetically altered animals. Hindlimb ischemia models offer the advantages of easier surgery with minimal mortality, easier assessment of angiogenesis and distal perfusion. The latter can be accomplished with peripheral blood pressure indices, skin limb perfusion measurement by color laser Doppler perfusion mapping or thermography between the ischemic and non-ischemic limbs and the use of imaging techniques not affected by cardiac motion (angiograms, magnetic resonance angiograms). Hindlimb ischemia is created by ligation or resection of the femoral artery. This model has most recently been validated in mice as well [45].

6.3 Routes of Growth Factor Administration

The best route of administration of growth factors or growth factor-encoding genes is not settled in either animal models or in clinical trials in humans. Almost every conceivable approach has been used, including intravenous or intracoronary bolus injections, prolonged infusions, intrapericardial instillation (via a catheter in the right atrium), subepicardial application of growth factor releasing substances, and direct intramyocardial injection. In addition, percutaneous angioplasty balloons have been used to deliver growth factor genes in patients with peripheral vascular disease, and growth factor genes may be injected via a fluoroscopically guided catheter into the endocardium [46]. The local intravascular delivery devices have the advantage of decreasing the risk of systemic side effects while at the same time obviating the need for thoracotomy. In addition, local administration may be able to deliver higher doses of growth factors without prior inactivation in the circulation where they are rapidly degraded [29]. If sustained release of a growth factor is desired, polymers can be used for subepicardial implantation or, in the form of microspheres, for intravascular application such as intracoronary injection [47]. The growth factor itself, the experimental or clinical condition to be investigated, cost, convenience and, above all, efficacy and side effects, are all factors that should be considered when deciding on the mode of administration.

6.4 Evaluation of Therapeutic Effect

Ideally, several methods of assessment should be used to measure the effect of angiogenic therapy. First, it is important to document increased perfusion and vascularization as evidence for angiogenesis/arteriogenesis. Then it also is important to show that this results in beneficial functional changes and clinical outcome. Outcome measures to be considered are listed in table 1.

Improved clinical outcome and regional wall motion or perfusion can result, at least theoretically, from small vessel angiogenesis, neoarteriogenesis, collateral recruitment, microvascular reactivity changes [1, 8, 10, 31] and possibly also from direct cell protective effects of growth factors (see below). The thoughtful use of several of these outcome measures may help to differentiate between those various mechanisms, as well as document any benefit.

Table 1. Outcome measures for therapeutic angiogenesis

Goal	Method
Quantification of vessels and vascular cell proliferation	Histology, histochemistry, radiological and nuclear magnetic angiography
Quantification of perfusion and microvascular function	Microsphere flow analysis, nuclear isotope perfusion techniques, nuclear magnetic resonance techniques, catheter-based hemodynamic measurements to assess coronary resistance/flow in response to endothelium-dependent receptor-mediated agonists
Quantification of left ventricular function	Echocardiography, nuclear isotope techniques, nuclear magnetic resonance imaging, ventriculography
Symptomatic/prognostic endpoints (mainly for human subjects)	Degree of angina, exercise tolerance, freedom of myocardial (re-)infarction, length of survival

6.5 Results of Animal Studies

6.5.1 Fibroblast Growth Factor-1

The first agent to be used in an attempt to stimulate myocardial neovascularization was FGF-1. In a report from 1991, up to 800 μg FGF-1 was applied via a polytetrafluoroethylene fiber or collagen I sponge to an internal mammary artery (IMA) pedicle that was placed over the left anterior descending artery (LAD) distal to an ameroid occluder in dogs [48]. Angiographically, there was no evidence of increased collateral formation from the IMA to the LAD in the treated group. A striking degree of smooth muscle cell hyperplasia in the small arteries and arterioles limited to areas of subendocardial infarction was observed by histological examination. In a follow-up study, 8 weeks of 2.4 mg continuous FGF-1 infusion with heparin into the IMA, which was implanted into ischemic myocardium, did not improve collateral flow into the ischemic area over that seen with heparin infusion alone [49]. Similarly, a 4-week FGF-1 infusion (30 μg/hour) with heparin into the left main coronary artery did not improve regional blood flow in the ischemic left circumflex artery area (after prior gradual occlusion and then ligation of the left circumflex artery) [50].

Another group of researchers, however, using a different preparation of human recombinant FGF-1, implanted collagen coated PTFE sponges with 1 μg of the growth factor between the aorta and myocardium in rats. After 9 weeks they demonstrated aorta to myocardium collaterals by digitized computed angiography and histology in 11 out of 12 FGF-treated rats, but none in the 6 control rats that had received sponges without FGF-1 [51].

Substituting one of the three cysteines of FGF-1 with a serine residue (S^{117}) decreases the tendency of FGF-1 to become inactivated by disulfide bond formation. Thereby its activity half-life increases from 15 minutes to 1.4 hours without heparin, and from about 24 hours to 240 hours in the stabilizing presence of heparin [10, 52]. Ten μg of this S^{117}-FGF-1 applied in a sustained-release polymer over the ischemic left circumflex artery territory in an ameroid pig model led to improved regional perfusion and improved global and regional wall motion [53].

Thus, there is somewhat conflicting evidence on the use of FGF-1 in therapeutic trials to increase neovascularization in animal models of myocardial ischemia. The differences could be related to the animal model used, as the negative studies were all done with dogs. Other possible reasons could relate to the preparation of the growth factor, the stability of which, as noted above, can be significantly improved by chemical modifications.

6.5.2 Fibroblast Growth Factor-2

An early positive study of FGF-2 was not only very encouraging but it also raised interesting questions [54]. Dogs received a 10 μg human recombinant FGF-2 infusion over 1 minute into the LCX artery 30 minutes and 6 hours after occluding the LAD and were found to have better left ventricular systolic function one week later. This improvement was associated with a reduced infarct size, a significant increase of small collateral arteries from the LCX artery towards the infarct on angiogram, and a significantly increased number of capillaries and arterioles in the infarct area. Hemodynamic changes were not noticed during the growth factor infusions, nor were there any differences on angiograms at 30 and 60 minutes after infarction. In a follow-up study, similar results were achieved with intrapericardial administration of 30 μg FGF-2 with heparin [55]. Heparin alone had a mild proangiogenic effect in both of these studies when compared to saline. FGF-2 in combination with heparin was the most effective regimen.

Since angiogenesis takes time (on the order of weeks) to occur while tissue necrosis due to coronary occlusion occurs very rapidly (hours), the reduction of infarct size in theses studies probably results from effects of

FGF-2 other than new vessel formation. Dilation of the extensive preexisting collaterals in dogs is one potential mechanism [8, 10]. Another possibility is a direct protective effect of FGF-2 on ischemic tissue. A rat ischemia-reperfusion study has supported this latter notion [56]. The possibility of a direct myocardial protective effect is further supported by similar evidence of reduced infarct size in a mouse stroke model. This benefit persisted even in the absence of an FGF-2 induced increase in regional cerebral blood flow in eNOS knock-out mice suggesting a blood flow independent mechanism [57]. In a porcine infarct study, microscopic beads were used to create infarcts and to administer FGF-2 at the same time. No preservation of left ventricular function with FGF-2 treatment could be observed echocardiographically, however, an increased number of capillaries in the infarct area were noted at 2 weeks [58].

In the setting of *chronic* myocardial ischemia, FGF-2 has been used with varying doses and modes of delivery [8, 10]. Ameroid ischemia models with an intracoronary bolus [59] or left atrial continuous [60] FGF-2 administration for up to 9 weeks demonstrated increased neovascularization and regional perfusion or at least recovery of perfusion in dogs. Much lower doses of FGF-2 (up to 100 μg) administered slowly to the ameroid-induced ischemic porcine myocardium via heparin-alginate pellets, were associated with increased perfusion, collateral formation, as well as some improvement of global and regional left ventricular function [61, 62].

In a rabbit hindlimb ischemia model, daily intramuscular injections of 3 μg FGF-2 for two weeks were reported to be very effective in increasing angiographic collateral vessel formation and capillary angiogenesis as well as in decreasing tissue necrosis [63]. Other investigators confirmed these results [64] and also showed that FGF-2 administration did not have any neovascularizing effect in non-ischemic limbs [65], a finding that has been supported by many more studies on growth factor therapy in ischemic vascular disease in general.

Thus, there is ample evidence that FGF-2 is a useful agent to increase neovascularization in a variety of animal models of acute and chronic ischemia. In addition, FGF-2 appears to have direct protective effects on ischemic tissues in acute infarction models.

6.5.3 Fibroblast Growth Factor-5

There are much less data on the therapeutic use of FGF-5. However, 2 weeks after intracoronary infusion of a recombinant adenovirus expressing the human FGF-5 gene, evidence for increased angiogenesis was seen in a porcine ameroid ischemia model. This resulted in improved perfusion and stress-induced regional left ventricular function, both of which were still demonstrable at 12 weeks [66]. This was the first published study of in-vivo gene transfer as a treatment of myocardial ischemia. Overall capillary density was not different in the groups; however, a mild increase in capillary number surrounding each myocardial fiber was more pronounced in the endocardium than in the epicardium both in ischemic and non-ischemic regions of the treated animals. The latter observation is in contrast to other studies of therapeutic growth factor administration in which evidence for angiogenesis was limited to ischemic tissues. Still, the feasibility of therapeutic angiogenesis by intracoronary FGF-5 gene administration with improvement of myocardial function was quite convincingly demonstrated in this study.

6.5.4 Vascular Endothelial Growth Factor

In rabbit hindlimb ischemia models $VEGF_{165}$ appears to work well. Daily administration of up to 1000 μg for 10 days into the ischemic muscle [67], even single intraarterial bolus administration of up to 1000 μg [68], as well as intravenous bolus administration of up to 5000 μg [69] resulted in improved perfusion, collateral formation and capillary angiogenesis. Evidence of angiogenesis remained limited to the area of ischemia [69]. Treatment with the 121-, 165- and 189-isoforms of the VEGF gene applied via a hydrogel peripheral angioplasty balloon to the iliac artery resulted in similar benefits in the rabbit hindlimb ischemia model [70]. Most recently, intraarterial administration of the VEGF-C protein by injection and of its gene by hydrogel-coated angioplasty balloon was shown to promote neovascularization in the same model [39].

In a dog ameroid model of chronic myocardial ischemia, 4 weeks of daily 45 μg intracoronary VEGF injections distal to the occluder resulted in improved regional perfusion during maximal vasodilation and an increased density of small vessels larger than 20 μm in diameter [71]. However, 7 days of daily left atrial VEGF infusions (720 μg per bolus) did not appear to have any significant benefit, in contrast to the case with FGF-2 (1.74 mg per bolus) which did improve maximal collateral flow in the same study [72]. It is also of interest that VEGF treated dogs exhibited a significant increase of neointima formation after iliofemoral artery balloon injury performed at the time of growth factor treatment initiation, while the FGF-2 treated animals did not show any such evidence [72]. This is surprising, as neointima formation, an important component of atherogenesis and restenosis, is usually related to the degree of local smooth muscle cell proliferation, and VEGF has been considered to be a pure endothelial cell mitogen. This has been the only reported study comparing two different growth factors directly for therapeutic myocardial angiogenesis.

In porcine ameroid models, 2 μg VEGF with 50 U heparin was continuously infused into the ischemic myocardium over 4 weeks via a minipump system and was associated with improved regional perfusion and capillary densities, as well as contractile function [10, 73]. Transmyocardial laser application (TMR) may be able to increase the rate of expression of intramyocardially injected genes [74]. The combination of TMR with intramyocardial injection of plasmids encoding $VEGF_{165}$ was associated with a significantly improved regional wall motion in a pig ameroid ischemia model, while the improvements in wall motion of the TMR or plasmid injection without TMR groups did not achieve statistical significance in this small study. Quantification of angiogenesis in this model has not been reported yet.

Thus, treatment with VEGF is clearly effective for the treatment of peripheral vascular ischemia in animal models and also appears to have the potential to be a useful agent for treatment of chronic myocardial ischemia. Whether the prolonged periods of administration of the protein via the intracoronary and perivascular routes used in the experiments so far are required for effect is not established, and it is likely that the methods of administering growth factors or their genes will need further refinement.

6.5.5 Other Agents

Other agents with potential angiogenic effects have been investigated as well. Promising results by single intraarterial injection of a novel growth factor, human recombinant hepatocyte growth factor (HGF), in a rabbit hindlimb ischemia model have been reported [75]. A hybrid gene of the hypoxia inducible factor 1α (HIF-1α) combined with the herpes simplex virus VP16 was found in a rabbit hindlimb ischemia model to increase $VEGF_{165}$ serum levels and to increase neovascularization in the ischemic limb [76]. Thus, other growth factors and gene therapy with transcription factors may offer us new options for the treatment of myocardial ischemia.

7. CLINICAL TRIALS IN CORONARY HEART DISEASE

Table 2 lists the clinical trials of therapeutic myocardial angiogenesis that we are aware of. Schumacher *et al.* reported the first trial of growth factor therapy for coronary heart disease in humans [77]. Patients (n=40) who were scheduled to undergo multiple vessel bypass surgery were randomized to injection of active or heat-denatured FGF-1 (0.01 mg/kg body weight). All of the patients had stenoses in the distal third of the LAD or at the origin of one of its branches in addition to the proximal lesion that was bypassed by the the internal mammary artery (IMA). The growth factor injections were targeted close to the vessel and distal to the anastomosis between (IMA) and the left anterior descending artery (LAD). Twelve weeks later, the IMA bypasses were imaged by selective transfemoral angiography and digital subtraction angiography. At the site of the injections and in the distal areas supplied by the LAD, a pronounced contrast accumulation could be seen only in the treated group of patients. Gray-scale analysis confirmed a significantly improved regional perfusion consistent with small vessel

Table 2. Clinical trials of therapeutic myocardial angiogenesis (modified with permission from Helisch and Ware [78])

Growth Factor	Delivery	Phase	Number of Patients active/placebo	Principle Investigator / Sponsor
FGF-1	Intramyocardial	I	20/20	Schumacher [77]
FGF-2	Intracoronary	I		Unger / NIH
FGF-2	Local (polymer)	I	16/8	Simons / NIH
FGF-2	Intracoronary	I	52/0	Chiron [79]
FGF-2	Intravenous	I	14/0	Chiron
FGF-2	Intracoronary	II		Chiron
FGF-4	Intracoronary (adenovirus)	I		Berlex
$VEGF_{165}$	Intracoronary	I	15/0	Genentech
$VEGF_{165}$	Intravenous	I	27/0	Genentech
$VEGF_{165}$	Intracoronary / intravenous	II	117	Genentech [80]
$VEGF_{165}$	Intramyocardial (plasmid)	I	24/0	Isner [81, 82]
$VEGF_{121}$	Intramyocardial (adenovirus)	I	21/0	Crystal [83]
HIF 1 α	Intramyocardial (adenovirus)	I		Genzyme

angiogenesis in the treated group. Clinical follow-up data on these patients are still pending.

Recently, the first double-blind, placebo-controlled phase I trial has been completed. In this study, 24 patients with one ungraftable vessel supplying ischemic but viable myocardium, were enrolled. They were randomized to receive ten 10 μg or 100 μg FGF-2 containing heparin-alginate microbeads or microbeads without growth factor. These were implanted during coronary bypass surgery in the subepicardial fat along the distribution of the non-bypassable vessel [84, 85]. Implantation of the heparin-alginate beads added only 2.8 minutes to the time of surgery. No increase of plasma FGF-2 levels or side effects attributable to the growth factor were seen. All patients, except for two control patients, were free of angina 90 days after surgery. Nuclear scans demonstrated improved target area perfusion in 2/5 controls, 2/5 patients of the 10 μg bFGF group, and 5/5 patients of the 100 μg FGF-2 group [85]. Global left ventricular function improved in all groups; however, regional wall motion by MRI analysis improved only in the patients treated with 100 μg FGF-2 microbeads. Thus, this trial is encouraging for the prospects of using local angiogenic therapy to promote more complete revascularization of patients undergoing bypass surgery.

In a phase I, open label, dose escalation trial, FGF-2 was given as a single 20 minute intracoronary infusion to 52 patients with severe coronary artery disease who were suboptimal candidates for any of the standard revascularization procedures. Doses of 0.33 to 48 μg/kg were administered. At the maximum dose, 2 out of 10 patients developed severe hypotension [79]. Severity of angina, exercise time, as well as regional perfusion and wall motion as assessed by MRI improved in the subjects.

Another phase I trial enrolled 24 patients with multivessel coronary heart disease who had failed conventional anti-anginal medical therapy to intramyocardial $VEGF_{165}$ plasmid DNA injections as sole therapy [81, 82]. In this trial 125 μg was injected in 4 aliquots via a small left anterior thoracotomy into the ischemic anterolateral left ventricular free wall. The patients treated this way tolerated the procedure well and reported reduced angina. Gene expression could be documented by a transient increase of serum VEGF levels. In 11 out of 12 studied patients reduced ischemia, compared to baseline, was seen by dobutamine single-photon emission computed tomography sestamibi imaging [82].

The feasibility of intramyocardial injections of a replication deficient adenovirus vector encoding $VEGF_{121}$ was demonstrated in a phase I study [83]. 15 patients were injected into non-bypassable ischemic myocardium during bypass surgery and another 6 patients without suitable target vessels received these as sole therapy through a minithoracotomy [83].

Recently, the preliminary 60-day results of the VIVA trial were reported, a phase 2 trial of intracoronary (20 minutes) $VEGF_{165}$ followed by three 4-hour intravenous infusions over 9 days. Patients in this trial (n=117), who had an area of viable underperfused myocardium, were assigned to either placebo, or $VEGF_{165}$ at either 17 or 50 ng/kg/min on days 3, 6 and 9. The growth factor was well tolerated but there was no improvement in the anginal class and quality of life criteria above that seen with placebo and no improvement in exercise time at day 60, which was the primary endpoint [80].

Other ongoing studies are evaluating the intravenous route of FGF-2 and $VEGF_{165}$ administration, intracoronary FGF-4 gene injection with an adenoviral vector, as well as the intramyocardial administration of the Hypoxia Inducible Factor (HIF)-1α gene with an adenoviral vector.

So far, these clinical trials support the short-term safety of therapeutic growth factor or growth factor gene administration for coronary heart disease. In addition, some appear to suggest beneficial short term effects on symptoms, perfusion and regional wall motion. It is important to realize, however, that only one small trial has been published that used a randomized, double-blind design. The largest and still ongoing randomized trial (VIVA), which has been presented in oral form, was negative for its primary endpoint of exercise tolerance at 60 days. Thus, it would be premature to conclude that angiogenic therapy is beneficial for patients with ischemic heart disease. More controlled trials with a sufficient number of patients are needed to test various forms of angiogenesis-promoting strategies with careful attention to their potential long term side effects.

8. CONCLUSIONS

The concept of treating ischemic vascular diseases by augmenting the naturally occurring revascularizing processes appears very attractive. Many studies in animals and more recently also in patients suggest the feasibility of treatment strategies that try to promote endogenous compensatory neovascularization processes. This may provide us with the opportunity to achieve more complete revascularization in patients undergoing bypass surgery or one of the percutaneous catheter interventions. It may even be able to replace some of our current therapeutic modalities with less invasive ones that may be more comfortable for the patient.

However, it appears that there still is a lot to be learned about these naturally occurring compensatory processes of angiogenesis, arteriogenesis and vasculogenesis, their regulatory mechanisms and their interactions. This should help us answer many open questions regarding angiogenic therapy. Which is the best growth factor or combination of growth factors, should other substances be considered as well, e.g. to mobilize endothelial progenitor cells? Whether to administer the growth factory protein or gene, the optimal mode of administration (intravenous, intraarterial, local intramyocardial, epicardial, pericardial) and packaging (naked DNA vs. vector, carrier substance with or without sustained release properties), the optimal dose and frequency of dosing; sole growth factor(-gene) therapy versus combined with transcutaneous intravascular procedures, bypass surgery, or laser revascularization. We do not have any data on true long term efficacy and prognosis. There is still concern about the long term side effects of proangiogenic therapies, as angiogenesis appears to be part of the pathological process of atherosclerotic plaque formation and plaque instability, and is a key component of proliferative retinopathies and some malignant tumors.

We are still at the beginning of a promising new approach to the treatment of ischemic vascular diseases.

REFERENCES

1. Schaper, W., Piek, J. J., Munoz-Chapuli, R., Wolf, C., and Ito, W., 1999, Collateral Circulation of the Heart. In *Angiogenesis and Cardiovascular Disease* (J. A. Ware and M. Simons, eds.), Oxford University Press, New York, Oxford, 159-198.
2. Risau, W., 1997, Mechanisms of angiogenesis. *Nature* **386**: 671-674.
3. Asahara, T., Masuda, H., Takahashi, T., Kalka, C., Pastore, C., Silver, M., Kearne, M., Magner, M., and Isner, J. M., 1999, Bone marrow origin of endothelial progenitor cells responsible for postnatal vasculogenesis in physiological and pathological neovascularization [In Process Citation]. *Circ Res* **85**: 221-228.
4. Asahara, T., Murohara, T., Sullivan, A., Silver, M., van der Zee, R., Li, T., Witzenbichler, B., Schatteman, G., and Isner, J. M., 1997, Isolation of putative progenitor endothelial cells for angiogenesis. *Science* **275**: 964-967.
5. Folkman, J., 1995, Clinical applications of research on angiogenesis. *N Engl J Med* **333**: 1757-1763.
6. Moulton, K. S., Heller, E., Konerding, M. A., Flynn, E., Palinski, W., and Folkman, J., 1999, Angiogenesis inhibitors endostatin or TNP-470 reduce intimal neovascularization and plaque growth in apolipoprotein E-deficient mice [see comments]. *Circulation* **99**: 1726-1732.

7. McCarthy, M. J., Loftus, I. M., Thompson, M. M., Jones, L., London, N. J., Bell, P. R., Naylor, A. R., and Brindle, N. P., 1999, Angiogenesis and the atherosclerotic carotid plaque: An association between symptomatology and plaque morphology. *J Vasc Surg* **30**: 261-268.
8. Ware, J. A., and Simons, M., 1997, Angiogenesis in ischemic heart disease. *Nat Med* **3**: 158-164.
9. Ware, J. A., 1999, Cellular Mechanisms of Angiogenesis. In *Angiogenesis and Cardiovascular Disease* (J. A. Ware and M. Simons, eds.), Oxford University Press, New York, Oxford, 30-59.
10. Simons, M., and Laham, R. J., 1999, Therapeutic Angiogenesis in Myocardial Ischemia. In *Angiogenesis and Cardiovascular Disease* (J. A. Ware and M. Simons, eds.), Oxford University Press, New York, Oxford, 289-320.
11. O'Reilly, M. S., Boehm, T., Shing, Y., Fukai, N., Vasios, G., Lane, W. S., Flynn, E., Birkhead, J. R., Olsen, B. R., and Folkman, J., 1997, Endostatin: an endogenous inhibitor of angiogenesis and tumor growth. *Cell* **88**: 277-285.
12. Westernacher, D., and Schaper, W., 1995, A novel heart derived inhibitor of vascular cell proliferation. Purification and biological activity. *J Mol Cell Cardiol* **27**: 1535-1543.
13. Engleka, K. A., and Maciag, T., 1999, Molecular mechanisms of the fibroblast growth factor family. In *Angiogenesis and Cardiovascular Disease* (J. A. Ware and M. Simons, eds.), Oxford University Press, New York, Oxford, 79-100.
14. Ferrara, N., and Gerber, H. P., 1999, The vascular endothelial growth factor family. In *Angiogenesis and Cardiovascular Disease* (J. A. Ware and M. Simons, eds.), Oxford University Press, New York, Oxford, 101-127.
15. Bunn, F. H., and Poyton, R. O., 1996, Oxygen sensing and molecular adaptation to hypoxia. *Physiological Reviews* **70**: 839-885.
16. Shweiki, D., Itin, A., Soffer, D., and Keshet, E., 1992, Vascular endothelial growth factor induced by hypoxia may mediate hypoxia-initiated angiogenesis. *Nature* **359**: 843-845.
17. Takagi, H., King, G. L., Ferrara, N., and Aiello, L. P., 1996, Hypoxia regulates vascular endothelial growth factor receptor KDR/Flk gene expression through adenosine A_2 receptors in retinal capillary endothelial cells. *Invest Ophthlmol Vis Sci* **37**: 1311-1321.
18. Maltepe, E., Schmidt, J. V., Baunoch, D., Bradfeld, C. A., and Simon, M. C., 1997, Abnormal angiogenesis and responses to glucose and oxygen deprivation in mice lacking the protein ARNT. *Nature* **386**: 403-407.
19. Stein, I., Neeman, M., Shweiki, D., Itin, A., and Keshet, E., 1995, Stabilization of vascular endothelial growth factor mRNA by hypoxia and hypoglycemia and coregulation with other ischemia-induced genes. *Mol Cell Biol* **15**: 5363-5368.
20. Wang, G. L., Jiang, B.-H., Rue, E. A., and Semenza, G. L., 1995, Hypoxia-inducible factor 1 is a basic-helix-loop-helix-PAS heterodimer regulated by cellular O_2 tension. *Proc Natl Acad Sci USA* **92**: 5510-5514.
21. Levy, A. P., Levy, N. S., Wegner, S., and Goldberg, M., 1995, Transcriptional regulation of the rat vascular endothelial growth factor gene by hypoxia. *J Biol Chem* **270**: 13333-13340.
22. Li, J., Brown, L. F., Hibberd, M. G., Grossman, J. D., Morgan, J. P., and Simons, M., 1996, VEGF, flk-1, and flt-1 expression in a rat myocardial infarction model of angiogenesis. *Am J Physiol* **270**: H1803-1811.

23. Li, J., Hampton, T., Morgan, J. P., and Simons, M., 1997, Stretch-induced VEGF expression in the heart. *J Clin Invest* **100**: 18-24.
24. Acevedo, A. D., Bowser, S. S., Gerritsen, M. E., and Bizios, R., 1993, Morphological and proliferative responses of endothelial cells to hydrostatic pressure: role of fibroblast growth factor. *J Cell Physiol* **157**: 603-614.
25. Schwartz, E. A., Bizios, R., Medow, M. S., and Gerritsen, M. E., 1999, Exposure of human vascular endothelial cells to sustained hydrostatic pressure stimulates proliferation. Involvement of the alphaV integrins. *Circ Res* **84**: 315-322.
26. Tardy, Y., Resnick, N., Nagel, T., Gimbrone, M. A., Jr., and Dewey, C. F., Jr., 1997, Shear stress gradients remodel endothelial monolayers in vitro via a cell proliferation-migration-loss cycle. *Arterioscler Thromb Vasc Biol* **17**: 3102-3106.
27. Arras, M., Ito, W. D., Scholz, D., Winkler, B., Schaper, J., and Schaper, W., 1998, Monocyte activation in angiogenesis and collateral growth in the rabbit hindlimb. *J Clin Invest* **101**: 40-50.
28. Coulier, F., Pontarotti, P., Roubin, R., Hartung, H., Goldfarb, M., and Birnbaum, D., 1997, Of worms and men: an evolutionary perspective on the fibroblast growth factor (FGF) and FGF receptor families. *J Mol Evol* **44**: 43-56.
29. Yu, C., English, A., and Edelman, E. R., 1999, Growth factor delivery strategies. In *Angiogenesis and Cardiovascular Disease* (J. A. Ware and M. Simons, eds.), Oxford University Press, New York, Oxford, 238-257.
30. Cuevas, P., Carceller, F., Ortega, S., Zazo, M., Nieto, I., and Gimenez-Gallego, G., 1991, Hypotensive activity of fibroblast growth factor. *Science* **254**: 1208-1210.
31. Sellke, F. W., and Harrison, D. G., 1999, Coronary Microcirculation and Angiogenesis. In *Angiogenesis and Cardiovascular Disease* (J. A. Ware and M. Simons, eds.), Oxford University Press, New York, Oxford, 258-288.
32. Ziche, M., Morbidelli, L., Choudhuri, R., Zhang, H. T., Donnini, S., Granger, H. J., and Bicknell, R., 1997, Nitric oxide synthase lies downstream from vascular endothelial growth factor-induced but not basic fibroblast growth factor-induced angiogenesis. *J Clin Invest* **99**: 2625-2634.
33. Vincenti, V., Cassano, C., Rocchi, M., and Persico, G., 1996, Assignment of the vascular endothelial growth factor gene to human chromosome 6p21.3. *Circulation* **93**: 1493-1495.
34. Keyt, B. A., Berleau, L. T., Nguyen, H. V., Chen, H., Heinsohn, H., Vandlen, R., and Ferrara, N., 1996, The carboxyl-terminal domain (111-165) of vascular endothelial growth factor is critical for its mitogenic potency. *J Biol Chem* **271**: 7788-7795.
35. Asahara, T., Takahashi, T., Kalka, C., Masuda, H., Chen, D., Silver, M., and Isner, J. M., 1998, A novel function of VEGF: mobilization of bone marrow derived endothelial progenitor cells. *Circulation* **98 (Suppl. I)**: I-605.
36. Kalka, C., Takahashi, T., Masuda, H., Pieczek, A., and Asahara, T., 1998, Intramuscular VEGF gene transfer mobilizes endothelial cell progenitor cells in human subjects. *Circulation* **98 (Suppl. I)**: I-1685.
37. Achen, M. G., Jeltsch, M., Kukk, E., Makinen, T., Vitali, A., Wilks, A. F., Alitalo, K., and Stacker, S. A., 1998, Vascular endothelial growth factor D (VEGF-D) is a ligand for the tyrosine kinases VEGF receptor 2 (Flk1) and VEGF receptor 3 (Flt4). *Proc Natl Acad Sci U S A* **95**: 548-553.

38. Meyer, M., Clauss, M., Lepple-Wienhues, A., Waltenberger, J., Augustin, H. G., Ziche, M., Lanz, C., Buttner, M., Rziha, H. J., and Dehio, C., 1999, A novel vascular endothelial growth factor encoded by Orf virus, VEGF- E, mediates angiogenesis via signalling through VEGFR-2 (KDR) but not VEGFR-1 (Flt-1) receptor tyrosine kinases. *Embo J* **18**: 363-374.
39. Witzenbichler, B., Asahara, T., Murohara, T., Silver, M., Spyridopoulos, I., Magner, M., Principe, N., Kearney, M., Hu, J. S., and Isner, J. M., 1998, Vascular endothelial growth factor-C (VEGF-C/VEGF-2) promotes angiogenesis in the setting of tissue ischemia. *Am J Pathol* **153**: 381-394.
40. Cao, Y., Linden, P., Shima, D., Browne, F., and Folkman, J., 1996, In vivo angiogenic activity and hypoxia induction of heterodimers of placenta growth factor/vascular endothelial growth factor. *J Clin Invest* **98**: 2507-2511.
41. Kurz, H., Wilting, J., Sandau, K., and Christ, B., 1998, Automated evaluation of angiogenic effects mediated by VEGF and PlGF homo- and heterodimers. *Microvasc Res* **55**: 92-102.
42. Ziche, M., Maglione, D., Ribatti, D., Morbidelli, L., Lago, C. T., Battisti, M., Paoletti, I., Barra, A., Tucci, M., Parise, G., Vincenti, V., Granger, H. J., Viglietto, G., and Persico, M. G., 1997, Placenta growth factor-1 is chemotactic, mitogenic, and angiogenic. *Lab Invest* **76**: 517-531.
43. Mezaros, J. G., Brunton, L. L., and Bloor, C. M., 1999, Animal Models of Angiogenesis in Cardiovascular Tissues. In *Angiogenesis and Cardiovascular Disease* (J. A. Ware and M. Simons, eds.), Oxford University Press, New York, Oxford, 213-237.
44. Michael, L. H., Entman, M. L., Hartley, C. J., Youker, K. A., Zhu, J., Hall, S. R., Hawkins, H. K., Berens, K., and Ballantyne, C. M., 1995, Myocardial ischemia and reperfusion: a murine model. *Am J Physiol* **269**: H2147-2154.
45. Couffinhal, T., Silver, M., Zheng, L. P., Kearney, M., Witzenbichler, B., and Isner, J. M., 1998, Mouse model of angiogenesis. *Am J Pathol* **152**: 1667-1679.
46. Sanborn, T. A., Tarazona, N., Deutsch, E., Lee, L., Hackett, N., El-Sawy, T., and Crystal, R. G., 1999, Percutaneous endocardial gene therapy: in vivo gene transfer and expression. *J Am Coll Cardiol* **33 (Suppl. A)**: 262A.
47. Arras, M., Mollnau, H., Strasser, R., Wenz, R., Ito, W. D., Schaper, J., and Schaper, W., 1998, The delivery of angiogenic factors to the heart by microsphere therapy. *Nat Biotechnol* **16**: 159-162.
48. Banai, S., Jaklitsch, M. T., Casscells, W., Shou, M., Shrivastav, S., Correa, R., Epstein, S. E., and Unger, E. F., 1991, Effects of acidic fibroblast growth factor on normal and ischemic myocardium. *Circ Res* **69**: 76-85.
49. Unger, E. F., Shou, M., Sheffield, C. D., Hodge, E., Jaye, M., and Epstein, S. E., 1993, Extracardiac to coronary anastomoses support regional left ventricular function in dogs. *Am J Physiol* **264**: H1567-1574.
50. Unger, E. F., Banai, S., Shou, M., Jaklitsch, M., Hodge, E., Correa, R., Jaye, M., and Epstein, S. E., 1993, A model to assess interventions to improve collateral blood flow: continuous administration of agents into the left coronary artery in dogs. *Cardiovasc Res* **27**: 785-791.
51. Schlaudraff, K., Schumacher, B., von Specht, B. U., Seitelberger, R., Schlosser, V., and Fasol, R., 1993, Growth of "new" coronary vascular structures by angiogenetic growth factors. *Eur J Cardiothorac Surg* **7**: 637-643.

52. Ortega, S., Schaeffer, M. T., Soderman, D., DiSalvo, J., Linemeyer, D. L., Gimenez-Gallego, G., and Thomas, K. A., 1991, Conversion of cysteine to serine residues alters the activity, stability, and heparin dependence of acidic fibroblast growth factor. *J Biol Chem* **266**: 5842-5846.
53. Lopez, J. J., Edelman, E. R., Stamler, A., Hibberd, M. G., Prasad, P., Thomas, K. A., DiSalvo, J., Caputo, R. P., Carrozza, J. P., Douglas, P. S., Sellke, F. W., and Simons, M., 1998, Angiogenic potential of perivascularly delivered aFGF in a porcine model of chronic myocardial ischemia. *Am J Physiol* **274**: H930-936.
54. Yanagisawa-Miwa, A., Uchida, Y., Nakamura, F., Tomaru, T., Kido, H., Kamijo, T., Sugimoto, T., Kaji, K., Utsuyama, M., Kurashima, C., and et al., 1992, Salvage of infarcted myocardium by angiogenic action of basic fibroblast growth factor. *Science* **257**: 1401-1403.
55. Uchida, Y., Yanagisawa-Miwa, A., Nakamura, F., Yamada, K., Tomaru, T., Kimura, K., and Morita, T., 1995, Angiogenic therapy of acute myocardial infarction by intrapericardial injection of basic fibroblast growth factor and heparin sulfate: an experimental study. *Am Heart J* **130**: 1182-1188.
56. Padua, R. R., Sethi, R., Dhalla, N. S., and Kardami, E., 1995, Basic fibroblast growth factor is cardioprotective in ischemia- reperfusion injury. *Mol Cell Biochem* **143**: 129-135.
57. Huang, Z., Chen, K., Huang, P. L., Finklestein, S. P., and Moskowitz, M. A., 1997, bFgf ameliorates focal ischemic injury by blood flow-independent mechanisms in eNos mutant mice. *Am J Physiol* **272**: H1401-1405.
58. Battler, A., Scheinowitz, M., Bor, A., Hasdai, D., Vered, Z., Di Segni, E., Varda-Bloom, N., Nass, D., Engelberg, S., Eldar, M., and et al., 1993, Intracoronary injection of basic fibroblast growth factor enhances angiogenesis in infarcted swine myocardium. *J Am Coll Cardiol* **22**: 2001-2006.
59. Unger, E. F., Banai, S., Shou, M., Lazarous, D. F., Jaklitsch, M. T., Scheinowitz, M., Correa, R., Klingbeil, C., and Epstein, S. E., 1994, Basic fibroblast growth factor enhances myocardial collateral flow in a canine model. *Am J Physiol* **266**: H1588-1595.
60. Lazarous, D. F., Scheinowitz, M., Shou, M., Hodge, E., Rajanayagam, S., Hunsberger, S., Robison, W. G., Jr., Stiber, J. A., Correa, R., Epstein, S. E., and et al., 1995, Effects of chronic systemic administration of basic fibroblast growth factor on collateral development in the canine heart. *Circulation* **91**: 145-153.
61. Harada, K., Grossman, W., Friedman, M., Edelman, E. R., Prasad, P. V., Keighley, C. S., Manning, W. J., Sellke, F. W., and Simons, M., 1994, Basic fibroblast growth factor improves myocardial function in chronically ischemic porcine hearts. *J Clin Invest* **94**: 623-630.
62. Lopez, J. J., Edelman, E. R., Stamler, A., Hibberd, M. G., Prasad, P., Caputo, R. P., Carrozza, J. P., Douglas, P. S., Sellke, F. W., and Simons, M., 1997, Basic fibroblast growth factor in a porcine model of chronic myocardial ischemia: a comparison of angiographic, echocardiographic and coronary flow parameters. *J Pharmacol Exp Ther* **282**: 385-390.
63. Baffour, R., Berman, J., Garb, J. L., Rhee, S. W., Kaufman, J., and Friedmann, P., 1992, Enhanced angiogenesis and growth of collaterals by in vivo administration of recombinant basic fibroblast growth factor in a rabbit model of acute lower limb ischemia: dose-response effect of basic fibroblast growth factor. *J Vasc Surg* **16**: 181-191.

64. Pu, L. Q., Sniderman, A. D., Brassard, R., Lachapelle, K. J., Graham, A. M., Lisbona, R., and Symes, J. F., 1993, Enhanced revascularization of the ischemic limb by angiogenic therapy. *Circulation* **88**: 208-215.
65. Pu, L. Q., Sniderman, A. D., Arekat, Z., Graham, A. M., Brassard, R., and Symes, J. F., 1993, Angiogenic growth factor and revascularization of the ischemic limb: evaluation in a rabbit model. *J Surg Res* **54**: 575-583.
66. Giordano, F. J., Ping, P., McKirnan, M. D., Nozaki, S., DeMaria, A. N., Dillmann, W. H., Mathieu-Costello, O., and Hammond, H. K., 1996, Intracoronary gene transfer of fibroblast growth factor-5 increases blood flow and contractile function in an ischemic region of the heart. *Nat Med* **2**: 534-539.
67. Takeshita, S., Pu, L. Q., Stein, L. A., Sniderman, A. D., Bunting, S., Ferrara, N., Isner, J. M., and Symes, J. F., 1994, Intramuscular administration of vascular endothelial growth factor induces dose-dependent collateral artery augmentation in a rabbit model of chronic limb ischemia. *Circulation* **90**: II 228-234.
68. Takeshita, S., Zheng, L. P., Brogi, E., Kearney, M., Pu, L. Q., Bunting, S., Ferrara, N., Symes, J. F., and Isner, J. M., 1994, Therapeutic angiogenesis. A single intraarterial bolus of vascular endothelial growth factor augments revascularization in a rabbit ischemic hind limb model. *J Clin Invest* **93**: 662-670.
69. Bauters, C., Asahara, T., Zheng, L. P., Takeshita, S., Bunting, S., Ferrara, N., Symes, J. F., and Isner, J. M., 1995, Site-specific therapeutic angiogenesis after systemic administration of vascular endothelial growth factor. *J Vasc Surg* **21**: 314-324.
70. Takeshita, S., Tsurumi, Y., Couffinahl, T., Asahara, T., Bauters, C., Symes, J., Ferrara, N., and Isner, J. M., 1996, Gene transfer of naked DNA encoding for three isoforms of vascular endothelial growth factor stimulates collateral development in vivo. *Lab Invest* **75**: 487-501.
71. Banai, S., Jaklitsch, M. T., Shou, M., Lazarous, D. F., Scheinowitz, M., Biro, S., Epstein, S. E., and Unger, E. F., 1994, Angiogenic-induced enhancement of collateral blood flow to ischemic myocardium by vascular endothelial growth factor in dogs. *Circulation* **89**: 2183-2189.
72. Lazarous, D. F., Shou, M., Scheinowitz, M., Hodge, E., Thirumurti, V., Kitsiou, A. N., Stiber, J. A., Lobo, A. D., Hunsberger, S., Guetta, E., Epstein, S. E., and Unger, E. F., 1996, Comparative effects of basic fibroblast growth factor and vascular endothelial growth factor on coronary collateral development and the arterial response to injury. *Circulation* **94**: 1074-1082.
73. Harada, K., Friedman, M., Lopez, J. J., Wang, S. Y., Li, J., Prasad, P. V., Pearlman, J. D., Edelman, E. R., Sellke, F. W., and Simons, M., 1996, Vascular endothelial growth factor administration in chronic myocardial ischemia. *Am J Physiol* **270**: H1791-1802.
74. Sayeed-Shah, U., Mann, M. J., Martin, J., Grachev, S., Reimold, S., Laurence, R., Dzau, V., and Cohn, L. H., 1998, Complete reversal of ischemic wall motion abnormalities by combined use of gene therapy with transmyocardial laser revascularization. *J Thorac Cardiovasc Surg* **116**: 763-769.
75. Nakamura, S., Moriguchi, A., Nakamura, Y., Mastsushita, H., Aoki, M., Tomita, N., and Nakamura, T., 1998, Therapeutic angiogenesis by a single intra-arterial injection of human hepatocyte growth factor (HGF) in rabbit ischemic hind limb model as "supplement cytokine therapy". *Circulation* **98 (Suppl. I)**: I-268.

76. Shyu, K.-G., Vincent, K. A., Luo, Y., Magner, M., Tio, R. A., Jiang, C., Akita, G. Y., Isner, J. M., and Gregory, R. J., 1998, Naked DNA encoding an hypoxia-inducible factor 1α (HIF-1α)/VP16 hybrid transcription factor enhances angiogenesis in rabbit hindlimb ischemia: an alternate method for therapeutic angiogenesis utilizing a transcriptional regulatory system. *Circulation* **98 (Suppl. I)**: I-68.
77. Schumacher, B., Pecher, P., von Specht, B. U., and Stegmann, T., 1998, Induction of neoangiogenesis in ischemic myocardium by human growth factors: first clinical results of a new treatment of coronary heart disease. *Circulation* **97**: 645-650.
78. Helisch, A., and Ware, J. A., 1999, Therapeutic angiogenesis in ischemic heart disease. *Thromb Haemostasis* **82**: 772-780.
79. Laham, R. J., Leimbach, M., Chronos, N. A., Vansant, J. P., Pearlman, J. D., Pettigrew, R., Guler, H.-P., Whitehouse, M. J., Hung, D., Baim, D. S., King III, S. B., and Simons, M., 1999, Intracoronary administration of recombinant fibroblast growth factor-2 (rFGF-2) in patients with severe coronary artery disease: results of phase I. *J Am Coll Cardiol* **33 (Suppl. A)**: 383A.
80. Henry, T. D., Annex, B. H., Azrin, M. A., McKendall, G. R., Willerson, J. T., Hendel, R. C., Giordano, F., Klein, R., Gibson, M., Berman, D. S., Luce, C. A., and McCluskey, E. R., 1999, Double blind, placebo controlled trial of recombinant human vascular endothelial growth factor - the VIVA trial. *J Am Coll Cardiol* **33 (Suppl. A)**: 384A.
81. Losordo, D. W., Vale, P. R., Symes, J. F., Dunnington, C. H., Esakof, D. D., Maysky, M., Ashare, A. B., Lathi, K., and Isner, J. M., 1998, Gene therapy for myocardial angiogenesis: initial clinical results with direct myocardial injection of phVEGF165 as sole therapy for myocardial ischemia. *Circulation* **98**: 2800-2804.
82. Vale, P. R., Losordo, D. W., Dunnington, C. H., Lathi, K., Esakof, D. D., Symes, J. F., and Isner, J. M., 1999, Direct intramyocardial injection of VEGF results in effective gene transfer for patients with chronic myocardial ischemia. *J Am Coll Cardiol* **33 (Suppl. A)**: 384A.
83. Rosengart, T. K., Lee, L. Y., Patel, S. R., Sanborn, T. A., Parikh, M., Bergman, G. W., Hachamovitch, R., Szulc, M., Kligfield, P. D., Okin, P. M., Hahn, R. T., Devereux, R. B., Post, M. R., Hackett, N. R., Foster, T., Grasso, T. M., Lesser, M. L., Isom, O. W., and Crystal, R. G., 1999, Angiogenesis gene therapy: phase I assessment of direct intramyocardial administration of an adenovirus vector expressing VEGF121 cDNA to individuals with clinically significant severe coronary artery disease. *Circulation* **100**: 468-474.
84. Sellke, F. W., Laham, R. J., Edelman, E. R., Pearlman, J. D., and Simons, M., 1998, Therapeutic angiogenesis with basic fibroblast growth factor: technique and early results. *Ann Thorac Surg* **65**: 1540-1544.
85. Laham, R. J., Sellke, F. W., Ware, J. A., Pearlman, J. D., Edelman, E. R., and Simons, M., 1999, Results of a randomized, double-blind, placebo-controlled study of local perivascular basic fibroblast growth factor (bFGF) treatment in patients undergoing coronary artery bypass surgery. *J Am Coll Cardiol* **33 (Suppl. A)**: 383A-384A.

Abstracts of Posters

INTEGRIN $\alpha_v\beta_3$ INVOLVEMENT IN THE MECHANISM OF ANGIOGENIC ACTION OF THROMBIN

Andriopoulou P., Tsopanoglou N.E., and Maragoudakis M.E.
University of Patras, Medical School, Dept. of Pharmacology, 261 10 Rio, Patras, GREECE

Brown et al. (1) have reported that thrombin upregulates the expression of a_vb_3 in smooth muscle cells.

In view of the pivotal role of $a_v\beta_3$ integrin in angiogenesis (2), we are exploring the possibility that thrombin may mediate its angiogenic action, at least in part, by upregulation of $a_v\beta_3$ in endothelial cells.

Preliminary results indicate that is $a_v\beta_3$ upregulated after exposure of endothelial cells to thrombin. In addition $a_v\beta_3$ peptide antagonists inhibit the thrombin-induced angiogenesis in the CAM. Furthermore, thrombin induces the secretion of the activated from of 72KD collagenase from endothelial cells, which is inhibited by the peptide antagonist to $a_v\beta_3$.

These results point to the involvement of $a_v\beta_3$ in the angiogenic action of thrombin.

REFERENCES

1. Brooks, P.C., Clark, R.A.F. and Cheresh, D.A. (1994) Science 264, 569-571.
2. Brown, S.L., Lundgren, C.H., Nordt, T. and Fujii, S. (1994) Cardiovasc. Research, 28, 1815-1820.

CLINICAL DETECTION OF METASTASIS ASSOCIATED TUMOR MICROVASCULARI-SATION IN UVEAL MELANOMA

Bechrakis N. E., Servetopoulou F., Anagnostopoulos I. and Foerster M.H.
Department of Ophthalmology and Pathology, Klinikum Benjamin Franklin, Free University of Berlin, Germany

Purpose: To identify Metastasis Associated Vascular Patterns (MAVP-according to Folberg) using indocyanine green angiography (ICG) and/or fluorescein angiography and to correlate them with histopathological and immunohistochemical detection of tumor microvascularisation in uveal melanomas.

Methods: 35 eyes with uveal melanomas were investigated using indocyanine green angiography with a scanning laser ophthalmoscope (n=15) and/or fluorescein angiography (n=34) prior to enucleation or local tumor resection. Vascular patterns thus visualised were compared to the histologicaly identified metastasis associated vascular patterns.

Results: indocyanine green angiography and histology revealed the same tumor vascular pattern in 14 out of 15 cases, whereas fluorescein angiography in 18 out of 34 cases.

Conclusion: Histologically described MAVP were detected in our cases with great accuracy using either fluorescein or indocyanine green angiography, allowing an evaluation of metastasis associated microvascularisation purely on clinical grounds.

Angiogenesis: From the Molecular to Integrative Pharmacology
Edited by Maragoudakis, Kluwer Academic / Plenum Publishers, New York, 2000

THE INFLUENCE OF LOW MOLECULAR WEIGHT AND UNFRACTIONED HEPARIN ON IN VITRO ANGIOGENESIS

Collen A, Smorenburg S.*, Koolwijk P., Buller H.* and Van Hinsbergh V.
*Gaubius LaboratoryTNO-PG, Leiden and *Dept. of Vascular Medicine, Academic Med. Ctr., Amsterdam, The Netherlands.*

Patients with malignant diseases frequently suffer from thromboembolic complications. Results of clinical trials have indicated that treatment of cancer patients with low-molecular weight heparin lead to a higher long-term survival than treatment with standard heparin, although recurrence of thromboembolic and bleeding complications was identical in both patient groups. Angiogenesis is an important factor in regulating tumour progression and metastasis and therefore, factors altering angiogenesis could influence long-term survival.

In a human *in vitro* angiogenesis model we investigated whether different heparin fractions would exert an effect on capillary-like tubular structure formation. We found that the addition of LMW heparin to the culture media of stimulated cells resulted in a larger capillary-like tubular structure foirmation than did unfractioned heparin. This was due to the upregulated expression and secretion of u-PA as well as the upregulated expression and surface localisation of its receptor, while the expression and secretion of PAI-1 remained unaltered. In addition to the cell-bound u-PA activity the structure of the fibrin network plays an important regulatory role in the capillary-like tubular structure formation. The presence of LMW-heparin during polymerisation of the fibrin matrix lead to a more transparent network than the presence of unfractioned heparin, which lead to a more opaque fibrin network. Endothelial cells cultured on top of transparent

Angiogenesis: From the Molecular to Integrative Pharmacology
Edited by Maragoudakis, Kluwer Academic / Plenum Publishers, New York, 2000

fibrin matrices migrated into the fibrin matrix and formed capillary-like structures. They showed a larger reduction in capillary-like tubular structure formation when they grew in the transparent (LMW-heparin containing) fibrin gels as compared to the more opaque (control and unfractionated heparin-containing) fibrin matrices.

In conclusion, we found a dual effect of LMW-heparin on *in vitro* angiogenesis, on one hand it stimulates angiogenesis by upregulation of the plasminogen activator activity and on the other hand it inhibits angiogenesis by rendering the fibrin matrix less permessive for angiogenesis. Both effects together resulted in a net inhibitory effect on *in vitro* angiogenesis of LMW heparin, while unfractioned heparin had no net negative effect.

ANGIOTENSIN II MODULATES THE ACTIVITY OF THE LARGE CONDUCTANCE, CALCIUM- AND VOLTAGE-ACTIVATED POTASSIUM CHANNEL IN SMOOTH MUSCLE CELLS FROM RAT MESENTERIC MICROVESSELS THROUGH AT2 RECEPTOR.

Christiana Dimitropoulou, Leslie Fuchs and Gerald Carrier
Medical College of Georgia, Augusta, GA 30912, U.S.A.

In cell-attached patches of smooth muscle cells from rat mesenteric microvessels, angiotensin enhanced the opening of a 120 ± 12pS K^+ channel. This channel was calcium and voltage dependent and was inhibited by 1mM tetraethylammonium (TEA), verifying this as the BK_{Ca} channel. In the presence of 100nM Angiotensin II (Ang II) the opening probability (+50mV) of the BK_{Ca} channel was significantly increased ($p<0.02$). In the presence of [Sar^1, Ile^8]- Angiotensin II, the effect of Ang II was completely antagonized. Additionally in inside out patches, Ang II had no effect on BK_{Ca} channel activity, indicating that Ang II did not directly interact with this channel. Pretreatment with Losartan (1mM, 30min) did not inhibit the stimulatory effects of Ang II (100nM, n=3) on BK_{Ca} channel activity, indicating that Ang II does not stimulate the BK_{Ca} activity via AT1 receptor mediated mechanism. Pretreatment with PD123,319, an Ang II type 2 (AT2) receptor inhibitor on Ang II (100nM) stimulatory effect on BK_{Ca}-channels, reversed the effect of Ang II, indicating that Ang II stimulates the BK_{Ca} activity via AT2 receptor mediated mechanism. In tension studies Ang II did not promote contraction of the arterial microvessels (150-280 Ìm diameter) but induced relaxation of arteries preconstricted with Endothelin 1. The vasodilatory effect remained after pretreatment of the vessels with Losartan (1mM), but was abolished when the vessels were pretreated with PD123,319. It is known that AT2 angiotensin receptors are involved in embryogenesis, but there is no indication of their physiological role after maturation. The present results suggest that AT2 plays a role in modulating the vascular tone after maturation.

ACKNOWLEDGMENT

This work was supported by American Heart Association.

REGULATION OF VEGF GENE EXPRESSION IN BROWN ADIPOCYTES

The Question Of The Identity Of The 2nd Messengers for the Adrenergic Stimulation

Fredriksson M., Lindquist J., Bronnikov G. and Nedergaard J.
The Wenner-Gren Institute, Stockholm University, SWEDEN

The aim of the present investigation was to clarify through which signalling mechanism the gene expression of the vascular endothelial growth factor (VEGF) is regulated in cultured brown adipocytes. The sympathetic neurotransmitter, norepinephrine (NE), an activator of brown adipose tissue heat production and growth, was a potent inducer of VEGF expression in brown adipocytes. Known inducers of VEGF expression in other cell types, such as the hypoxia-mimicking agent cobalt, as well as serum and phorbol ester, were effective also on brown adipocytes. Concerning each of these inducers, the effect of NE was additive, implying a separate pathway for the NE-induced VEGF expression. The NE effect was mediated by b-adrenoceptors, with a lack of influence by the α_1-adrenoceptor signalling pathway. The capacity of cAMP to induce VEGF expression was saturated with forskolin treatment and the high forskolin-induced cAMP levels were not altered by NE cotreatment. However, the effect of NE on VEGF expression was additive to the effect of forskolin, as was the effect of the β-adrenoceptor agonist isoprenaline. Thus, the regulation of VEGF expression by NE does apparently not adhere to simple cAMP-mediation kinetics.

A number of inhibitors of intracellular signalling factors are being used in attempts to identify downstream components of the NE signalling pathway inducing VEGF expression. However, the NE effect was not

Angiogenesis: From the Molecular to Integrative Pharmacology
Edited by Maragoudakis, Kluwer Academic / Plenum Publishers, New York, 2000

inhibited by cycloheximide, suggesting that synthesis of new proteinaceous factors was not needed for the mediation of the signal.

It is evident that the mechanisms involved in regulating VEGF gene expression in brown adipocytes are complex and further studies are required to reach further understanding in this matter.

REGULATION OF *IN VITRO* ANGIOGENESIS AT HYPOXIC CULTURE CONDITIONS

Kroon M.E., Koolwijk P., van der Vecht B., and Van Hinsbergh V.
Gaubius Laboratory TNO-PG, Leiden, The Netherlands.

Hypoxia is a common feature of pathological conditions and stimulates the formation of neovessels. In this study an in vitro angiogenesis model was used consisting of human microvascular endothelial cells (hMVEC) cultured on a three-dimensional fibrin matrix to investigate the effects of hypoxia on capillary-like tube formation. hMVEC stimulated with basic fibroblast growth factor (bFGF) and tumour necrosis factor-α (TNFα) formed 3 to 4 times more tubular structures when cultured at hypoxic conditions (1% O2 in the surrounding air) compared to hMVEC cultured at normoxic conditions (20% O_2). We previously showed that this increase in tube formation could partly be explained by a higher proteolytical activity of the cells; u-PAR as well as u-PA/u-PAR could not be mimicked by $CoCl_2$, $NiCl_2$ or DFO, which indicates that a heme-protein is not involved in the oxygen sensing pathways of hMVECs.

Beside an upregulation of proteolytical activity, hypoxia is known to have a number of other pro-angiogenic effects. For example, vascular endothelial growth factor (VEGF) is known to be upregulated at hypoxic conditions and is often indicated as the explanation for increased angiogenesis during hypoxia. We investigated if VEGF could be involved in tube formation at hypoxic culture conditions in our model. VEGF could not be detected by ELISA and Northern blot at both normoxic and hypoxic conditions. By RT-PCR it was possible to detect VEGF mRNA in both

Angiogenesis: From the Molecular to Integrative Pharmacology
Edited by Maragoudakis, Kluwer Academic / Plenum Publishers, New York, 2000

normoxic as hypoxic hMVEC. This was, however, a faint signal which indicates that the amount of VEGGF mRNA was extremely low. Furthermore, the addition of soluble VEGFR-1 and antibodies against VEGFR-2 could not prevent the increased tube formation at hypoxic culture conditions.

These data indicate that in our model *in vitro* angiogenesis model a higher proteolytical activity is the main explanation for an increased formation of capillary-like tubular structures at hypoxic culture conditions.

INHIBITION OF ANGIOGENESIS BY BLOCKERS OF VOLUME-REGULATED ANION CHANNELS

Vangelis G. Manolopoulos[1,2], Sandra Liekens[3], Pieter Koolwijk[4], Thomas Voets[2], Erna Peters[4], Guy Droogmans[2], Peter I. Lelkes[5], Erik De Clercq[3], and Bernd Nilius[2]

[1]*Lab. of Pharmacology, Democritus University of Thrace Medical School, 68100 Alexandroupolis, GREECE;* [2]*Lab. of Physiology, Katholieke Universiteit Leuven Medical School, 3000 Leuven, BELGIUM;* [3]*Rega Institute for Medical Research, Katholieke Universiteit Leuven, 3000 Leuven, BELGIUM;* [4]*Gaubius Laboratory TNO-PG, 2333 CK Leiden, THE NETHERLANDS;* [5]*Lab. of Cell Biology, University of Wisconsin Medical School, Milwaukee, WI 53201, USA*

Osmotic cell swelling activates an outwardly rectifying Cl^- current, $I_{Cl,swell}$, in endothelial cells, which is mediated by volume-regulated anion channels (VRAC). We have identified several compounds that potently block the endothelial VRAC, and we have shown that serum-induced proliferation of endothelial cells is arrested in the presence of such blockers. In the present study, we used four chemically diverse blockers of VRAC (NPPB, the phenol derivative mibefradil, and the antiestrogens tamoxifen and clomiphene), to assess the possible involvement of VRAC in angiogenesis. Four models of angiogenesis were used: the matrigel and the fibrin gel assays of growth factor-induced tube formation *in vitro*, the rat aorta-ring assay of spontaneous microvessel formation *ex vivo*, and the chick chorioallantoic membrane (CAM) assay of neovascularization

Angiogenesis: From the Molecular to Integrative Pharmacology
Edited by Maragoudakis, Kluwer Academic / Plenum Publishers, New York, 2000

in vivo. Mibefradil (20 μM), NPPB (100 μM), tamoxifen (20 μM), and clomiphene (20 μM), inhibited tube formation by rat microvascular endothelial cells plated on matrigel by 42.9 ± 8.8%, 25.3 ± 10.4%, 32.2 ± 4.5%, and 20 ± 5.8%, respectively ($p<0.05$). Also, NPPB (50-100 μM) and mibefradil (10-30 μM), significantly inhibited bFGF (10 ng/ml) + TNFα (2.5 ng/ml)-stimulated microvessel formation by human microvascular endothelial cells plated on a fibrin gel by 30-70%. Furthermore, NPPB, mibefradil, and clomiphene, concentration-dependently inhibited spontaneous microvessel formation in the rat aorta-ring assay and neovascularization in the developing chick CAM. These results indicate that VRAC blockers are potent inhibitors of angiogenesis. Further, our results suggest that anion channels may play a role in angiogenesis and thus their potential should be explored for possible use as therapeutic tools in cancer and other angiogenesis-dependent pathophysiological processes.

IDENTIFICATION OF ANGIOSTATIN-BINDING PROTEINS

Borris Troyanovsky, Tanya Levchenko and Lars Holmgren
Center for Genomics Research, Karolinska Institute, S-171 77 Stockholm, SWEDEN

Angiostatin, a proteolytic fragment of plasminogen, is a potent inhibitor of angiogenesis and tumor growth in mice. Our aim was to identify the molecules on the surface of the endothelial cells, which bind angiostatin. We have employed the yeast two-hybrid system to identify potential angiostatin binding proteins. A human placenta library was used to screen for molecules that interact with the Kringles 1-4 of angiostatin. We identified three independent positive clones that contained the same angiostatin binding sequence. Sequence analysis revealed no homology to any known genes. We have cloned the full-length cDNA encoding angiostatin-binding protein (ABP-1) into a eukaryotic expression vector. This construct was transfected into non-responsible cells/immortalized endothelial cells, fibroblasts. We have shown that angiostatin binds to Hela cells transfected with ABP1 but not to cells transfected with the vector alone. Immunostainings of NIH-3T3 cells transfected with ABP1 with antibodies against the angiostatin binding domain shows that the protein is localized on the cell surface. Furthermore, in migrating cells, ABP1 colocalizes with focal adhesion kinase. This is of interest as angiostatin has been shown to inhibit endothelial cell migration and angiostatin induces FAK activity in these cells. Studies of endogenous ABP1 in HUVEC cells shows that the protein is predominantly localized in focal adhesion complexes. Functional studies are performed to investigate whether ABP1 functions as an angiostatin receptor.

NEUROPEPTIDE Y (NPY) AND VASCULAR REMODELING

Zofia Zukowska-Grojec and Derrick S. Grant[1]
Dept. Physiology & Biophysics, Georgetown Univ. Med. Center, Washington, DC, and [1]Cardeza Found. Hem. Res., Jefferson Univ., Philadelphia, PA, USA

NPY is a pleiotropic central and peripheral sympathetic co-transmitter/neuromodulator acting via multiple Gi/o-coupled receptors (Rs: Y1-Y6). In the cardiovascular system, NPY is released during stress, exercise and myocardial ischemia and causes vasoconstriction via Y1 receptors (Rs). Recently, we discovered that, at concentrations below vasoconstrictive, NPY is potently mitogenic for vascular smooth muscle (VSMCs) and endothelial cells (Ecs), and chemotactic and angiogenic *in vitro* , *ex vivo* in the rat aortic sprouting assay, and *in vivo* in a Matrigel mouse model. Using R antagonists and antisense oligonucleotides we found that NPY's actions are bimodal (subpM-nM), mediated by multiple Rs: Y1, Y2 and Y5, and regulated by endothelial peptidase, dipeptidyl peptidase IV. Endothelium also possesses its own autocrine NPY system. Since VSMCs and ECs possess no or low abundance of NPY Rs, NPY's actions require their induction/up-regulation - a process mediated by NPY itself, steroids, ·-adrenergic R agonists and fibroblast growth factor - acting as permissive factors and augmenting NPY's actions. *In vivo*, in the rat carotid artery angioplasty model, injury up-regulated expression of the NPY peptide and its Rs: Y1, Y2 and Y5 in the vessel wall. Angioplasty-induced neointimal thickening was inhibited by local application of Y1 R antagonist (H409/22 1mg/pellet/14 days) but

Angiogenesis: From the Molecular to Integrative Pharmacology
Edited by Maragoudakis, Kluwer Academic / Plenum Publishers, New York, 2000

markedly augmented by NPY (1 μg/pellet/14 days). In a rat angiogenesis model, occlusion of the femoral artery increased NPY outflow from the ischemic versus non-ischemic limb, decreased NPY vascular content and induced/up-regulated NPY Rs, Y1, Y2 and/or Y5; these effects were prevented by ipsilateral lumbar sympathectomy. Local administration of NPY (1 μg/pellet/21 days, in the popliteal) up-regulated the expression of these Rs in calf skeletal muscle and restored vessel density (latex-perfused) of the ischemic limb (calf muscles and foot pad) to that of the non-ischemic one. Thus, NPY, a sympathetic nerve-derived growth factor, stimulates ischemia-driven angiogenesis and neointimal proliferation via activation of multiple Gi/o-coupled Rs; the action requires the presence of the permissive factors: hormones, stress mediators and growth factors. Our studies suggest that NPY-modeled drugs may be useful prevention or treatment of angioplasty-induced restenosis (antagonists) and induction of angiogenesis in the ischemic peripheral and heart disease (agonists).

ACKNOWLEDGMENTS

The work was supported by USPHS HL 55310 and Astra Hassle grants.

MODULATION OF ENDOTHELIAL WOUND HEALING: CELL SPREADING VERSUS MIGRATION.

S. Wilson, A.K. Keenan.
Department of Pharmacology, University College Dublin, Belfield, Dublin 4, Ireland.

Re-establishment of a functional endothelial barrier after physical or inflammatory injury is of major importance in vascular wound healing. Among the mediators thought to contribute to the inflammatory injury associated with the pathogenesis of atherosclerosis is the LDL constituent lysophosphatidylcholine (LysoPC)[1]. The aim of this study was to assess effects of LysoPC on wound healing of human umbilical vein endothelial cell (HUVEC) monolayers exposed to inflammatory injury.

HUVEC grown on porous membrane supports were exposed to H_2O_2 and injury was assessed as increased monolayer permeability to trypan blue-albumin (TB-BSA). Wound healing was measured as the recovery of barrier function 24 h later. Modulation of wound healing was assessed following exposure of cells to LysoPC during the 24 h recovery period. In parallel experiments changes in cytoskeletal actin disposition following injury and recovery were visualised using rhodamine phalloidin staining. In addition, the ability of LysoPC to inhibit angiogenesis was assessed using the matrigel assay: tube formation by HUVEC grown on matrigel was measured in the presence or absence of LysoPC. Finally, effects of LysoPC on cell viability were investigated using the MTT assay. The significance of differences between treatments was assessed by ANOVA followed by an appropriate post test.

A 20 min exposure to 0.2 mM H_2O_2 significantly increased percentage transfer of TB-BSA across the monolayer above control values (from 5.87 ± 0.4 to 20.26 ± 0.4%, $p<0.0001$, n=4). After 24 h, transfer across monolayers previously treated with H_2O_2 was significantly reduced to 10.12 ± 1.42% ($p<0.0001$ w.r.t. the level immediately post injury). A 24 h post-injury treatment with 40 _M LysoPC did not affect the recovery from injury. Likewise, cytoskeletal actin changes during the recovery process were unaffected by post-injury treatment with 0.? _M LysoPC. In the matrigel assay, treatment with 8 _M LysoPC for an 18 h period significantly reduced tube formation to 47.1 ± 10% ($p<0.0005$, n=6) of control. At the higher concentration of 40 _M, LysoPC further significantly reduced tube formation to 6.9 ± 3% ($p<0.0001$) of control. The vehicle control, 0.05% methanol/chloroform had no significant effect on tube formation. Finally, cell viability was unaffected by the LysoPC treatments.

Since cell migration plays a key role in tube formation in the angiogenesis assay, it can be inferred that LysoPC inhibits such migration. This is in agreement with previous evidence that LysoPC can block cell migration in response to physical injury[2]. In contrast, cell spreading is the primary determinant of wound healing in the permeability model and the present results would suggest that LysoPC does not modulate this process. This is supported by the lack of effect of LysoPC on the cytoskeletal actin changes associated with cell spreading during the recovery process. It can be concluded therefore that the cell spreading characteristic of wound healing following inflammatory injury is resistant to inhibition by LysoPC.

ACKNOWLEDGMENT

This work was supported by the Irish Heart Foundation.

REFERENCES

1. Busse R, Fleming I. Endothelial dysfunction in atherosclerosis. J Vasc Res 1996; 33: 657-661.

2. Murugesan G, Fox PL. Role of lysophosphatidylcholine in the inhibition of endothelial cell motility by oxidized low density lipoprotein. J Clin Invest 1996; 97: 2736-2744.

1. Kanthou
2. Wedge
3. Banerjee
4. Kroon
5. Zukowska
6. Hassessian
7. Sang
8. Andriopoulou
9. Grant-Williamson
10. Giannopoulou
11. Koolwijk
12. Collins
13. Papadimitriou
14. Collen
15. Breir
16. Egginton
17. Tziora
18. Marmara
19. Dimitropoulou
20. Keenan
21. Thompson
22. Gregoriou
23. Manolopoulos
24. Haudenschild
25. Catravas
26. Maragoudakis
27. Hussain
28. Grant
29. Hellerqvist
30. Kaklamanis
31. Gubish
32. Koukourakis
33. Vinores
34. Giatromanolaki
35. Wagner
36. Fotsis
37. Behrakis
39. Wong
40. Li
41. Gagliardi
42. R. Kumar
43. Fredriksson

Participants' photo

PARTICIPANTS

ANDRIOPOULOU, PARASKEVI
Dept. of Pharmacology,
Univ. of Patras Medical School,
261 10 Rio, Patras, GREECE

BANERJEE, DIPAK
University of Puerto Rico, Medical
Sciences Campus, School of Medicine
Dept. of Biochemistry, PO Box
365067 San Juan, PR 00936-5067,
U.S.A.

BEHRAKIS, NICK
Dept. of Ophthalmology
Benjamin Franklin Clinic
Freie Universität Berlin GERMANY

BOICE, JUDITH
Merck Res. Labs.
PO Box 2000, RY50-105
Rahway, NJ 07065, U.S.A.

BREIER, GEORG
Max-Planck Institut for Physiological
and Clinical Research, Parkstr.1
D-61231 Bad Nauheim GERMANY

CATRAVAS, JOHN
Medical College of Georgia,
Department of Pharmacol. & Toxicol.,
Augusta, GA 30912-2300, U.S.A.

COLLINS, DELWOOD
Dept. of Biochemistry and Human
Preproductive Endocrinology
University of Kentucky
204 Health Sciences Res. Bldg.
Lexington, KY 40536-0305, U.S.A.

DIMITROPOULOU, CHRISTIANA
Medical College of Georgia,
Department of Pharmacol. & Toxicology,
Augusta, GA 30912-2300, U.S.A.

EGGINTON, STUART
Department of Physiology University
Birmingham Medical School Vincent
Drive, Edgbaston
Birmingham B15 2TT,
UNITED KINGDOM

FOTSIS, THEODORE
Laboratory of Biological Chemistry
Medical School, Univ. of Ioannina
45110 Ioannina GREECE

FREDRIKSSON, MAGNUS
The Wenner-Gren Institute
Dept. of Physiology, Arrhenius
Laboratories F3, Stockholm University
S-106 91 Stockholm SWEDEN

GAGLARDI, ANTONIO
Division of Human Reproductive
Endocrinology,
Dept. of Obs./Gynecology
University of Kentucky
204 Health Sciences Res. Bldg.
Lexington, KY 40536-0305, U.S.A.

COLLEN, ANEMIE
Vascular/Conn Tissue Research
Gaubius Laboratory TNO-PG
Zernikerdreef 9,
Leiden, N-2333 CK NETHERLANDS

GIANNINI, GIUSEPPE
SIGMA-TAU
Via Pondine Km 30,400
Pomezia 00040 ITALY

GIANNOPOULOU, EFI
University of Patras
School of Pharmacy
Dept. of Mol. Pharmacology
261 10 Rio, Patras GREECE

GIATROMANOLAKI, ALEX.
Tumour & Angiogenesis Research Gr.
18 Dimokratias Avenue
Heraklion 71306, Crete GREECE

GRANT, DERRICK
College of Medicine/RM812
Thomas Jefferson University
1015 Walnut St. /Curtis Bldg.
Philadelphia, PA 19107, U.S.A.

GUBISH, EDWARD
Regulatory & Clin. Development
EntreMed Inc.
9610 Medical Center Dr. Ste. 200
Rockville MD 20850 U.S.A

HASSESSIAN, HAR. MINAS
Ophthalmology Department
Guy-Bernier Research Center
Hopital Maisonneuve Rosemont
5415, boul. de l' Assomption,
Montreal CANADA H1T 2M4

HAUDENSCHILD, CHRISTIAN
Holland Laboratories
American Red Cross
15601 Crabbs Branch Way
Rockville, MD 20855 U.S.A.

HELLERQVIST, CARL
Vanderbilt University
23RD AT Pierce/634 MRBI
Nashville, TN 37212, U.S.A

HORSMAN, MICHAEL
Danish Cancer Society
Dept. of Exper. Clinical Oncology
Aarhus University Hospital
Norrebrogade 44, Bldg. 5
DK-8000 Aarhus C DENMARK

HUSSAIN , ZAMIRAL
University of California,
Department of Surgery,
Parnassus Ave., Room HSE 839,
San Francisco, U.S.A.

KAKLAMANIS, LOUCAS
Onasion Surgery Center
Dept. of Pathology
Leof. Sigrou 356, Athens 176 74
GREECE

KANTHOU, CHRYSO
Tumour Microcirculation Group
Gray Laboratory Cancer Research Trust,
PO Box 100
Mount Vernon Hospital, Northwood
Middlesex HA6 2JR
UNITED KINGDOM

KEENAN, ALAN
National University of Ireland
Dept. of Pharmacology
Belfield, Dublin 4 IRELAND

KONERDING, MORITZ
Institute of Anatomy, University of
Mainz, Johannes Gütenberg, Beckerweg
13, 55099 Mainz, GERMANY

KOOLWIJK, PIETER
Vascular/Conn Tissue Research
Gaubius Laboratory TNO-PG
Zernikerdreef 9, P.O. Box 2301 Leiden,
2333 CK NETHERLANDS

KROON, MAR.
Vascular/Conn Tissue Research
Gaubius Laboratory TNO-PG
Zernikerdreef 9,
Leiden, 2333 CK NETHERLANDS

KOUKOURAKIS, MICHAEL
Tumour & Angiogenesis Res. Group
18 Dimokratias Avenue
Heraklion 71306, Crete GREECE

KUMAR, CHANDRA
Dept. of Tumor Biology, K-15-4
(4600) Schering Research Institute
2015, Galloping Hill Road
Kenilworth NJ 07033 U.S.A.

KUMAR, RAKESH
Dept. of Cancer Biology
Glaxo Wellcome R & D
Five Moore Drive, P.O. Box 13398
Research Triangle Park
North Carolina 27709 U.S.A.

LELKES, PETER
University of Wisconsin, Lab of Cell
Biology, Department of Medicine,
Sinai Samaritan Medical Center,
Mount Sinai Campus, 950 North
Twelfth Str., Box 342, Milwaukee,
WI 53201-0342 U.S.A.

LI VINCENT
The Angiogenesis Foundation, Inc.,
124 Mount Auburn Street, 207 N,
Cambridge, MA 02138, U.S.A.

LI, WILLIAM
The Angiogenesis Foundation, Inc.,
124 Mount Auburn Street, 207 N,
Cambridge, MA 02138, U.S.A.

LIEKENS, SANDRA
Rega Institute for Med. Res.
Katholieke University of Leuven
Minderbroedersstraat 10
B-3000 Leuven, BELGIUM

MARAGOUDAKIS, MICHAEL
University of Patras
Medical School
Dept. of Pharmacology
261 10 Rio, Patras GREECE

MANOLOPOULOS, VANGELIS
Laboratory of Pharmacology,
Medical School, Democritus University
of Thrace, I. Kaviri 6, 68100
Alexandroupolis, GREECE

MARRON, MARIE
University of Leicester, Cardiovascular
Res. Institute, Dept. of Surgery
Robert Kilpatrick Bldg., P.O. Box 65
Leicester LE2 7LX,
UNITED KINGDOM

MORBIDELLI, LUCIA
University of Florence
Dept. of Preclinica and Clin. Pharmac.
Viale Pieraccini 6,
50139 Florence, ITALY

NAGAI, S.
Tokyo Metropolitan Komagome Hospital
3-18-22, Honkomagome, Bunkyo-ku,
Tokyo, 113, JAPAN

PAPADIMITRIOU, EVANGELIA
University of Patras
School of Pharmacy
Dept. of Mol. Pharmacology
261 10 Rio, Patras GREECE

PARDANAUD, LUC
Institute of Embryology
CNRS and College of France
49bis avenue de la Belle Gabrielle
94736 Nogent-sur-Marne FRANCE

PISANO, CLAUDIO
Sigma-Tau S.p.A.
Via Pentina Km 30,400
00040 Pomezia-Roma ITALY

POMPLIANO, DAVID
Antimicrobial and Angiogenesis Res.
DuPont Pharmaceuticals
Experimental Station E400/4257
Rt. 141 and Henry Clay Rd.
Wilmington, DE 19880 U.S.A.

PRESTA, MARCO
University of Brescia, Department of Science, Biomedicine and Biotechnology, Via Valsabbina 19, I-25123 Brescia, ITALY

SANG, QINGXIANG AMY
Dept. of Chemistry,
Bldg. DLC, Room 262
Florida State University
Tallahassee, Florida 32306-4390, U.S.A.

SATO, YASUFUMI
Dept. of Vascular Biology
Institute of Development, Aging and Cancer, Tohoku University
4-1 Seiryomachi Aoba-ku
Sendai 980-77, JAPAN

SIVRIDIS, EFTHIMIOS
Democritus University of Thrace
Medical School, Dept. of Pathology
P.O. Box 128,
Alexandroupolis 68100, GREECE

THOMPSON, DOUGLAS
University of Aberdeen Med. School, Dept. of Pathology
Aberdeen Royal Infirmary,
Aberdeen AB25 2ZD
UNITED KINGDOM

TOI, MASAKAZU
Tokyo Metropolitan Komagome Hospital, 3-18-22, Honkomagome, Bunkyo-ku Tokyo, 113 JAPAN

TROYANOVSKY, BORIS
Karolinska Instiutet (MTC)
Doctorsringen 13, Box 280
Stockholm, S 171 77 SWEDEN

TSAKAYANNIS, DIMITRIOS
Angiogenesis Foundation
Marathonodromon 113
154 52 Athens GREECE

TSOPANOGLOU, NIKOS
University of Patras Medical School
Dept. of Pharmacology
261 10 Rio, Patras GREECE

WAGNER, SILVIA
Forshungberelch Abt für
Allgomeinchlrurgie Zentrum
Medizinischer Forshun
Arbeitsgruppe Wundheilung,
Waldhörnlestr. 22, D-72072 Tübingen, GERMANY

WARE, ANTONY
Albert Einstein College of Medicine of Yeshiva Univ., Dept. of Medicine/ Division of Cardiology, 1300 Morris Park Avenue Forch G46, Bronx, NY 10461, U.S.A.

WEDGE, STEVE
Cancer & Infection Res. Dept., ZENECA Pharmac. Mereside, Alderley Park, Macclesfield, SK10 4TG,
UNITED KINGDOM

WEICH, HERBERT
Dept. of Gene Regulation and Differentiation, Molecular Biotechnology, National Res. Center for Biotechnology, Mascheroder Weg 1B, D-38124 Braunschweig, GERMANY

WONG, KING-PING
California State University, School of Natural Sciences, 2555 East San Ramon M/S NS90, Fresno, CA 93740-8034, U.S.A.

ZUKOWSKA-GROJEC, ZOFIA
Dept. of Physiology and Biophysics, Georgetown University Medical Ctr, 3900 Reservoir Rd., NW,
Washington, DC 20007, U.S.A.

INDEX